ENZYME INDUCTION

BASIC LIFE SCIENCES

Alexander Hollaender, General Editor

Biology Division
Oak Ridge National Laboratory
and The University of Tennessee
Knoxville
and Associated Universities, Inc.
Washington, D.C.

A Continuation Order Plan is available for this series. A continuation order will
bring delivery of each new volume immediately upon publication. Volumes are billed
only upon actual shipment. For further information please contact the publisher.

ENZYME INDUCTION

Edited by

DENNIS V. PARKE

University of Surrey
Guildford, Surrey, England

PLENUM PRESS · LONDON AND NEW YORK

Library of Congress Catalog Card Number 74-32541
ISBN 0-306-36506-5

© 1975 Plenum Press, London
A Division of Plenum Publishing Company, Ltd.
4a Lower John Street, London W1R 3PD, England
Telephone 01-437 1408

U.S. edition published by Plenum Press, New York,
A Division of Plenum Publishing Corporation
227 West 17th Street, New York, N.Y. 10011

Printed in Northern Ireland at The Universities Press, Belfast

Foreword

Our present concepts of the regulation of enzyme activity in the cell have been largely based on the extensive body of work which has been carried out with micro-organisms. A distinction between constitutive and adaptive enzymes had already been made well before World War II and work on enzyme adaptation, both in yeast and bacteria, was done by several workers, especially Marjorie Stephenson and her group in Cambridge in the 1930s. In studies starting about 1947 Stanier demonstrated that the oxidation of aromatic compounds by species of Pseudomonas involved the coordinate and sequential induction of a group of enzymes concerned in the orderly catabolism of a substrate which acted as the inducer. The investigations of Umbarger and of Pardee, both in 1956, established the principle, which is now firmly established for almost all anabolic reaction chains, that the first 'committed' step in a biosynthetic pathway is sensitive to feedback control by the final product of the particular reaction sequence. This control can be exercised in two ways. It can either act on the rate of formation of the enzyme or it can affect the activity of the latter without altering the concentration of the enzyme. The idea that regulatory enzymes are inhibited or activated by substances not resembling their substrate in their chemical structures was first put forward by Gerhardt and Pardee in the early part of 1963, but the allosteric concept in its most general form was clearly and elegantly advanced in the classical paper by Monod, Changeux and Jacob in that same year. The relative merits of the concepts of the French workers and those of Koshland have been the subject of continuous debate over the last few years.

The classical model of Jacob and Monod which explains repression of enzyme formation in terms of an interaction of a specific repressor protein with a specific part of the DNA—the operon region or operator DNA—has stood, at least in its general outline, the test of time.

Several of these repressors have now been isolated and many of the inducers in microbial systems must be presumed to act at the level of the operator DNA.

However, there is now ample evidence that control of enzyme production also occurs at levels other than DNA. Thus it appears that specific aminoacyl-tRNA molecules may affect the levels of the enzymes concerned with the synthesis of the amino acids for which they act as acceptors. Secondly, the amino acids can affect the levels of the aminoacyl-tRNA synthetases. Evidence is now accumulating that both mechanisms operate in bacterial systems and it seems likely that they may also be important in eucaryotic cells. In any case it is clear that regulation of the rate of synthesis of a specific enzyme protein can occur at various points of the total process starting from the transcription of the first message and ending in the release of the peptide from the ribosome.

In animals the possibilities of regulation are somewhat greater than in microbes. First of all eucaryotic cells have some internal compartments with permeability barriers and this has of course been studied particularly well with respect to electron transfer between the cytosol and the mitochondrion. Special shuttle systems are needed to transfer certain intermediates from one compartment to the other, an example being the carnitine system which is used to transfer acetyl groups from the mitochondrion to the cytosol. In connection with at least one of the contributions to this volume it may be relevant to mention that it has been suggested that the effect of haemin, in inhibiting the activity of the ALA synthetase, can be explained by postulating that it interferes with the translocation of the enzyme from the cytosol to the mitochondrion. The enzyme which is presumed to be formed by the cytoplasmic ribosomes must be transferred to the mitochondrion where it combines with the succinyl CoA which is synthesized by the tricarboxylic acid cycle. Another complication is that in a multicellular organism the product of one cell can affect the metabolism of another. Thus hormones and other substances have general and often specific effects on the formation of enzymes and also on their activities.

We have learnt during the last years that proteins, even soluble proteins in the cytosol of the same cells, have life spans varying from a few hours to several days. The concentration of the enzyme at any one time is the result of both the rate of synthesis and the rate of degradation. Any agent, may it be diet, hormone or drug, can affect either the rate of appearance or disappearance of the enzyme, or it may affect the structure of the cell or the concentration of a necessary cofactor or inhibitor of the enzyme. This field of enzyme regulation in animal cells is expanding at a fast rate and at the present time speculation which is largely based on bacterial models sometimes runs too much ahead of experimental facts. A field in which our knowledge is particularly scarce is that of degradation of enzymes. We have no definite evidence that lysosomes are involved, nor do we know whether specific enzymes are involved in the breakdown of certain enzymes. We cannot even be sure that inactivation of an enzyme involves simple hydrolysis to amino acids.

Knowledge of enzyme regulation is obviously essential for a full understanding of cellular differentiation, metabolism of the whole animal and the effect of drugs and other xenobiotics on the organism.

A. Neuberger

Preface

The incentive for putting together this volume came initially from discussions with various colleagues concerning the mechanisms of biological regulation and their possible roles in health and disease. This resulted in the organisation of a colloquium on the subject of Enzyme Induction, under the auspices of the Biochemical Society, which was held at the University of Surrey in June 1972. As the proceedings of this meeting were published only in brief synopses several of the participants agreed with a suggestion to publish a fuller account of their papers and, to make the scene more comprehensive, we persuaded a few other collaborators to join with us. The result is a monograph directed primarily to a consideration of the importance of enzyme induction to man, but relating this phenomenon to those areas where knowledge is more explicit and detailed, namely the simpler unicellular micro-organisms.

The revelation of the mechanisms of transcription and translation of genetic information into the synthesis of functional and structural proteins—the genetic code, and the concept of enzyme regulation by genomal repression and derepression, the Jacob-Monod operon hypothesis—have led to considerable speculation as to the relevance of these hypotheses, developed from studies with micro-organisms, to biological control in multicellular organisms, and especially to man. Biological phenomena of higher living organisms, such as tissue differentiation, perinatal development, the regulation of cellular metabolism and mechanisms of hormonal control, adaptation to dietary and environmental changes such as exposure to drugs and toxic chemicals, the immune response, and the development of malignancy and possibly aging, are all probably dependent, in some degree, on the regulation of *de novo* enzyme synthesis. The various contributors to this book have reviewed the possible roles of enzyme induction in some of these phenomena, including accounts of their own recent investigations, with the aim of focusing on the importance of this mechanism of biological regulation in higher organisms, particularly man.

In recording my thanks to the many who have contributed to this publication it is both a pleasure and a privilege to acknowledge the considerable encouragement and wise counsel received from Sir Charles Dodds, F.R.S. and Professor Albert Neuberger, F.R.S. who, as Joint Chairmen of the original Biochemical Society Colloquium, so appropriately placed in perspective the various contributions to that meeting. Professor Neuberger has contributed the Foreword which gives essentially the same views he expressed at the Colloquium, and I am most grateful to him for all his kind help and advice. Tragically, the recent death of Sir Charles Dodds prevented him from carrying out a similar intention and it is in gratitude of his considerable help and encouragement that this book is dedicated to his memory.

To the many friends and colleagues who accepted the invitation to participate in this venture goes my sincere appreciation of their scholarship and their dedication to the demanding and ill-rewarded task of writing a scientific review. I hope that seeing their own particular field of research in the context of the wider scene may in itself prove rewarding to them. To the publishers my grateful thanks for their constant advice and patience, and lastly, to my personal assistant, Mrs. Margaret Whatley, my deep appreciation for the endless tasks she has undertaken so competently and selflessly, and which I so often seemed to take for granted.

To the reader, whether biologist or clinician, I hope that these insights into the molecular mechanisms of biological regulation and their importance to man, in health and disease, may be of interest and of some value in his deeper understanding of the mysteries of life.

D. V. Parke

Contents

Chapter 1

Enzyme Induction in Microbial Organisms

Alan Wiseman
*Department of Biochemistry, University of Surrey,
Guildford, Surrey, U.K.*

1.1. INTRODUCTION

The *de novo* biosynthesis of certain enzymes of a microbial cell involved in catabolism of, for example, disaccharides, are subject to a switch-on mechanism elicited by a corresponding component of the surrounding medium. Such a component is by definition an inducer for the corresponding enzymes, which classically are responsible for the uptake and catabolism respectively of an appropriate inducer, whose metabolism by the cell is thereby initiated. For example, maltose induces maltase in yeast while lactose induces β-galactosidase in *Escherichia coli*.

Rapid induction of enzymes able to break down foodstuffs appearing in the environment of the micro-organism is clearly of great ecological advantage. This induction process effects a change in phenotype allowing further production of energy required for metabolism and/or growth. It is existing genetic potentiality that is revealed by the switch-on action of the inducer, and the induced enzyme is produced at about the same rate in all the cells of the culture. Even without enzyme induction by a potential substrate, it is possible to observe somewhat similar changes in the content of such enzymes due to populational adaptation in cultures, based on mutation and selection of mutants able to make these enzymes, and grow therefore on the added substrate. Populational adaptation should be viewed as an alternative to enzyme induction, although selection of cell lines able to produce appropriate induced enzymes needed for growth, is feasible in both microbial and animal cell cultures (and tissues).

In a non-growing microbial culture, enzyme induction can be observed, with addition of inducer, after a short induction period. In an exponentially-growing culture, moreover, the addition of a suitable inducer rapidly elicits maximum production of the induced enzyme(s), although the time taken for the enzyme level (specific activity) to reach its plateau will depend on the mean generation time of the organism (see Fig. 1.1).

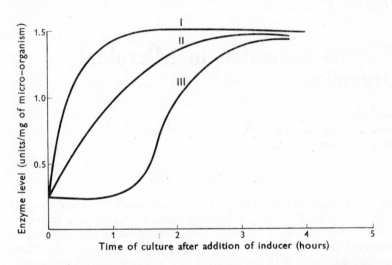

Fig. 1.1. Effect of addition of inducer on induced enzyme production, relative to cell weight, under three conditions of incubation of a micro-organism

   I. Addition of inducer to a rapidly growing exponentially-growing culture causes almost immediate switch-on of production of the appropriate induced enzyme. The enzyme level in units/mg of organism (specific activity), which parallels the differential rate of synthesis in growing cultures, rapidly reaches its maximal constant value. Here the enzyme is being produced in constant ratio to the rest of the cellular protein.

  II. Attainment of maximal enzyme level is slower in more slowly growing cultures, because it is reached after about three mean generation times have passed.

 III. In essentially non-growing cultures, or otherwise stationary phase cultures, a considerable lag (induction period) is possible in reaching full rate of enzyme biosynthesis. Eventually, the same fully switched-on level per cell might be reached, perhaps rapidly (i.e. expressed as specific activity, as the differential rate expression is not meaningful due to lack of growth), if the full synthetic capability can be realized before enzyme biosynthesis prematurely ceases for other reasons, which here are usually the lack of amino-acids if not supplied.

Inducible enzyme systems in microbial organisms display many features of microbiological and biochemical interest. Some of these general features, including definitions of commonly used terms, are considered in sections below as an introduction to the detailed description of progress with special cases. The most intensely studied case is the induction of β-galactosidase (EC 3.2.1.23) in *E. coli* controlled by the *lac* operon, discussed in Section 1.2.

De novo Synthesis

There is no doubt, in the microbial systems investigated, that the major induced enzyme (usually a hydrolase) is formed *de novo*. Enzyme formation from amino-acids occurs, rather than from some inactive peptide or protein precursor existing prior to addition of the inducer to the culture (Monod, 1956). Thus, incorporation of labelled amino-acids into the enzyme (subsequently isolated), occurs with the addition of inducer immediately after transfer of the culture to this labelled medium. Structural analogues of amino-acids prevent the appearance of functional induced enzymes.

The overall rate of enzyme biosynthesis is controlled by the rates of transcription and translation of messenger RNA, which is relatively short-lived in many bacteria—see p. 11 (review books: Wiseman, 1965; Watson, 1970).

Basal Enzyme and Induction Ratio

Even in the absence of inducer it is usually possible to detect some activity of the inducible enzymes in the micro-organism. This is the basal (un-induced) level of the enzyme and there is considerable evidence that basal and induced enzyme are biochemically identical and made by transcription of the identical structural genes. Clearly a slow or occasional transcription of these genes does occur, which is accelerated by the switch-on effect of the inducer. The induced activity divided by the basal activity of a particular enzyme is known as the induction ratio of the inducer used. In rapidly growing cultures this ratio also refers to the relative rates of biosynthesis of that enzyme observed in the culture. In stationary-phase cultures, enzyme biosynthesis may cease while the enzyme activity may remain high for some time. Here the enzyme is not diluted out, on a per gram dry weight basis, due to lack of growth.

It is a common misconception that the induction ratio for all microbial inducible enzymes is extremely high; one thousand-fold usually. Investigators achieving up to about ten-fold induction in rat liver of tryptophan pyrrolase induced by tryptophan (*via* cortisol action), or of cytochrome p-450 induced by drugs, might reasonably feel that a two orders of magnitude difference from microbial systems may require an entirely different mechanism (see other chapters). In fact, the induction ratio is highly dependent on the system and the particular inducer employed and even for β-galactasidase of *E. coli* the range quoted is from 232 for lactose to, admittedly, as high as 1830 for *n*-propyl galactoside (induction profile tabulated in review by Richmond, 1968). The range with penicillinase of *Staphylococcus aureus* is about 30–50 (Richmond *et al.*, 1964), and it may be as low as 3–8 for penicillin β-lactamases in Gram-ve bacteria (Smith, 1963). Induction ratios in eukaryotic cells such as yeasts may be even lower, depending on the yeast strain used. An induction ratio for maltase by maltose of about 5 was observed in stationary phase brewers' yeast, *Saccharomyces cerevisiae* (Harris and Millin, 1963), and was reduced to about 2.5 in our studies on α-glucosidase induction by maltose in exponentially-growing brewers' yeast (A. Wiseman and T-K. Lim, unpublished). Induction ratio is of course dependent on basal level, and that may be higher in some organisms than others. Such studies are of prime importance in controlling industrial fermentations such as brewing.

Gratuitous Induction

Certain substances related structurally to the natural inducer expected in a system may be much better inducers, whether they serve as a productive substrate of the inducible enzyme or not. For example, yeast β-glucosidase is induced, rather poorly, by the expected substrate cellobiose (induction ratio 13). Methyl-β-glucoside, however, is a much better inducer of the enzyme (induction ratio 500), and in fact is also hydrolysed much faster than cellobiose (Halvorson, 1960). Nevertheless, thiomethyl-β-glucoside is not hydrolysed while being an excellent inducer. Such induction, obtained "free" of hydrolysis of inducer is termed gratuitous induction. Explanations of inducer effect could not therefore depend on concepts of substrate breakdown, or even on substrate affinity, as induction is achieved at much lower concentrations of inducer than are needed to half-saturate at least the active site of the enzyme. Very high affinities are involved in the action of normal or gratuitous inducers, involving direct complexing with the genome DNA (see on).

Simple kinetic studies on enzyme induction usually involve the use of such gratuitous inducers. Under these conditions, enzyme induction can be studied without a progressively decreasing inducer concentration due to its cleavage by the induced enzyme. Also, by employing some other energy source (but not one simple enough to cause catabolite repression of the inducible enzyme—see p. 10), the production of energy for enzyme biosynthesis is not rate-limited by the supply of the inducible enzyme otherwise needed for example, to produce the required glucose from maltose.

Kinetics of Induction

With excess gratuitous inducer in exponentially-growing cultures of bacteria or yeast many induced enzymes, e.g. β-galactosidase (*E. coli*) and α-glucosidase (*S. cerevisiae*) can be shown to have their biosynthesis switched on at full rate almost immediately by the addition of this inducer (Duerksen and Halvorson, 1959). The level of the enzyme rises rapidly in the culture exactly in step with the new protein (and total cell material) being made by cell growth. The rate of increase of enzyme activity (or weight) in batch culture equals the rate of increase in total dry weight of cell material in the culture, so that this ratio is constant and is usually referred to as the "differential rate of synthesis" of that inducible enzyme. (Expressed graphically, the differential plot (see Fig. 1.2) is useful in showing intersecting straight

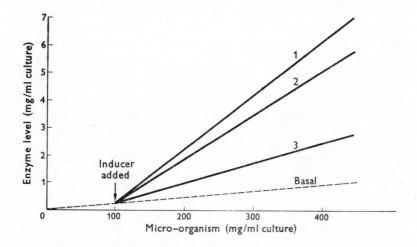

Fig. 1.2. Differential rates of synthesis of an induced enzyme in a rapidly growing microbial culture—calculation of induction ratio (IR) for each of three inducers

From the slope of the appropriate line, the differential rate of synthesis expressed as a percentage of microbial weight, is approximately:

Basal	0.2%	
Inducer 1	1.4%	IR 7
Inducer 2	1.2%	IR 6
Inducer 3	0.5%	IR 2.5

Switch-on of enzyme synthesis can be almost immediate on addition of the inducer, so that the enzyme is immediately (or soon) being produced in constant ratio to the rest of the cellular protein in the growing culture. The differential rate of synthesis, and the specific activity (units/mg of protein) of the induced enzyme is constant for each inducer.

lines, e.g. under more complex conditions of culture.) This value could be, say, 1% so that during exponential growth, e.g. fully 1% of the weight of the new cell material is due to this newly synthesized induced enzyme, in any interval of time. Differential rate of synthesis is expressed by some authors as percentage enzyme activity of the culture (or per ml of culture) over total protein in the culture (or per ml of culture), ignoring the low basal level of inducible enzyme and the low culture density at the time of addition of inducer. Of course, if expressed as enzyme activity per mg of cells, this is the usual mode of expression of the specific activity value for the enzyme in the culture. This term is used for non-growing cultures (see Fig. 1.1).

It might be noted here that some authors also used the term, differential, to refer to a change in induced enzyme level with respect to time, that is referring to the rate of enzyme biosynthesis as such and not to its constant ratio in the culture. Also, induced enzyme specific activities may be given as activity per mg dry or wet weight of cells, or even as activity per ml of culture. Another feature that makes comparisons difficult is that the activity of intracellular enzymes usually can be measured only after some technique for disruption of the cell (Wiseman, 1969). The various popular methods for disruption of bacteria and of yeasts are often harsh and lead to destruction of some of the enzyme to be assayed. Also, assay-increment in homogenates must be clearly distinguished from enzyme solubilization in techniques where supernatant fractions only are assayed for enzyme activity (Wiseman and Jones, 1971). The sub-cellular location of the enzyme is of prime importance in controlling its accessibility to substrate, for cell wall enzymes such as yeast invertase are always fully available for assay, although insoluble.

Sequential Induction

The classic example is in *Pseudomonas fluorescens* where successive steps occur in the oxidation of the indole ring of tryptophan to β-keto-adipic acid. Here, seven enzymes are induced, each in turn by the intermediary metabolite, the potential substrate, produced by the enzyme previous in the metabolic pathway. Switch-on of biosynthesis of the required enzyme occurs when the appropriate concentration of potential substrate is reached, to effect overall the efficient degradation of tryptophan. This induction by intermediates in the pathway allows growth on any such intermediate without wasteful production of earlier enzymes of this complex pathway.

Nevertheless, recent work on *Pseudomonas* species have revealed more complicated control mechanisms, which allows these organisms to grow on a variety of aromatic chemicals, e.g. D-mandelate in *Ps. putida*. One of the intermediary metabolites here, *cis,cis*-muconate was shown to be the inducer of the enzyme that produces it. The rest of the pathway follows the usual sequential induction, however (see review by Ornston, 1971). Other work reported in this area on the induction of aliphatic amidases of *Ps. aeruginosa* (Clarke, 1970), is of considerable industrial interest in relation to enzyme production.

Deadaptation and Crypticity

Rapid destruction of the induced enzyme can be achieved by incubation of induced cells in buffer solution, i.e. without growth being possible, in the absence of inducer. In growing cultures, removal of inducer will switch off the appropriate messenger RNA production so that enzyme biosynthesis on preformed messenger RNA will decrease at a rate dependent on the stability of this template in that organism (also, further growth will dilute out the enzyme in the culture). An analogous effect can be achieved by using an inhibitor of messenger RNA synthesis such as actinomycin D, although its use with microbial cultures is limited by the inability of this inhibitor to enter the intact cell. With cells incubated in buffer solution, the enzyme itself may be rapidly destroyed, not just diluted out, by little understood energy-dependent degradation processes—probably in addition to decay of the messenger RNA required for further biosynthesis of the enzyme. For example, if maltose-grown yeast is incubated aerobically in phosphate buffer solution containing 3% glucose, the maltase activity is lost within two hours (Halvorson, 1960). Under these conditions the maltose active-transport system, maltose permease (normally inducible with maltase, an example of coordinate induction— see pp. 8–9 is lost more rapidly than the maltase. Maltose cannot be used by a cell lacking the maltose permease even though ample maltase, now cryptic (hidden), can be found inside the cell after disruption. Bacterial permeases, including β-galactoside permease, have been reviewed in detail (Cohen, 1968). Recent work has noted the kinetics of β-galactoside permease synthesis in *E. coli* (West and Stein, 1973)—see p. 15.

Co-ordinate Induction: Jacob–Monod Explanation

The switch-on mechanism of inducers caused the appearance of two or more enzymes in constant molar ratio usually equal, in many systems (the activity measured will depend on respective turnover numbers). Maltase and maltose permease are induced together by maltose in yeast. In the *lac* operon of *E. coli* with inducers of β-galactosidase, β-galactoside permease and the enzyme of essentially unknown function, galactoside transacetylase, are also produced.

The latter case was the famous classic work of Jacob and Monod who showed, using now established techniques of microbial genetics, that there was negative control of expression of the contiguous

structural genes (in the operon, consisting of a cistron for each enzyme—
see Fig. 1.3), concerned with the biosynthesis of these enzymes. (See
definitive review by Monod *et al.*, 1962; also the more recent review
with the emphasis on microbial genetics by Richmond, 1968.) This
well-known model (Fig. 1.3) is widely accepted, and certainly applies
to many bacterial systems, but not to all of these. With negative
control, the operon is normally switched off by the action of a protein
repressor molecule acting at the operator. Positive control, however,
has been suggested more recently for the operon that controls the
biosynthesis of enzymes in *E. coli* responsible for the pathway of
arabinose degradation. This system is discussed on p. 16.

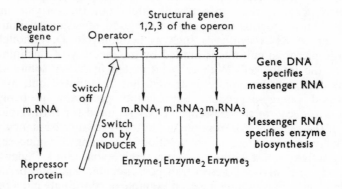

Fig. 1.3. Basic "Jacob–Monod" hypothesis for negative control of induced
biosynthesis of some enzymes in bacteria (*E. coli*)
        The inducer removes the repressor protein from the operator gene
        causing switch-on of structural genes of the operon, coding *via* messenger
        RNA for the biosynthesis of the corresponding enzymes subject to co-
        ordinate induction by that inducer.

The regulation of induced transcription of the appropriate genes
is a topic that includes much of the vast area of studies in protein
biosynthesis in microbial organisms and others.
        Most of the well-characterized operons in bacteria are each under
the negative control of a protein regulator molecule (repressor),
which switches off the operator by combining with the DNA at this
regulatory point. The repressor protein is synthesized in the usual way
on a messenger RNA template (in polysomal location) made by a
distinct regulatory gene. The mode of action of this gene does not
require it to be closely linked to the operon it represses, whereas
the operon itself consists of the contiguous segments of DNA forming,
in linear sequence, one region called the promoter, next the operator

region and then the appropriate linked structural genes that show co-ordinate induction (Figs 1.3 and 1.4). It is the combination of inducer with this protein repressor that removes the latter from its DNA-binding site on the operator, presumably by eliciting a conformational change that lowers its affinity for this site—as can be shown *in vitro*. Separate loss by point mutation has provided evidence, in the *lac* operon of *E. coli*, for a separate promoter region. It is this region that is responsible for the rate of initiation of transcription of messenger RNA by RNA polymerase (containing a sigma factor),

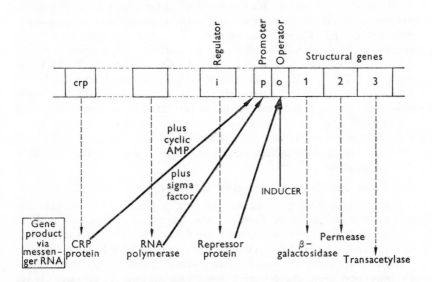

Fig. 1.4. Regulation of transcription of the *lac* operon of *E. coli*
    The genes marked 1, 2, 3, *o*, *p* and probably *i* are contiguous in this organism. Additional control (positive) by cyclic AMP plus its receptor protein (CRP) is of particular interest.

a concept that has extended the original Jacob–Monod theory. In this *E. coli* system the operator region only recognizes and binds the repressor protein molecule (see below). More recently, the essential role of cyclic AMP, and its binding protein, in promotor effectiveness has been recognized and studied *in vitro*. Full details are given in Section 1.2.

Structure of the *lac* Repressor (see p. 15)

    This protein has been isolated from *E. coli* and is recognized by its affinity for a typical inducer (gratuitous) of the *lac* operon, isopropyl-β-D-thiogalactoside (Gilbert and Muller-Hill, 1966). It has a molecular weight of 152,000 and is composed of four subunits, each binding one molecule of inducer, probably *in situ* while on the DNA. The repressor protein is synthesized constitutively, i.e. it is not subject to any other repressor protein action on its own messenger RNA synthesis—a relatively simple system in this microbial organism.

End-product Repression of Enzymes of Biosynthetic Pathways

The Jacob–Monod theory of negative control of induction also includes an explanation for the usual repression of all the enzymes of a synthetic pathway by the end-product of that synthetic pathway, such as an amino-acid. Biosynthesis of these enzymes is prevented because the end-product activates a latent repressor that subsequently switches off the appropriate operon. A classic example here is the repression of the synthetic pathways for certain amino-acids in *E. coli* by addition of these amino-acids (see Cohen, 1968; also Epstein and Beckwith, 1968). This phenomenon is quite distinct from feed-back inhibition of these enzymes, where the amino-acid causes a rapid halt to the action of the first enzyme in the pathway, without preventing its biosynthesis.

Derepression of enzyme biosynthesis caused by removal of an end-product or its controlling influence, should not be equated with induction, as no inducer is involved.

Catabolite Repression of Inducible Enzymes

Certain very simple substrates, especially glucose, will prevent the co-ordinate induction of known inducible enzymes, such as maltase and maltose permease in yeast or the *lac* operon products in *E. coli*, by the usually highly effective inducers (Paigen and Williams, 1970).

This "glucose-effect" was for many years considered to be unrelated to induction and repression phenomena discussed on pp. 7–10, despite extensive investigation often in relation to diauxic growth observed on glucose–disaccharide mixtures. Diauxic growth of yeast is of particular importance in brewing. Here catabolite repression by glucose causes the glucose present in the brewers' wort to be used up before induction of maltose permease and maltase can occur to use the large quantity of maltose present. Two stages of yeast growth, diauxie, are therefore observed here.

Recent work has shown that glucose lowers the level of cyclic AMP in the cell. This reduces the rate of initiation of transcription of certain operons, *via* the promotor region, that make messenger RNA for the production of inducible enzymes (see Section 1.2).

Transcription of Genes, Factors and Inhibitors

RNA polymerase haloenzyme of *E. coli* has been fractionated by chromatography on phosphocellulose into apoenzyme and another protein (eluted), named sigma factor (Burgess *et al.*, 1969). This factor transiently associates with RNA polymerase and confers on it an ability to discriminate between promoter sites of operons, so that positive control is exerted over messenger RNA synthesis. No doubt a variety of such sigma factors are involved in regulation of microbial enzyme biosynthesis. Many sigma factors are thought to be involved in induction of enzymes and proteins in viral infection of a bacterial cell. Another factor, rho, directs the termination of the action of RNA polymerase after completion of appropriate operons (Roberts, 1969).

Most studies on inhibition of transcription have used actinomycin D (combines with DNA), and use of this inhibitor is usually assumed to block transcription in all organisms, including animals. Growing *E. coli* does not take up this compound, but *tris*-EDTA treated cells can (Leive, 1965) and so will yeast protoplasts (Van Wijk *et al.*, 1969). Recently, rifampicin has been shown to be a competitive inhibitor of RNA polymerase for its promoter binding site (see review by White, 1970)—see p. 19 for another new inhibitor.

Translation of Messenger RNA—Factors and Inhibitors

Several enzyme factors required for translation of messenger RNA, including its initiation, have been discovered in *E. coli*. These are usually referred to as f1, f2, f3. (See review by Wiseman, 1971b.)

A large number of inhibitors have been found to inhibit overall translation in microbial organisms, by affecting one or more steps, e.g. transfer (of amino-acid from its transfer RNA complex), chain elongation or translocation of ribosomes along the messenger RNA. For example, studies have been reported on tetracyclines (Cerna *et al.*, 1969), pactamycin (Cohen *et al.*, 1969), streptomycin (Ozaki *et al.*, 1969)—all of which mainly inhibit transfer. Chain elongation, mediated by peptidyl transferases, is inhibited by puromycin, chloramphenicol, sparsomycin (Monro *et al.*, 1969) and lincomycin; and translocation by the steroidal antibiotic fusidic acid. Actidione (cycloheximide) is often used to inhibit cytoplasmic protein synthesis in yeasts.

All of these compounds inhibit protein biosynthesis in animals or animal systems. Thus, cycloheximide inhibits several animal systems (Davies and Garren, 1968), and so does puromycin. Chloramphenicol is of particular interest, for although bacteria are much more sensitive to its effects at low concentrations, this antibiotic inhibits mainly mitochondrial protein synthesis and induced protein synthesis (including antibody synthesis in lymph node tissue culture) in eukaryotes (Beard *et al.*, 1969), including yeast (A. Wiseman and T-K Lim, in the press).

The effect of such inhibitors in animal systems is generally non-specific, with toxic side-effects, making strict interpretation by analogy with microbial studies hazardous.

1.2. CONTROL OF THE *LAC* OPERON IN ESCHERICHIA COLI

Much of the recent interest in this area has centred around the role of cyclic AMP (and its receptor protein) as an additional stimulator of the *lac* operon; and especially its ability to remove catabolite repression by glucose. The development of these studies in *E. coli* is considered in the sections that follow (for an earlier review see Perlman and Pastan, 1971).

Effect of Cyclic AMP *in vivo*

The presence of cyclic 3',5'-AMP (at about 10^{-4} M maximum) had been reported some years ago in *E. coli* (Makman and Sutherland, 1965). Addition of glucose to these cells caused a rapid fall in cyclic AMP level. Later studies noted that cyclic AMP appeared to be without effect on β-galactosidase synthesis in normal growing *E. coli*, probably due to failure to enter the cells. However, with *tris*-EDTA-treated non-growing cells it was possible to restore the differential rate (calculated on rate of protein biosynthesis, by uptake of labelled amino-acid, see Section 1.4), of induced β-galactosidase synthesis to control level despite the presence of glucose (10^{-2} M), which otherwise caused 68% catabolite repression (Perlman and Pastan, 1968). In this work the gratuitous inducer, isopropyl-β-D-thiogalactoside was used at high concentration (5×10^{-4} M), with cyclic AMP at 10^{-3} M for full effect. Tryptophanase induction, by tryptophan, was also stimulated in the presence of catabolite repressors, but here a lower concentration of cyclic AMP was required (see p. 14).

A very similar result was reported soon afterwards using normal growing *E. coli* (Ullmann and Monod, 1968). Under these conditions removal of two-thirds of the catabolite repression due to glucose was achieved by 5×10^{-3} M cyclic AMP, using isopropyl-β-D-thiogalactoside (5×10^{-4} M) as inducer of β-galactosidase. Similar studies agreed with these findings, while work with mutant strains showed that at least the main site of action of cyclic AMP was the *lac* operon promoter, where it presumably increased the rate of transcription, to form messenger RNA (Pastan and Perlman, 1968; Perlman et al., 1969). More general studies extended this work to show that glucose repression of many enzymes of *E. coli* could be decreased by cyclic AMP, and furthermore that this applied to β-galactosidase biosynthesis in many bacterial species (De Crombrugghe et al., 1969; also Carpenter and Sells, 1973). In agreement with this work, cyclic AMP-degrading enzyme (Monard et al., 1969) and adenylate cyclase (Tao and Lipmann, 1969; Ide, 1969) have been isolated from *E. coli*. Furthermore, glucose exhaustion has been shown to derepress adenylate cyclase activity in intact *E. coli*, using a novel *in vivo* assay for the enzyme (Peterkofsky and Gazdar, 1973). Many attempts have been made to reproduce this effect of cyclic AMP in existing crude *in vitro* systems from *E. coli* that will respond to inducer. Non-specific effects, however, and problems of assay of specific *lac* messenger RNA made on *lac* DNA, led to many difficulties—see below.

Studies With *in vitro* Systems, from *E. coli* and Associated Transducing Bacteriophages

Attempts to show stimulation of biosynthesis of *lac* operon messenger RNA *in vitro* must rely on the estimation of this messenger RNA. The best way of recognizing it is probably to note the increase, due to induction, in phenol-purified tritiated RNA (using ^{3}H-ATP and other nucleoside triphosphates with *lac* DNA template). This is assayed by its hybridization ability to the correct single strand prepared from *lac* DNA. Pure *lac* DNA, however, has been mainly unavailable, it seems, although its isolation from transducing phages has been reported (Shapiro *et al.*, 1969).

Transducing phages have been isolated that carry the *lac* operon of *E. coli*. Thus, the phages λ plac5 and ϕ80plac1 contain the *E. coli lac* operon but no other bacterial genes in common (Shapiro *et al.*, 1969). Furthermore, there is no homology of DNA sequence in these phages that might be mistaken for *lac* after transcription, so they can be used *in vitro* as if they contain only *lac* DNA. One problem, "read-through" transcription of *lac* operon plus other operons too, can be reduced using the termination and release factor (rho factor: Roberts, 1969). It is of course necessary to include RNA polymerase in the *in vitro* system, with its sigma-factor fraction added separately if desired, but excluding extraneous initiation factors and ribosomes associated with translation of messenger RNA. The detection of *lac* messenger RNA made in such systems was achieved by using DNA from the other phage, the one not used to make the messenger RNA. This DNA was subjected to a strand separation process and then immobilized on nitrocellulose filters prior to hybridization with the tritiated RNA extract (Arditti *et al.*, 1970; Eron *et al.*, 1971). These workers reported that the *lac* messenger RNA synthesized constituted a small fraction of the tritiated RNA produced in their *in vitro* system (evidence also comes from deletion mutants). Nevertheless, considerable stimulation (about five-fold) was observed by addition of cyclic AMP plus cyclic AMP receptor protein (abbreviated to CAP or CRP usually). The authors commented on the unexpected failure of promoter deletion mutants of the phage used for transcription to alter their results.

Similar techniques have been employed by other groups of workers (Varmus *et al.*, 1970a, 1970b; De Crombrugghe *et al.*, 1971). The latter group used a mutant phage, λh80dlacp[5], which is a much improved template for induced β-galactosidase synthesis in complete systems *in vitro*. The synthesis of this enzyme, in a complete transcription-translation system, is greatly increased by the addition of cyclic AMP (plus its receptor protein) and is repressed when *lac* repressor protein is added. The maximum stimulation of *lac* messenger RNA synthesis by cyclic AMP + CRP was about 20-fold, with half-maximal effect of cyclic AMP at the low level, compared to *in vivo* stimulation, of 2.5×10^{-6} M; reflecting the dissociation constant of the cyclic AMP–CRP complex. Up to 18 % of the RNA made *in vitro* was shown to be the correct strand of *lac* messenger RNA. It has been recently noted, however, that not all apparently undergraded messenger RNA, detectable by hybridization techniques, is necessarily functional (Puga *et al.*, 1973).

Evidence is also presented (De Crombrugghe *et al.*, 1971) that cyclic AMP plus CRP are required for binding of RNA polymerase to the *lac* promoter (in fact, CRP binds tightly to *lac* DNA). This conclusion was derived from studies using a competitive inhibitor for RNA polymerase (rifampicin—also known for its antiviral activity). It is further suggested here that the formation of a complex between CRP and *lac* DNA promoter region is probably required for the binding of RNA polymerase to another part of this *lac* DNA promoter region. A similar conclusion was reached by other workers (Eron and Block, 1971).

Another report on this topic had concluded, however, that cyclic AMP reversed catabolite repression *in vivo* by an unblocking effect on translation of *lac* messenger RNA (Aboud and Burger, 1970). This was in agreement with a similar conclusion for β-galactosidase induction in *E. coli* (Yudkin and Moses, 1969) and for tryptophanase induction by tryptophan in *E. coli in vivo* (Pastan and Perlman, 1969b). Evidence for such conclusions is usually derived from the use of inhibitors of translation and transcription *in vivo*, which can present difficulties of interpretation. Nevertheless, transcription and translation are tightly coupled processes in bacteria, and both effects may occur, with one predominating in some systems.

Purification of Cyclic AMP-Receptor Protein (CRP)

Partial purifications by the usual biochemical techniques were reported (Zubay *et al.*, 1970; Emmer *et al.*, 1970—purification $\times 100$; extensive purification, Riggs *et al.*, 1971; full purification to homogeneity by Anderson *et al.*, 1971).

The latter authors report that purified CRP has a molecular weight of about 45,000, a value based on sedimentation equilibrium data, and is composed of two apparently identical subunits. Isoelectric focusing data showed the protein to have an isoelectric point of 9.12.

CRP must have a specific attachment site that recognizes a particular deoxyribonucleotide sequence characteristic of a part of the promoter region of all operons that show stimulation by CRP-cyclic AMP. There is no unequivocal evidence for binding of CRP or cyclic AMP to RNA-polymerase itself.

Interaction Between *lac* Repressor Protein and *lac* DNA Operator (see p. 9)

It has now been suggested that *lac* repressor protein has an N-terminal sequence of about fifty amino-acids that bind directly to *lac* operator DNA (Adler *et al.*, 1972). Possible sequences and models have been constructed by these authors. Structural determination by X-ray crystallography, on a suitable crystal, is yet to come.

Presumably, the differing induction ratios for various inducers reflect their particular affinity for the *lac* repressor protein, and each is able to maximally remove a characteristic proportion of this blocking repressor.

Fig. 1.4, page 9, presents a diagrammatic summary of macro-molecular interactions at the *lac* operon.

1.3. SOME ADVANCES WITH OTHER MICROBIAL SYSTEMS

Permease Systems (see p. 7)

The lactose-transporting active-transport system has recently been isolated and characterized from *Staphylococcus aureus*. This organism takes up lactose, and other sugars, by action of a phosphotransferase in the cell membrane. This enzyme system catalyses the transfer of a phosphoryl group from phosphoenol pyruvate to β-galactosides with the formation of galactoside-6-phosphate (Simoni *et al.*, 1973; Simoni and Roseman, 1973). The phosphotransferase system was shown to require four protein components. One of the phosphocarrier proteins involved was purified to homogeneity. The system was inducible by lactose or galactose (McClatchy and Rosenblum, 1963), although the level of only two of the four components was increased. Earlier studies on β-galactoside permeases of *E. coli* have been thoroughly reviewed (Cohen, 1968) and a recent paper has dealt with the kinetics of this system (West and Stein, 1973).

The *gal* Operon

In *E. coli*, the three enzymes responsible for substrate inducible fermentation of galactose, are galactokinase, galactose-1-phosphate uridyltransferase and UDP-glucose epimerase. This is another example of co-ordinate induction, with the structural genes for the production of these enzymes forming one operon (Buttin, 1963). The *gal* repressor protein has now been isolated and subjected to a four-step purification procedure (Nakanishi *et al.*, 1973). Like the *E. coli lac* system, the *gal* system requires cyclic AMP and cyclic AMP receptor protein. Also, the use of appropriate bacteriophage has led to isolation of most of the *gal* operon with phage (λp gal 8) DNA, a system suitable for *in vitro* production of *gal* messenger RNA (detected by standard hybridization techniques). The inducer, galactose, was shown to overcome the specific switch-off action of *gal* repressor, on the *gal* operator, with half-maximum effect at 0.5 mM.

Positive Control Systems

One apparent exception to the classic Jacob–Monod mode of negative control of an inducible system, is found with the arabinose operon of *E. coli*.

The first three enzymes for arabinose utilization here are an epimerase, an isomerase, and a kinase. There is also a linked regulator gene (ara C) which if deleted causes the non-functioning of the operon, and also of an unlinked arabinose permease.

The usual tests were done with constructed partial diploid organisms carrying two sets of these genes (see review by Clarke, 1972). Evidence based on data from such microbial genetics experiments suggested that the wild-type organism produced an activator protein (p2) required for the initiation of the *ara* operon. It has been suggested that the regulator gene produces a normal repressor protein (p1) but arabinose, the inducer, instead of just lifting its repression at the operator, converts it to an activator protein (p2) (Englesberg *et al.*, 1969). This scheme would require the *ara* operon to be under the dual, and opposing, control of a single gene through a first product and its derivative.

Such a positive control system is now not so surprising in view of the positive control seen at the promoter for initiation of transcription of the operon by cyclic AMP receptor protein (switched on by cyclic AMP). Catabolite repression of the *ara* operon is indeed lifted by cyclic AMP *in vivo*, and with *in vitro* (θ80dara DNA phage) systems (Zubay *et al.*, 1971; Greenblatt and Schleif, 1971).

Enzymes of Industrial Interest

A large number of amylases and proteases have been produced from bacteria and fungi for industrial uses, ranging from brewing and baking to enzyme washing powers (review by Wiseman, 1971a; Wiseman, 1973). For example, effective inducers of α-amylase biosynthesis in *Aspergillus oryzae* include starch, glycogen and maltotriose (Banks *et al.*, 1967). Much of the interest in industrial production is in the selection of a mutant that produces the enzyme with exactly the desired characteristics, preferably constitutively. Substrate-induction conditions are usually used anyway, for example in the growth of bakers' yeast, intended for production of a usually constitutive invertase; this is carried out on molasses (sucrose). Nevertheless, large increases ($\times 30$) in invertase levels have been claimed using this substrate-induction under large-scale conditions. Much better invertase induction is achieved in *Pullularia pullulans* by using sucrose monopalmitate as the inducer rather than sucrose (80-fold better). The former is slowly hydrolysed to sucrose, and catabolite repression by build up of metabolites of sucrose is avoided. Similar considerations apply to induction of cellulase or dextranase (Reese *et al.*, 1969).

Many other cases of industrial interest have been studied in some detail, for example the substrate induction of pectinases in *A. oryzae* and other filamentious fungi (Nyiri, 1968; Zetelaki-Horvath and Bekassay-Molnar, 1973). Another most interesting system is the induction of glucose isomerase in *Streptomyces albus* by xylose (Takasaki *et al.*, 1969). Glucose isomerase is of especial interest because of its ability to convert glucose to fructose and so effectively produce invert sugar (by 50% conversion), which is sweeter than glucose. The control of fermentation by "inducers" has been recently subjected to a most interesting review (Perlman, 1973). A mathematical treatment of industrial induction and repression of lipase has been published (Van Dedem and Moo-Young, 1973).

α-Glucosidase (Maltase) in Yeast

The induction in yeast of maltase and maltose permease by maltose has been discussed briefly above (see pp. 1, 4 and 7). Figure 1.5 relates to recent findings on the effect of the glucose concentration, in catabolite repression of 4% maltose induced α-glucosidase biosynthesis in non-growing *S. cerevisiae* (A. Wiseman and T-K. Lim, in the press). Rapid induction is seen in the presence of 4% maltose plus 0.1% or 1% glucose, but with a longer induction period for the latter. None occurs with 4% maltose plus 2% or 5% glucose. No induction was observed with 4% maltose or 0.1% glucose alone. Surprisingly, at 4% maltose plus 10% glucose a time-dependent loss of enzyme is observed. Such loss of enzyme has been reported previously in deadaptation conditions (using no maltose) in the presence of 3% glucose (see p. 7). Little is known about the cause and control of enzyme degradation *in vivo*.

An interesting study had been reported using protoplasts of *S. carlsbergensis*, which are amenable to uptake of inhibitors such as actinomycin D (Van Wijk *et al.*, 1969). These workers investigated the cause of catabolite repression by glucose in a maltose-induced, essentially non-growing, suspension of protoplasts. Maltase (α-glucosidase) was assayed here too by using the chromogenic substrate, *p*-nitrophenyl-α-glucoside (PNPG). The ability of maltose to induce α-glucosidase was controlled by the concentration of glucose present.

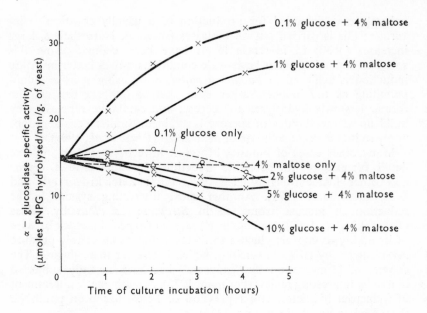

Fig. 1.5. Induction and catabolite-repression, of α-glucosidase caused by
maltose plus glucose in non-growing *S. cerevisiae* (N.C.Y.C. No. 240)
(A. Wiseman and T-K. Lim, *Biochem. Soc. Trans.*, in the press)

The yeast was grown in complex medium containing 1% glucose,
washed thoroughly and resuspended in phosphate buffer pH 6.5, as
indicated. α-Glucosidase was assayed, after ultrasonic disruption of
yeast, by the *p*-nitrophenyl-α-glucoside (PNPG) technique (Wiseman and
Jones, 1971). Induction with 4% maltose required the presence of 0.1%
glucose. Catabolite repression by glucose at 2% and above prevented
the rise in enzyme level. In fact, 10% glucose (plus 4% maltose) clearly
caused a deadaptation effect with genuine loss of the enzyme, despite
the presence of the inducer (maltose).

They reported that 1% glucose caused catabolite repression, but that
0.1% allowed much improved induction by maltose. This effect was
considered to be due to derepression of the appropriate operon(s).
The rate of change of α-glucosidase activity with time, referred to
here as the differential curve when plotted (compare differential rate
in growing organisms, Section 1.4), became zero after about 2 hours,
so that in this system the rate of α-glucosidase biosynthesis increases
relatively slowly to its fully switched-on rate, even though a supply of
amino acids is provided.

As in rapidly growing cultures, stationary phase or otherwise non-
growing cells do not preferentially synthesize the induced enzyme. A

range of enzymes, characteristic of the organism under those conditions, is made, however, with the addition of the switched on induced enzyme. It is thought that amino-acids are provided for enzyme biosynthesis, in the absence of nitrogen compounds normally required for growth, by degradation of cellular protein (Richards and Hinshelwood, 1961), although eventually this source can be exhausted (Strange, 1961). Nevertheless, induction of maltose permease but not maltase was observed prior to addition of an amino-acid supply (Harris and Millin, 1963).

It is of interest that glucose itself at 0.1% was able to derepress α-glucosidase synthesis (Van Wijk et al., 1969), although this synthesis soon ceased due to the removal from the medium of all the glucose, now the sole carbon-source in this experiment. Actinomycin D is able to enter these protoplasts and presumably prevent gene transcription to form messenger RNA. When added soon after the maltose, actinomycin D prevented the appearance of α-glucosidase but had little effect later when presumably a stable messenger RNA had already synthesized. Here the system was sensitive to inhibitors of translation of messenger RNA, such as puromycin or cycloheximide (see pp. 10–11). Glucose at 0.1% was shown to derepress α-glucosidase synthesis despite the inhibition of messenger RNA synthesis by actinomycin D. It is possible that derepression involves the initiation of translation of some existing messenger RNA for α-glucosidase. Glucose at 1% prevented both transcription and translation, each to different extents. These data on transcription and translation effects of maltose and glucose in yeast illustrate the complexities of interpretation even in microbial systems where manipulation of media can be readily made. It is not yet clear, however, whether the specificity of inhibitors such as actinomycin D can be relied upon even in these relatively simple microbial systems. A new inhibitor seems more promising—lomofungin from *Streptomyces lomodensis* (Fraser *et al.*, 1973). This inhibitor is able to rapidly enter cells of *Schizosaccharomyces pombe* and causes rapid inhibition of ribosomal RNA and poly-disperse RNA synthesis.

More recent studies by this group (Van Wijk and Konijn, 1971; see also permease study De Kroon and Koningsberger, 1970, see p. 15) have involved, once again, cyclic AMP (see pp. 12–13). Under conditions of catabolite repression by 2% glucose in *S. carlsbergensis*, intracellular levels of cyclic AMP were lowered six-fold. Subsequent adaptation to maltose utilization was achieved in a medium containing 2% maltose plus 0.1% glucose, and supplemented to 1% with a casein hydrolysate. This caused the appearance (after a lag phase of about 1 hour) of high maltose fermenting ability in 2–4 hours (see Fig. 1.1), with a parallel restoration of cyclic AMP level. Cyclic AMP (plus CRP protein) is likely to stimulate transcription, and possibly translation too (see Section 1.2). No mention was made of the effect of addition of cyclic AMP to yeast or yeast protoplasts under conditions of glucose repression. Cyclic AMP will easily enter only damaged *E. coli* (see p. 12) and yeast too is usually impermeable to negatively charged molecular species. Studies are in progress here on that question, although we have observed no lifting of catabolite repression by 10 mM cyclic AMP, or 10 mM dibutyryl-cyclic AMP in the presence of 2% glucose plus 4% maltose in a non-growing yeast culture. Removal of glucose repression of α-glucosidase synthesis was acheived in yeast protoplasts however (A. Wiseman and T-K Lim, in the press.

Induction of Cytochrome P-450 in Micro-Organisms

Cytochrome P-450, especially of liver microsomal systems, has been subjected to extensive study in relation to its drug metabolizing ability (see Chapter 8). Only one microbial system has aroused much interest to date, that of *Pseudomonas putida*. Growth of this organism on D-(+)-camphor occurs after induction of cytochrome P-450$_{cam}$ (P-448 in fact, Peterson and Griffin, 1972). This type of cytochrome was found as one component of the inducible methylene hydroxylase system that catalyses the hydroxylation of camphor to its exo-5-alcohol, followed by further degradative steps (Hedegaard and Gunsalus, 1965; Katagiri *et al.*, 1968; Gunsalus *et al.*, 1967). The early enzymes of the camphor oxidation metabolic pathway are induced by a variety of bicyclic mono-terpenes, some gratuitously. The principal structural requirement for such induction is for the two 5-membered rings of bornane to be present in the molecule. The very limited specificity of isolated cytochrome P-450$_{cam}$ (Gibson *et al*, 1970), is in distinct contrast to the wider specificity of cytochrome P-450 from animal systems. *In vivo* too, inducibility of liver cytochrome P-450 is seen with a wide range of drugs, anutrients and carcinogens. There is considerable interest in the use of the relatively low-molecular weight cytochrome P-450$_{cam}$ as a model compound, and relatively large-scale isolation and purification of it has been achieved (Peterson, 1971) and it has been crystallized in stable form as its substrate complex with D-camphor (Yu and Gunsalus, 1970).

Cytochrome P-450 has been detected in several other microbial species under some conditions of growth. The role of this enzyme is, then, usually completely unknown. It is in substrate inducible systems, however, that its function as a terminal oxidase can be investigated. Such studies have been done in yeasts and bacteria growing on hydrocarbons.

Some interest has been shown in the cytochrome P-450 concerned in tetradecane metabolism by *Candida tropicalis*, although under similar conditions of growth, *C. lipolytica* does not appear to produce this enzyme (A. Wiseman and C. McCloud, unpublished). Thus, cytochrome P-450 was isolated from *C. tropicalis* grown on tetradecane with high oxygenation and stirring rate (Lebeault *et al.*, 1971; similar system reported by Gallo *et al.*, 1971). A soluble enzyme system containing cytochrome P-450 as one component was subsequently isolated and shown to be capable of oxidizing *n*-alkanes to the corresponding primary alcohol. The system was also able to oxidize fatty acids to their ω-hydroxyl derivative, being especially active in the oxidation of lauric acid. It was also able to achieve the *N*-demethylation of e.g. aminopyrine, hexobarbital and ethyl morphine. Like the liver system this eukaryotic system also requires NADPH. The *Ps. putida* system, and that from *Corynebacterium* sp., uses NADH as the cofactor in the electron transport system. Strains of *Corynebacterium* grown on octane as sole carbon source have been reported to contain cytochrome P-450 (Cardini and Jurtshuk, 1968). Studies in inducibility *per se* have rarely been reported, however.

1.4. CONCLUSION

The recent recognition of the action of cyclic AMP in microbial systems (see pp. 12–15) has once again brought together concepts developed by investigators each in his own field—animal or microbial. This convergence of ideas has happened many times in the past, but usually with the discoveries in microbial systems being extrapolated, however tentatively, to the much more complex and difficult animal systems. Over the last fifteen years, almost without exception, the concepts derived from microbial studies on protein and enzyme biosynthesis have been incorporated into the understanding of animal systems—often after modifications. Messenger RNA, for example, discovered as a short-lived species after viral infection of *E. coli* (Volkin and Astrachan, 1956; 1968), was also detected in normal micro-organisms (Ycas and Vincent, 1960—in yeast). Later, the concept of inherent instability gave way to ideas of template stability (Pitot *et al.*, 1965).

Cyclic AMP seems to stimulate only a relatively few genes (say 1%), even in *E. coli* where ten enzymes have been listed recently (review by Pastan and Perlman, 1970). In other Gram-ve bacteria (Gram + ve data awaited) β-galactosidase only was listed, except for an additional galactokinase in *Salmonella typhimurium*. In all cases, glucose repression caused lowered intracellular cyclic AMP levels, and extra cyclic AMP lifted the repression. Glucose transport into microbial cells is a partially energy-dependent process coupled usually to a hexokinase (glucokinase)-ATP system (yeast) or a phosphenol-pyruvate-phosphotransferase system (*E. coli*). Glucose uptake by *E. coli* has been reported to be slightly increased by cyclic AMP (Moses and Sharp, 1970).

Transport of disaccharides is also a partially energy-dependent process and it is not obvious why glucose transport particularly should lower cyclic AMP levels. Some evidence is available that simple binding of glucose (without phosphorylation) to its transport system causes efflux of cyclic AMP from cells of *E. coli* (Pastan and Perlman, 1969a).

In animal cells, cyclic AMP increases the activity of certain protein kinases that can, for example, cause the activation of phosphorylase kinase or the phosphorylation of histones (see review by Robinson *et al.*, 1968). Cyclic AMP and its receptor protein in micro-organisms might also be involved in such phosphorylation reactions, but this question remains unanswered as no phosphorylation appears to be involved, at least in bacteria.

The difficulty in developing a unified concept of enzyme induction in microbial organisms and in cells of higher organisms is associated with their environment and nature. Microbial cells have evolved with mechanisms that adjust their growth rate to the maximal with whatever potential substrate is present, and the cyclic AMP supply is adequate, except under glucose-repression. Cells of higher organisms usually maintain a stable environment involving cellular interaction, and tissue interaction controlled by hormones that can modify the rates of transcription or possibly translation. Cyclic AMP is formed from ATP by activation of membrane adenylate cyclase when hormones such as thyrotrophin or adrenocorticotrophin bind to their target tissue cells. These cells then produce thyroxine or steroid hormone, respectively.

Nevertheless, consideration of cells as prokaryotic or eukaryotic, means that microbial organisms such as yeasts are profitably considered along with cells of higher organisms, for some aspects of enzyme induction. Certainly the stability of messenger RNA noted in yeast (pp. 17–20) is comparable with that noted in liver (Pitot *et al.*, 1965). Also, the presence of a discrete nucleus raises the usual problems of transport of messenger RNA from nucleus to cytoplasm—unlike bacteria where simultaneous transcription and translation of poly-cistronic messenger RNA is usual. Mammalian genes, especially, exist in several copies it seems, with enormous apparent redundancy of information. In yeast, too, it is usual to have up to six genes for α-glucosidase synthesis, for example.

Few studies of enzyme induction in higher organisms can be compared directly with microbial results. Induction of cytochrome P-450 and also of tryptophan pyrrolase are essentially detoxication reactions effecting the removal (usually) of harmful substances. Interferon induction by viral nucleic acid, reverse transcriptase induction, and induction of other proteins and enzymes are associated with viral attack of animal cells (see review by Gallow, 1971). These systems are hardly comparable in effect with substrate-induced growth caused by the inducer in microbial organisms. In addition, hormonal and developmental effects appear *in vivo*, and over the long periods usually studied induction is certainly not the simple switch-on effect (Jacob–Monod operon model) with constant and maximal differential rate of synthesis seen for certain enzymes in microbial cultures.

Nevertheless, some analogies are possible. Steroid hormones, for example, have been shown to be gratuitous inducers of enzyme biosynthesis by increase of transcription (and possibly translation–Tomkins *et al.*, 1969).

Recent studies have added further complications still. There seems to be little doubt that translation of messenger RNA, as well as transcription, is stimulated by cyclic AMP in the case of *E. coli* tryptophanase (Pastan and Perlman, 1970). Moreover in studies of isocitrate lyase induction by growth of *Chorella* sp. on acetate, catabolite repression by glucose could not be reversed by cyclic AMP nor by the more easily permeable form, dibutyryl cyclic AMP. Messenger RNA is present, but is not translated under conditions of glucose repression—the synthesis of isocitrate lyase stops immediately on addition of glucose (McCullough and John, 1972). One alternative effect claimed for cyclic AMP, which might explain its stimulatory effect on *E. coli* tryptophanase *via* a translation effect, is through its combination with *E. coli* G factor (translocase). Inhibition of the GTPase activity and ribosome-associated degradation of messenger RNA could raise the level of this template molecule and so of the observed rate of enzyme synthesis (Kuwano and Schlessinger, 1970). Now a cyclic AMP–protein complex has also been claimed to control the release of tyrosine aminotransferase from polysomes (Donovan and Oliver, 1972), while other work has shown that cyclic GMP antagonises the effects of cyclic AMP (Kram and Tomkins, 1973).

Much remains to be discovered about the effect of turnover of both messenger RNA and the enzyme itself in regulation of enzyme biosynthesis. Yet, new factors are discovered at frequent intervals. For example, a low molecular weight protein factor (π) increased the effectiveness on all genes of RNA polymerase in yeast (DiManio *et al.*, 1972; see also Travers, 1973). Also, induction of λ phage proteins in *E. coli* is helped by D protein, which binds strongly to DNA. D protein is of low molecular weight and was reported to have histone-like features (Ghosh and Echols, 1972). On the contrary, histones in animal and plant cells have been the subject of extensive study as possible repressors, perhaps after specific phosphorylation, of gene action—their essential role is undecided also.

REFERENCES

Aboud, M. and Burger, M. (1970). *Biochem. Biophys. Res. Commun.*, **38**, 1023–1032
Adler, K., Beyreuther, K., Fanning, E., Geisler, N., Gronenborn, B., Klemm, A., Muller-Hill, B., Pfahl, M. and Schmitz, A. (1972). *Nature (Lond.).*, **237**, 322–327

Anderson, W. B., Schneider, A. B., Emmer, M., Perlman, R. L. and Pastan, I. (1971). *J. Biol. Chem.*, **246**, 5929–5937

Arditti, R. R., Eron, L., Zubay, G., Tocchini-Valentini, G., Connaway, S. and Beckwith, J. (1970). *Cold Spring Harbor Symposia on Quantitative Biology*, **25**, 437–442

Banks, G. T., Binns, F. and Cutcliffe, R. L. (1967). *Progress in Industrial Microbiology*, **6**, 95–139

Beard, N. S., Armentrout, S. A. and Weisberger, A. S. (1969). *Pharmacol. Rev.*, **21**, 213–245

Burgess, R. A., Travers, A. A., Dunn, J. J. and Bautz, E. K. F. (1969). *Nature (Lond.)*, **221**, 43–46

Buttin, G. (1963). *J. Molec. Biol.*, **7**, 183–205

Cardini, G. and Jurtshuk, P. (1968). *J. Biol. Chem.*, **243**, 6071–6072

Carpenter, G. and Sells, B. H. (1973). *Biochim. Biophys. Acta*, **287**, 322–329

Cerna, J., Rychlik, I. and Pulkrabek, P. (1969). *European J. Biochem.*, **9**, 27–35

Clarke, P. H. (1970). *Advan. Microb. Physiol.*, **4**, 179–222

Clarke, P. H. (1972). *Sci. Prog. Oxf.*, **60**, 245–258

Cohen, G. N. (1968). *The Regulation of Cell Metabolism*, Holt, Rinehart and Winston; N.Y.

Cohen, L. B., Herner, A. E. and Goldberg, I. H. (1969). *Biochemistry*, **8**, 1312–1326

Davies, W. W. and Garren, L. D. (1968). *J. Biol. Chem.*, **243**, 5153–5157

De Crombrugghe, B., Chen, B., Anderson, W., Nissley, P., Gottesman, M. and Pastan, I. (1971). *Nature, New Biology*, **231**, 139–142

De Crombrugghe, B., Perlmann, R. L., Varmus, H. E. and Pastan, I. (1969). *J. Biol. Chem.*, **244**, 5828–5835

De Kroon, R. A. and Koningsberger, V. V. (1970). *Biochim. Biophys. Acta*, **204**, 590–609

DiManio, E., Hollenberg, C. P. and Hall, B. D. (1972). *Proc. Natl. Acad. Sci., U.S.*, **69**, 2818–2822

Donovan, G. and Oliver, I. T. (1972). *Biochemie*, **11**, 3904–3910

Duerksen, J. C. and Halvorson, H. (1959). *Biochim. Biophys. Acta*, **36**, 47–55

Emmer, M., De Crombrugghe, B., Pastan, I. and Perlman, R. (1970). *Proc. Natl. Acad. Sci., U.S.*, **66**, 480–487

Englesberg, E., Sheppard, D., Squires, C. and Meronk, F. (1969). *J. Molec. Biol.*, **43**, 281–298

Epstein, W. and Beckwith, J. R. (1968). *Ann. Revs. Biochem.*, **37**, 411–436

Eron, L., Arditti, R., Zubay, G., Connaway, S. and Beckwith, J. R. (1971). *Proc. Natl. Acad. Sci., U.S.*, **68**, 215–218

Eron, L. and Block, R. (1971). *Proc. Natl. Acad. Sci., U.S.*, **68**, 1828–1832

Fraser, R. S. S., Creanor, J. and Mitchison, J. M. (1973) *Nature (Lond.)*, **244**, 222–224

Gallo, M., Bertrand, J. C. and Azoulay, E. (1971). *FEBS letts.*, **19**, 45–49

Gallow, R. C. (1971). *Nature (Lond.)*, **234**, 194–198

Ghosh, S. and Echols, H. (1972). *Proc. Natl. Acad. Sci., U.S.*, **69**, 3660–3664

Gibson, D. T., Cardini, G. E., Meseles, F. C. and Kallio, R. F. (1970). *Biochemistry*, **9**, 1631–1635

Gilbert, W. and Muller-Hill, B. (1966). *Proc. Natl. Acad. Sci., U.S.*, **56**, 1891–1898

Greenblatt, J. and Schlief, R. (1971). *Nature, New Biol.*, **233**, 166–170

Gunsalus, I. C., Bertland, A. U. and Jacobson, L. A. (1967). *Arch. Microbiol.*, **59**, 113–122

Halvorson, H. O. (1960). In *Advances in Enzymology*, **22**, 99–156 (Nord, F. F., ed.) Interscience; N.Y.

Harris, G. and Millin, D. J. (1963). *Biochem. J.*, **88**, 89–95

Hedegaard, J. and Gunsalus, I. C. (1965). *J. Biol. Chem.*, **240**, 4038–4043

Ide, M. (1969). *Biochem. Biophys. Res. Commun.*, **36**, 84–92

Katagiri, M., Ganguli, B. N. and Gunsalus, I. C. (1968). *J. Biol. Chem.*, **243**, 3543–3546

Kram, R. and Tomkins, G. M. (1973). *Proc. Natl. Acad. Sci.*, *U.S.*, **70**, 1659–1663

Kuwano, M. and Schlessinger, D. (1970). *Proc. Natl. Acad. Sci.*, *U.S.*, **66**, 146–152

Lebeault, J-M., Lode, E. T. and Coon, M. J. (1971). *Biochem. Biophys. Res. Commun.*, **42**, 413–419

Leive, L. (1965). *Biochem. Biophys. Res. Commun.*, **18**, 13–17

McClatchy, J. K. and Rosenblum, E. D. (1963). *J. Bacteriol.*, **86**, 1211–1215

McCullough, W. and John, P. C. L. (1972). *Nature (Lond.)*, **239**, 402–405

Makman, R. S. and Sutherland, E. W. (1965). *J. Biol. Chem.*, **240**, 1309–1314

Monard, D., Janecek, J. and Rickenberg, H. V. (1969). *Biochem Biophys. Res. Commun.*, **35**, 584–591

Monod, J. (1956). In *Enzymes; Units of Biological Structure and Function*, p. 7 (Gaebler, O. H., ed.), Academic Press; N.Y.

Monod, J., Jacob, F. and Gros, F. (1962). In *The Structure and Biosynthesis of Macromolecules*, pp. 104–120, Biochemical Society Symposium, No. 21, Cambridge University Press

Monro, R. E., Celma, M. L. and Vazquez, D. (1969). *Nature*, **222**, 356–358

Moses, V. and Sharp, P. B. (1970). *Biochem. J.*, **118**, 481–489

Nakanishi, S., Adhya, S., Gottesman, M. E. and Pastan, I. (1973). *Proc. Natl. Acad. Sci.*, *U.S.*, **70**, 334–338

Nyiri, L. (1968). *Process Biochem.*, **3**, 27–30

Ornston, L. N. (1971). *Bacteriol. Rev.*, **35**, 87–116

Ozaki, M., Mizushima, S. and Nomura, M. (1969). *Nature (Lond.)*, **222**, 333–339

Paigen, K. and Williams, B. (1970). *Advan. Microb. Physiol.*, **4**, 252–324

Pastan, I. and Perlman, R. (1968). *Proc. Natl. Acad. Sci.*, *U.S.*, **61**, 1336–1342

Pastan, I. and Perlman, R. (1969a). *J. Biol. Chem.*, **244**, 5836–5842

Pastan, I. and Perlman, R. (1969b). *J. Biol. Chem.*, **244**, 2226–2232

Pastan, I. and Perlman, R. (1970). *Science*, **169**, 339–344

Perlman, R. (1973). *Process Biochemistry*, **8(7)**, 18–20

Perlman, R. and Pastan, I. (1968). *Biochem. Biophys. Res. Commun.*, **30**, 656–664

Perlman, R. L., De Crombrugghe, B. and Pastan, I. (1969). *Nature (Lond.)*, **223**, 810–812

Perlman, R. L. and Pastan, I. (1971). In *Current Topics in Cellular Regulation*, **3**, 117–134 (Horecker, B. L. and Stadtman, E. R., eds.), Academic Press; N.Y.

Peterkofsky, A. and Gazdar, C. (1973). *Proc. Natl. Acad. Sci.*, *U.S.*, **70**, 2149–2152

Peterson, J. A. (1971). *Arch. Biochem. Biophys.*, **144**, 678–693

Peterson, J. A. and Griffin, B. W. (1972). *Arch. Biochem. Biophys.*, **151**, 427–433

Pitot, H. C., Peraino, C. P., Lamar, C. and Kennan, A. L. (1965). *Proc. Natl. Acad. Sci.*, *U.S.*, **54**, 845–851

Puga, A., Borras, M-T., Tessman, E. S. and Tessman, I. (1973). *Proc. Natl. Acad. Sci.*, *U.S.*, **70**, 2171–2175

Reese, E. T., Lola, J. E. and Parrish, F. W. (1969). *J. Bacteriol.*, **100**, 1151–1154

Richards, N. and Hinshelwood, C. (1961). *Proc. Roy. Soc. B.*, **154**, 463–477

Richmond, M. H. (1968). *Essays in Biochemistry*, **4**, 106–154 (Campbell, P. N. and Greville, G. D., eds.), Academic Press

Richmond, M. H., Parker, M. T., Jevon, M. P. and John, M. (1964). *Lancet*, **1**, 293–296

Riggs, A. D., Reiness, G. and Zubay, G. (1971). *Proc. Natl. Acad. Sci.*, *U.S.*, **68**, 1222–1225

Roberts, J. W. (1969). *Nature*, **224**, 1168–1174

Robinson, G. A., Butcher, R. W. and Sutherland, E. W. (1968). *Ann. Rev. Biochem.*, **37**, 149–174

Shapiro, J. L., Machattie, L., Eron, G., Ihler, K., Ippen, K. and Beckwith, J. (1969). *Nature (Lond.)*, **224**, 768–774

Simoni, R. D. and Roseman, S. (1973). *J. Biol. Chem.*, **248**, 966–976
Simoni, R. D., Nakazawa, T., Hays, T. B. and Roseman, S. (1973). *J. Biol. Chem.*, **248**, 932–940
Smith, J. T. (1963). *J. Gen. Microbiol.*, **30**, 299–306
Strange, R. E. (1961). *Nature (Lond.)*, **191**, 1272–1273
Takasaki, Y., Kosugi, Y. and Kanbayashi, A. (1969). In *Fermentation Advances* pp. 561–589 (Perlman, D., ed.), Academic Press; N.Y.
Tao, M. and Lipmann, F. (1969). *Proc. Natl. Acad. Sci., U.S.*, **63**, 86–92
Tomkins, G. M., Gelehrter, T. D., Granner, D., Martin, D., Samuels, H. and Thompson, E. B. (1969). *Science*, **166**, 1474–1480
Travers, A. (1973). *Nature (Lond.)*, **244**, 15–18
Ullmann, A. and Monod, J. (1968). *FEBS letts.*, **2**, 57–60
Van Dedem, G. and Moo-Young, M. (1973). *Biotech. Bioeng.*, **15**, 419–439
Van Wijk, R. and Konijn, T. M. (1971). *FEBS lrtts.*, **13**, 184–186
Van Wijk, R., Ouwehand, J., Van Den Bos, T. and Koningsberger, V. V. (1969) *Biochem. Biophys. Acta*, **186**, 178–191
Varmus, H. E., Perlman, R. L. and Pastan, I. (1970a). *J. Biol. Chem.*, **245**, 2259–2267
Varmus, H. E., Perlman, R. L. and Pastan, I. (1970b). *J. Biol. Chem.*, **245**, 6366–6372
Volkin, E. and Astrachan, L. (1956). *Virology*, **2**, 149–161
Volkin, E. and Astrachan, L. (1958). *Biochim. Biophys. Acta*, **29**, 536–544
Watson, J. D. (1970). *Molecular Biology of the Gene*, 2nd edn., W. A. Benjamin, Inc.; N.Y.
West, I. C. and Stein, W. D. (1973). *Biochim. Biophys. Acta*, **308**, 161–167
White, R. (1970). *New Scientist*, **48**, 546–548
Wiseman, A. (1965). *Organization for Protein Biosynthesis*, Blackwell Scientific Publications; Oxford
Wiseman, A. (1969). *Process Biochemistry*, **4(5)**, 63–65
Wiseman, A. (1971a). *Process Biochemistry*, **6(8)**, 27–29
Wiseman, A. (1971b). *Biologist*, **17**, 217–220
Wiseman, A. (1973). *J. appl. Chem. Biotechnol.*, **23**, 159
Wiseman, A. and Jones, P. R. (1971). *J. Appl. Chem. Biotechnol.*, **2**, 26–28
Ycas, M. and Vincent, W. S. (1960). *Proc. Natl. Acad. Sci., U.S.*, **46**, 804–811
Yu, C-A. and Gunsalus, I. C. (1970). *Biochem. Biophys. Res. Commun.*, **40**, 1423–1428
Yudkin, M. D. and Moses, V. (1969). *Biochem. J.* **113**, 423–428
Zetelaki-Horvath, K. and Bekassay-Molnar, E. (1973). *Biotech. Bioeng.*, **15**, 163–179
Zubay, G., Gielow, L. and Englesberg, E. (1971). *Nature, New Biol.*, **233**, 164–165
Zubay, G., Schwartz, D. and Beckwith, J. (1970). *Proc. Natl. Acad. Sci., U.S.*, **66**, 104–110

Chapter 2

Enzyme Induction in the Process of Development

J. W. T. Dickerson and T. K. Basu
Department of Biochemistry, University of Surrey,
Guildford, England.
and
The Marie Curie Memorial Foundation,
The Chart, Oxted, Surrey, England.

2.1. INTRODUCTION

The maintenance of homeostasis is necessary for normal growth. Before birth, the mother bears much of the responsibility for maintaining the constancy of the internal environment. Nutrients cross the placenta to the foetus, and waste products pass in the opposite direction. The placenta also acts as a "barrier", protecting the foetus from exposure to many potentially harmful non-nutrient substances.

When the foetus leaves the uterus there is an abrupt change both in the way in which nutrients enter the body and the form in which they are presented. Normally, and for a period of time which differs for different species, the newborn animal receives milk. This must be processed in the hitherto largely untried stomach and intestine, and from it the nutrients required for maintenance, growth and development are absorbed. Metabolic pathways in the liver change as an adaptation to the changing demands made upon it. Anutrients can freely enter the growing animal, and the neonate develops its own mechanisms for excreting them. The nervous system also undergoes rapid development.

Smaller and more gradual changes in the environment also occur during weaning when there is a change in the nature of the food.

Adaptation to these environmental changes involves changes in the activities of many different enzymes, and this chapter describes the role of induction in their regulation. The review has been restricted to the small intestine, liver and brain, and "induction" has been interpreted as an increase in enzyme activity caused by a change in substrate concentrations or by hormonal action. In only a few instances has the work cited fulfilled the requirements of the more precise mechanistic definition, that is, has been shown to be mediated by the *de novo* synthesis of protein. Parts of the subject have been reviewed previously, and these publications have been drawn upon in an attempt to present a co-ordinated view of the subject and to illustrate some of its implications in relation to paediatrics.

The relationships of mammalian hormones and enzyme concentrations in adult animals has been reviewed by Pitot and Yatvin (1973), and although only passing reference is made in the present review to changes in isoenzymes, isoenzyme changes in ontogeny have been reviewed by Masters and Holmes (1972).

2.2. THE SMALL INTESTINE

The changes, during development, in the enzymes of the mucosal cells of the small intestine are in part a response to the change in the nature of its contents. The foetus swallows amniotic fluid, and this has been shown for a number of species including rabbit, guinea pig, pig, sheep and monkey. Studies with radio-opaque material injected into the amniotic sac (McLain, 1963) have shown that the human foetus has started to swallow amniotic fluid by the sixth month of gestation, and that the motility of the intestine and the amount of fluid swallowed increases with the progress of gestation. After birth the food of the young mammal, for varying periods of time, is maternal milk which in almost all species contains rather more fat and less carbohydrate than the adult diet. Moreover, the carbohydrate present in milk is lactose, whereas the bulk of the carbohydrate present in the adult diet is starch. There are also differences in the nature of the proteins present in milk and in the mixed adult diet.

The development of some of the mucosal enzymes is of direct relevance to the problem of the dietary intolerance of certain foods and nutrients, and for this reason has excited considerable interest in the past decade or so and a number of excellent reviews of the subject are already available. That by Koldovský (1972) contains some 350 references. The present review will be selective rather than exhaustive, and will seek to summarize the present state of knowledge on the substrate and hormonal control of enzyme activity.

The Disaccharidases

Lactose is hydrolysed by a β-galactosidase located in the microvilli of the mucosal cells (Alpers, 1969) and normally referred to as "lactase". Another isoenzyme, acid β-galactosidase, occurs in the lysosomes. There is also an α-galactosidase present in some tissues including the small intestine (Jumawan *et al.*, 1972). The acid hydrolases will be discussed later.

Lactase activity is present in the foetal intestinal mucosa of man, rabbit and rat. In man (Auricchio *et al.*, 1965) and rabbit (Doell and Kretchmer, 1962) it has been shown that the activity increases before birth and falls afterwards. There is evidence that, for different species, the rate of change after birth is related to the state of maturity at birth and to the suckling period. Thus in the guinea pig, which can be weaned at birth, the activity shows the smallest change. In the cat, the most rapid fall in activity coincides with the weaning period at 4 to 7 weeks, and in the rat it occurs between 20 and 25 days after birth (Koldovský and Sunshine, 1970). During this period of rapid decline in lactase activity there is no difference between the activity in the jejunum and ileum, whereas at birth and in the adult, the activity in the jejunum is greater than that in the ileum. Thus, at these ages it seems likely that lactose is hydrolysed in specific regions of the small intestine, whereas during suckling, hydrolysis occurs along the entire length (Newey, 1967).

Doell and Kretchmer (1964) correlated the pre-natal rise in lactase activity in the rabbit with the rise in the lactose concentration in maternal blood. However, the enzyme appears not to be induced by substrate, for the rise occurs pre-natally after surgical removal of the mammary glands, and the fall postnatally, though occurring at the time of weaning, also appears to be independent of the presence of substrate at physiological concentrations (Sriratanaban *et al.*, 1971). The decrease is delayed by adrenalectomy in the rat (Koldovský, 1972).

Sucrase activity is low in the intestinal mucosa of suckling rats, but at 9 to 15 days of age it can be induced precociously by artificial feeding (Herbst and Sunshine, 1969). The normal increase is delayed by adrenalectomy and restored by hydrocortisone.

The Acid Hydrolases

The mucosal cells of the small intestine contain a group of acid hydrolases whose development has been studied extensively. These include acid phosphatase, β-glucuronidase, α- and β-galactosidase, arylsulphatase and cathepsin D. All these enzymes show a similar pattern of development in the rat with a peak specific activity at about 10 days of age, followed by a decline during the gradual weaning process (Jumawan *et al.*, 1972; Raychaudhuri and Desai, 1972). The activity of each of them is repressed by cortisone. The action of cortisone in repressing the activity of acid β-galactosidase indicates clearly the dependence of this action on the age of the animal, for it is without effect in the 8-day old animal and can only be evoked in the middle of the suckling period (Koldovský, 1972). The rise in the specific activity of arylsulphatase activity in the neonatal rat has also been reported by Danovitch and Laster (1969) and may be induced by an aromatic sulphate ester present in rat milk. The concomitant change in activity of these enzymes suggests that they have a common role complementary to each other and to the overall function of the small intestine (Raychaudhuri and Desai, 1972).

Alkaline Phosphatase

The specific activity of alkaline phosphatase changes little during the postnatal development of the rat jejunum, whereas that of the ileum rises sharply to a peak again at about 10 days of age with a rapid fall to 20 days when the activity is not much higher than in the ileum (Pelichová et al., 1967). Cortisone induces the activity of this enzyme in the foetal rat (Ross and Goldsmith, 1955). In the mouse, phosphatase activity normally begins to increase rapidly at 14 days of age (Moog, 1953), although cortisone will induce an increase 6 days earlier and ACTH will similarly induce 2 days earlier. Between 8 and 14 days after birth the intestinal epithelium responds well to cortisone but shows no response after the enzyme has reached its maximum activity on the 18th day. Studies with different steroids (Moog and Thomas, 1955) suggest that the basic step in the accumulation of phosphatase is sensitive to cortico-steroids of the 11-oxy, 17-hydroxy type.

Other Enzymes

Changes in other enzymes in the developing small intestine of rat may be related to changes in the diet at weaning. Thus, Hahn and Skala (1971) reported that glycerophosphate dehydrogenase (EC 1.1.18) and malic enzyme (EC 1.1.40) were always highest in the proximal third of the small intestine, and that malic enzyme attained its maximum activity on the 20th day. Pyruvate kinase (EC 2.7.1.40), on the other hand, increased after the 20th day. The effect of diet, hormones and drugs on the activities of some glycolytic enzymes in adult man has been reviewed by Rosensweig et al., (1971). Folic acid, though neither a substrate nor a coenzyme of the glycolytic enzymes, causes a prompt rise in their activity in the jejunum. These enzymes in the adult also respond to sex hormones and to certain drugs, but there appears to be no information about the possible inductive role of these substances in the developing organ.

2.3. THE LIVER

Energy Metabolism

Glycogenesis

Maternal glucose is the main source of energy for the developing foetus. Triglycerides do not cross the placental barrier in any species that has so far been investigated (Robertson and Sprecher, 1968), and can be oxidized only to a limited extent in foetuses of man (Hahn *et al.*, 1964) and the rat (Drahota *et al.*, 1964). During the later stages of foetal growth three types of energy reserve are deposited in the foetus. Thus, triglycerides are synthesized from glucose and deposited in the liver and fat depots; the amounts of fat deposited in this way varies considerably from one species to another, being approximately 15% of body weight in a full-term human foetus and about 1% of body weight in the newborn rat (Widdowson, 1950). A further type of lipid storage is that which takes place in brown adipose tissue; this again is more abundant in some species than others, and seems to have a particular function in relation to post-partum thermo-regulation (Hull, 1966). It has also been known for many years that foetal tissues are rich in glycogen (for reviews see Shelley, 1961; Dawes and Shelley, 1968). These glycogen stores decrease rapidly after birth, and it seems clear that they represent a readily available source of energy during the early neonatal period. The relatively high glycogen content of cardiac muscle enables the neonate to withstand a considerable degree of anoxia (Dawes and Shelley, 1968). The newborn human infant could probably live on its own carbohydrate stores for three to four hours after birth, but before these are exhausted the concentration of free fatty acids in the plasma rises and reaches a peak in 6 to 24 hours after birth (Heim and Hull, 1966).

The newborn animal is then normally fed a milk diet. Of necessity, the metabolic adaptation to this diet has been studied most fully in the laboratory rat, and in this species the milk is essentially a high-fat, high-protein, low-carbohydrate diet (Dymsza *et al.*, 1964). In the laboratory, rats begin to wean themselves from about the 14th day of postnatal life. During this weaning period, the consumption of the high-fat milk diet decreases and that of a high-protein, high-carbohydrate solid diet increases.

The enzymes associated with the metabolic changes of the developing organism have been the subject of considerable investigation. Those concerned with carbohydrate metabolism have been reviewed by Walker (1971). The factors that control the deposition of glycogen in the foetus are complex, though it is clear that the enzymes concerned with glycogen synthesis occur before those concerned with its degradation (Dawkins, 1963). Experiments in which rat foetuses were decapitated *in utero* have shown that there is a correlation between the activity of glycogen synthetase (EC 2.4.1.11) and glycogen deposition (Jacquot and Kretchmer, 1964). Broadly speaking, there are two types of cells, the hepatocytes and the haemopoietic elements, in the liver at this time of life. The rise in total enzyme activity is, however, considerably greater than the increase in the volume of hepatocytes and must therefore represent a true intracellular change (Greengard *et al.*, 1972).

The thyroid and pituitary-adrenal axis begin to function in the rat at around the 18th day of gestation and studies in which rat and rabbit foetuses were decapitated (Jost, 1961) provided evidence for the involvement of the corresponding hormones in the induction of glycogen synthetase. The formation of glycogen synthetase is induced by the glucocorticoids. The concentration of corticosterone in the plasma of foetal rats has been reported to be much higher than in maternal plasma at the 19th day of gestation, is approximately equal to the maternal value from the 20th day to term, and then rises to 3.5 times the maternal value within 5 hours of delivery (Holt and Oliver, 1968b). The administration of exogenous cortisol to foetal rats increases the accumulation of glycogen in the foetal liver (Greengard and Dewey, 1970).

When thyroxine is injected into the rat foetus at the 19th day of gestation it induces a premature rise in the activity of reduced nicotinamide adenine dinucleotide phosphate (NADPH)-cytochrome c reductase (EC 1.6.99.2) and glucose-6-phosphatase (EC 3.1.3.9) (Greengard, 1969). This latter enzyme is of particular interest since it can also be induced prematurely by the administration of glucagon, though not by cortisol (Greengard and Dewey, 1967). Greengard has indeed suggested that the biphasic appearance of glucose-6-phosphatase can be explained if thyroxine is the normal inducer during foetal development and glucagon is responsible for the secondary rise after birth (Greengard, 1973). Porcine pancreatic glucagon does not pass the placenta in man even with a ten-fold rise in maternal levels (Johnston et al., 1972). However, Assan and Boillot (1971) determined the glucagon content of the pancreas and the glucagon-like activity of the gut and stomach of 65 human foetuses of 10 to 25 weeks gestation by radio-immunoassay techniques, and reported that the glucagon concentration in the pancreas increases with age. At 25 weeks, the concentration in the pancreas was considerably higher than in the adult organ extracted and assayed by the same procedure; however, the foetal plasma concentrations were within the normal adult range. Glucagon-like material is present during the last few days of gestation in the rat (Orci et al., 1969).

Glucagon causes the release of insulin from the pancreas in the full-term human baby (Milner and Wright, 1967), but glucose is a relatively poor stimulus to insulin secretion in the premature and full-term infant (Baird and Farquhar, 1962). However, in the premature infant both glucagon and a mixture of nine essential amino acids are potent stimuli of insulin release (Grasso et al., 1968). In vitro studies have shown that the β-cells of the pancreas are functional in the human foetus between 14 and 24 weeks gestation (Milner, 1971). As at the later stages of foetal development, glucose does not act as a stimulant and tolbutamide is ineffective in causing the release of the hormone. Stimuli that are effective fall into three groups, namely, the amino acids, arginine and leucine; the ionic stimuli, potassium, ouabain and barium; and stimuli which act by raising intracellular cyclic AMP concentrations—glucagon, theophylline and dibutyryl cyclic AMP. It seems to be a matter of conjecture as to which of these various stimuli may be active in the foetus.

Few studies have been made of the induction of enzyme activity in the foetus in species other than the rat. As already pointed out, however, high concentrations of glycogen are found in the liver of other species at birth and in the guinea-pig, as in the rat, the increase is paralleled by a rise in the activity of glycogen synthetase (Lea and Walker, 1964). In man, the concentration begins to increase between 90 and 100 days of gestation (Capková and Jirácek, 1968). In spite of the fact that the guinea-pig is born in a much more developed condition, glucose-6-phosphatase has been reported to be absent from the liver (Nemeth, 1954). Glucose-6-phosphatase activity has, however, been reported to be present in human foetal liver from the 22th week of gestation.

Glycogenolysis

After birth the newborn animal is dependent for energy, until it is fed, upon the mobilization of its own body stores. In the rat, there appears to be a lag of 1 to 2 hours after birth before significant amounts of glucose are mobilized from the glycogen stores, and this may partly explain the pronounced hypoglycaemia after parturition (Snell and Walker, 1973a). The concentrations of liver glycogen and plasma D-glucose have been measured in neonatal rats delivered by caesarian section at time intervals up to 3 hours after treatment of the neonatal rats with glucagon or cortisol, with and without dibutyryl cyclic AMP (Snell and Walker, 1973b). Glycogenolysis was promoted by glucagon or dibutyryl cyclic AMP in the third hour after birth but not at earlier times. However, cortisol plus dibutyryl cyclic AMP promoted glycogenolysis in the second hour after birth, but no hormone combination was effective during the first hour. In adult rats, hepatic glycogen mobilization is caused by the activation by glucagon of α-glucan phosphorylase (Sutherland and Rall, 1960; Hers *et al.*, 1970), an enzyme that is present in newborn rat liver in about the same concentration as in the adult organ. The apparent lack of phosphorylase activity may in part be attributed to the high plasma glucose concentration in the newborn rat immediately at birth (Snell and Walker, 1973b). Adrenalectomy blocks the stimulation of phosphorylase activity and glycogenolysis by cyclic AMP in adult rat liver (Schaeffer *et al.*, 1969a), skeletal muscle (Schaeffer *et al.*, 1969b), and heart muscle (Miller *et al.*, 1971). It is known that there is a relative adrenal corticosteroid insufficiency at birth, with plasma corticosterone values lower at this time than at 18 to 20 days of gestation, and that they rise to the pre-natal level at 2 hours post-partum (Holt and Oliver, 1968b). It is to be noted that in newborn rats, in contrast to adults, corticosterone values do not rise in response to stress (Schapiro et al., 1962; Levine et al., 1967).

Gluconeogenesis

Despite the lack of effect of glucagon, cortisol or dibutyryl cyclic AMP on glycogenolysis during the first 2 hours after birth, these agents do promote a rise in plasma glucose concentration which is blocked by the simultaneous administration of actinomycin D. This is in contrast to the effect after 2 hours when actinomycin D has no effect (Snell and Walker, 1973b). It is clear, therefore, that the hormone-stimulated rise in plasma glucose at 1 hour after birth results from the induction of one or more of the enzymes involved in gluconeogenesis. The rate-limiting enzyme in this pathway at birth is probably phospho-pyruvate carboxylase (EC 4.1.1.32) (Ballard and Hanson, 1967; Yeung and Oliver, 1967; Philippidis and Ballard, 1969; Ballard, 1971), and the fact that glucagon and dibutyryl cyclic AMP induce the activity of this enzyme *in utero* makes it likely that this pathway is responsible for the observed increase in plasma glucose concentration. The later rise in glucose concentration that occurs after giving dibutyryl cyclic AMP is not affected by actinomycin D and this is consistent with it being due to glycogenolysis. The increase in phosphorylase activity responsible for the glycogenolysis is an activation phenomenon that is insensitive to actinomycin D or puromycin (Sutherland and Rall, 1960; Cake and Oliver, 1969).

The synthesis of phosphopyruvate carboxylase can be induced prematurely by the injection of glucagon or adrenaline into rat foetuses *in utero* (Yeung and Oliver, 1968a). The action of these hormones is mediated by cyclic 3', 5'-AMP and injection of this substance causes a similar precocious development of phosphopyruvate carboxylase (Yeung and Oliver, 1968b). The effect of cyclic 3', 5'-AMP is approximately doubled by the simultaneous injection of caffeine.

It seems likely that amino acids are the main source of the glucose formed by gluconeogenesis in the neonatal period, for all the amino acids that can be converted into carbohydrate are metabolized via oxaloacetate and the phosphopyruvate step. This suggestion is supported by the finding that the activities of several of the transaminases and serine dehydratase are induced at this time (*vide infra*). Another possible precursor is glycerol derived from triglyceride metabolism, and Walker (1971) considers that glycerol may play a major role in the increased potential for gluconeogenesis during the neonatal period.

The newborn piglet, apparently, is not capable of gluconeogenesis (Swiatek *et al.*, 1968), in spite of the fact that the activity of phospho-pyruvate carboxylase is quite substantial (Mersmann, 1971). This clearly suggests that in the pig, in contrast to the rat, phospho-pyruvate carboxylase is not the limiting enzyme for gluconeogenesis. The newborn piglet depends almost exclusively on its carbohydrate reserves as a source of energy, and it is of interest in this connection that the skeletal muscle of the newborn piglet contains up to 8% of glycogen (McCance and Widdowson, 1959). Compared with this energy reserve, the amount of body fat is small (Widdowson, 1950) and only small amounts of free fatty acids are mobilized under starving conditions (Gentz *et al.*, 1970).

Neonatal hypoglycaemia is a frequent occurrence in many species, and is a problem in clinical paediatrics (Cornblath, 1968). The reason for the problem may well be a delayed induction of the enzymes involved in gluconeogenesis (Shelley and Nelligan, 1966).

Carbohydrate utilization

Carbohydrates are used by the body as a direct source of energy, for the synthesis of glycogen, and for conversion to lipid. The role of enzyme induction in the development of the ability of the organism to synthesize glycogen has been discussed. The metabolism of glucose passed from the mother to the foetus *in utero*, and of other hexoses, such as fructose and galactose, absorbed from the gastrointestinal tract after birth, requires the participation of a large number of enzymes. The change in the activities of many of these enzymes during develop-ment have been studied in the past decade or so. For obvious reasons, much of the work has been done on rat liver. The following review will therefore mainly concern the changes in this organ.

The liver contains a number of hexokinases that are involved in the first intracellular step in the utilization of hexoses. There are three isoenzyme forms of hexokinase (EC 2.7.1.1) and these act on a broad spectrum of substrates. Hexokinase activity is high in foetal liver and decreases rapidly to a low constant level after birth (Burch *et al.*, 1963). After birth hexokinase activity is replaced by enzymes which appear to be substrate-induced and which therefore develop at different times in the life of the animal as it is exposed to different carbo-hydrates.

When the newborn animal is exposed to galactose, derived from the dissaccharide lactose in milk, hepatic galactokinase (EC 2.7.1.6) develops and increases to a peak of activity within a few days after birth (Cuatrecases and Segal, 1965; Walker *et al.*, 1965/66). The increase in enzyme activity is paralleled by an increase in the ability to incorporate ^{14}C-galactose into glycogen in rat liver slices. There is also an increased capacity to metabolize galactose in the newborn lamb (Ballard and Oliver, 1965). Galactokinase activity shows no change, however, over a wide range of pre-natal and postnatal ages in the guinea-pig (Walker and Khan, 1968). The advanced development of the guinea-pig at birth may, in part, explain the lack of change in galactokinase activity, since in the rat the adaptive response decreases with increasing age, and the rate of decrease is only retarded slightly by feeding a diet containing 40% galactose (Cuatrecases and Segal, 1965).

Ketohexokinase (EC 2.7.1.3) activity is not present in foetal liver of a number of species (rat, rabbit and guinea-pig) but increases between birth and 7 to 10 days of age (Walker, 1963). Glucokinase (EC 2.7.1.2) is absent from foetal livers of all species so far examined; the increase after birth is induced by the transition at weaning from the high-fat, low-carbohydrate diet to an essentially high-carbohydrate low-fat diet (Walker and Holland, 1965; Walker and Eaton, 1967).

The activities of the various enzymes involved in the metabolism of carbohydrate rise and fall in a complex manner during the development of rat liver (Burch et al., 1963; Schaub et al., 1972) and kidney (Burch et al., 1971). Two of the key enzymes in the process of glycolysis are phosphofructokinase (EC 2.7.1.11), which controls the rate at which glucose is broken down to the two trioses, glyceraldehyde-3-phosphate and dihydroxyacetone phosphate, and pyruvate kinase (EC 2.7.1.40) which controls the rate of production of pyruvate. Changes in the activities of these enzymes have been described in rat liver (Burch et al., 1963; Hommes and Wilmink, 1968) and in various tissues in the human foetus (Hahn and Skala, 1970). In vitro studies of foetal liver obtained from legal abortions have shown that pyruvate kinase activity is induced by incubation of cultures for 5 hours in the presence of dibutyryl cyclic AMP. In the rat, the total activity of pyruvate kinase remains at the neonate level for about 20 days postnatal and then increases markedly over a few days (Taylor et al., 1967; Vernon and Walker, 1968a; 1968b) when the rat is weaned on to a high-carbohydrate diet. Premature weaning on to a high-fat or high-protein diet prevents the normal increase in activity (Vernon and Walker, 1968b). The response to a high-carbohydrate diet is also age dependent, for it occurs at 18 to 21 days after birth but not at 13 to 15 days. It seems likely that the response of hepatic pyruvate kinase to the carbohydrate-rich diet is controlled by insulin. Two isoenzyme forms of the enzyme called L (liver) and M (muscle) are found in adult rat liver and only one of these, the L form, is stimulated by increasing the carbohydrate content of the diet and by insulin (Tanaka et al., 1967). Whether this is the form stimulated in the developing animal does not appear to be known, although the ratio of L and M isoenzymes is different in foetal liver (Tanaka et al., 1967; Tepper and Hommes, 1970) with type M predominating, as it also does initially in liver undergoing regeneration after partial hepatectomy.

Fat metabolism

The mammalian foetus is able to oxidize fatty acids only very slowly. In the rat, carnitine and acylcarnitine transferase are absent from the heart at birth, and the mechanisms for fatty acid oxidation do not develop until 10 days after parturition (Wittels and Bressler, 1965). After birth there is a sharp rise in the concentration of fatty acids in the plasma and these investigators have suggested that the circulating fatty acids induce the synthesis of enzymes involved in their oxidation. There seems to be no evidence, however, that this is so. In the adult, fatty acid oxidation is thought to be limited to mitochondria and it could be that the explanation lies in the immature mitochondria (Greenfield and Boell, 1968), since a feature of this immaturity could be the low activity of certain enzyme systems. However, a high-carbohydrate diet such as that received by the foetus, is well known to promote fatty acid synthesis rather than oxidation.

The rise in plasma fatty acids after birth occurs in both rat and man, but for different reasons. In the rat, which has low body stores of fat, it is due to the commencement of feeding a high-fat diet, whereas in the human baby, with its high body stores of fat, it is due to the mobilization of stored triglycerides.

The nature of the diet seems to be the decisive factor in determining the activity of enzymes involved in fatty acid synthesis in the liver. In the mouse (Smith and Abraham, 1970) and rat (Volpe and Kishimoto, 1972) the activity of the multienzyme complex, fatty acid synthetase, falls during the suckling period, then rises sharply at weaning; moreover, this rise can be induced precociously by early weaning. A rise in activity also occurs after hatching in the chick (Ryder, 1970). The increase in liver fatty acid synthetase at weaning occurs in response to a low-fat diet and is due to a marked acceleration of enzyme synthesis (Volpe *et al.*, 1973) which can be prevented by actinomycin D (Smith and Abraham, 1970). The changes in activity of fatty acid synthetase in the brain are in contrast to those in the liver, for in brain the activity is highest in foetal life and the specific activity gradually falls, due to increased degradation, as the animal matures (Volpe and Kishimoto, 1972); these changes are unaffected by diet (Volpe *et al.*, 1973).

Premature weaning at 15 days has been found to cause an immediate rise in the activity of hepatic malic enzyme (EC 1.1.1.40), whereas the activities of glucose 6-PO_4 dehydrogenase (EC 1.1.1.49) and ATP citrate lyase (EC 4.1.3.8) did not rise until 4 to 5 days later (Lockwood *et al.*, 1970). Glucagon, growth hormone, and cortisol had no effect on the activities of these enzymes, but thyroxine increased the activity of malic enzyme some 20-fold, and that of ATP citrate lyase two-fold.

Amino acid metabolism

Tyrosine α-ketoglutarate aminotransferase (EC 2.6.1.5)

The first step in the metabolism of tyrosine in the homogentisic acid pathway involves transamination to p-hydroxyphenylpyruvate. The activity of the enzyme involved, tyrosine aminotransferase, can be induced in adult rat liver by the administration of either tyrosine or cortisol (Lin and Knox, 1957). The "inductive" effect of the amino acid is abolished by adrenalectomy, and occurs only in the presence of small amounts of the glucocorticoid. During foetal life the activity of the enzyme is very low in both the rat and man (Sereni *et al.*, 1959). In the rat, the activity rises rapidly during the first twelve hours after birth, reaching a peak value which is about twice the value in the adult. This postnatal rise is not induced by injection of tyrosine (Kenney and Flora, 1961), but is induced by a number of hormones. It represents a significant increase of specific enzyme protein synthesis, and can be partially or wholly prevented by the administration at birth of inhibitors of protein synthesis, such as puromycin or actinomycin D (Holt and Oliver, 1968a).

The exact mechanisms whereby various hormones act, and interact, in induction of tyrosine aminotransferase activity during the early postnatal period, have not been worked out. However, it seems clear that corticosteroids (Barnabei and Sereni, 1964; Gelehrter and Tomkins, 1967; Granner *et al.*, 1968) act at a transcriptional level, whereas glucagon and catecholamines (Wicks, 1968a) act at a translational level, with their action mediated by an increased synthesis of cyclic 3′,5′-AMP (Wicks, 1968b).

The normal postnatal rise in tyrosine aminotransferase activity is completely obliterated by adrenalectomy. Cortisol is completely effective in restoring the synthesis of the enzyme at this time of life, though it has little effect on the activity of the enzyme if given pre-natally (Sereni *et al.*, 1959; Greengard and Dewey, 1967). Glucagon, on the other hand, induces a sharp rise in enzyme activity in the foetus (Greengard and Dewey, 1967), and experiments with 13-day foetal explants have yielded similar results for these two hormones (Räihä *et al.*, 1971). It is possible that growth hormone may play a role in repressing the inductive action of cortisol (Kenney, 1967).

Since glucagon acts by increasing the synthesis of cyclic 3′,5′-AMP it acts by affecting the activity of adenyl cyclase, and it has been suggested, therefore, that the timing of the competency of glucagon to induce enzyme activity is related to the developmental pattern of adenyl cyclase (Hommes and Beere, 1971). In the liver, increase in the activity of adrenyl cyclase occurs four days before birth and the adult level is reached at birth.

Transient hypoglycaemia is a normal concomitant of birth in the rat (Yeung and Oliver, 1968b), in which natural delivery of an entire litter may take up to two hours, and it has been suggested (Greengard and Dewey, 1967) that the hypoglycaemia triggers a release of glucagon which then induces the activity of tyrosine aminotransferase, as well as that of serine dehydratase (EC 4.2.1.13). The same workers also showed that the rise in the activity of these enzymes could be prevented by the administration of glucose. The role of food deprivation in the induction of these enzymes has been confirmed (Reynolds and Potter, 1971).

The role of hormones in the induction of tyrosine aminotransferase is very complex. In the adult, immunochemical and isotopic studies by Holten and Kenney (1967) have shown that insulin, as well as glucagon, causes an increase in the rate of synthesis of the enzyme, which is comparable to that caused by cortisol. They found that the effects of the pancreatic hormones were not additive to each other, but were additive to the induction by cortisol. Induction by the steroid hormone is, however, more extensive than that by the protein hormones and operates by a different mechanism. It is also possible that the different hormones stimulated different forms of the enzyme (Holt and Oliver, 1969).

The developmental pattern of tyrosine aminotransferase is of some clinical importance to the paediatrician, for in babies of low birthweight there is a close association between protein intake, ascorbic acid availability and plasma tyrosine levels (Light et al., 1966). The elevated plasma tyrosine levels are related to gestational age rather than to bodyweight (Rizzardini and Abeliuk, 1971; Menkes et al., 1972) and to a lower IQ at 7 to 8 years of age (Menkes et al., 1972). This is one reason why it is unwise to give high-protein diets to premature babies (Snyderman, 1971). The possible benefit of pre-natal induction of tyrosine aminotransferase (Kirby and Hahn, 1973) remains to be explored.

Although there are now recognized dangers of giving too much tyrosine to premature babies due to the underdevelopment of enzyme systems responsible for its metabolism, tyrosine is nevertheless an essential amino acid for the premature baby (Snyderman, 1971). The concentration of its precursor, phenylalanine, may also be high in the plasma of premature babies (Räihä, 1973). It is tempting to attribute this to underdevelopment of the enzyme phenylalanine hydroxylase (EC 1.99.1.2), the absence of which leads to phenylketonuria. However, the system in the liver that catalyses the conversion of phenylalanine to tyrosine is complex and consists of several enzymes and coenzymes, and whilst a low activity of phenylalanine hydroxylase may be one factor, there is evidence to suggest that there may also be a deficiency of a pteridine cofactor (Friedman and Kaufman, 1971; Räihä, 1973).

Alanine aminotransferases

The mitochondria and cytosol of rat liver contain isoenzymic forms of L-alanine-glyoxylate aminotransferase and L-alanine-2-oxoglutarate aminotransferase (EC 2.6.1.2). The activity of mitochondrial L-alanine-glyoxylate aminotransferase begins to rise before birth and reaches a peak at about the time the animal is weaned (Snell and Walker, 1972). The similarity between this pattern of development and that of a number of enzymes involved in gluconeogenesis (Vernon and Walker, 1968a) suggests that the mitochondrial glyoxylate aminotransferase may be involved in this process. This possibility is supported by the finding that the enzyme is stimulated by glucagon in the adult (Snell and Walker, 1972) and that this hormone also stimulates the activities of gluconeogenesis enzymes (Exton et al., 1970). The administration of glucagon, dibutyryl cyclic AMP and thyroxine to foetal rats three days before birth increased the activity of the mitochondrial aminotransferase (Snell and Walker, 1972), but glucagon and cyclic AMP had no effect on the cytosol enzyme, although this enzyme was stimulated by thyroxine, and also by cortisol. After birth cortisol had no effect on the cytosol enzyme except when given simultaneously with casein. Casein also acted synergistically with cortisol in increasing the activity of both mitochondrial and cytosol L-alanine-2-oxoglutarate aminotransferase.

Thus, before birth when the foetus is in a nutritionally constant environment, the alanine aminotransferases are induced by hormones alone, whereas after birth changing patterns of nutrition are additional factors which act synergistically with the hormones.

Tryptophan oxygenase (pyrrolase) (EC 1.13.1.12)

This enzyme catalyses the conversion of tryptophan to formylkynurenine. In guinea-pig liver, activity begins to appear at the end of gestation and rises rapidly to adult levels 24 hours after birth (Nemeth and Nachmias, 1958). In the rabbit, the activity also rises to the adult level during the first 24 hours after birth, whereas in the rat, the level remains low until 12 to 15 days post partum (Nemeth, 1954; Auerbach and Weisman, 1959) when it rises sharply to reach the adult level within 24 hours. Premature delivery of rabbit foetuses stimulates early development of the enzyme (Nemeth, 1959), and birth has a similar stimulatory effect on the enzyme in the guinea-pig (Nemeth, 1954). Development is not induced by the administration of corticosteroids or tryptophan to foetal rats, nor by tryptophan and ACTH to foetal guinea-pigs (Nemeth and Nachmias, 1958). However, both cortisol (Schimke, 1966) and tryptophan (Knox and Auerbach, 1955; Goldstein et al., 1962) stimulate the rate of synthesis of the enzyme in the adult, but they do so by different mechanisms

(Greengard et al., 1963). Thus, whereas puromycin (an inhibitor of protein synthesis) inhibits the cortisol-induced increase of the enzyme, actinomycin (an inhibitor of RNA synthesis) inhibits the rise induced by cortisol but does not influence that due to substrate. The same workers found that development of the enzyme up to 23 days of age in the rat was independent of adrenocorticoid control and could proceed in spite of significant inhibition of RNA synthesis.

Insulin and testosterone also induce activity of this enzyme in the adult rat (Schor and Frieden, 1958).

Unlike the development of tyrosine aminotransferase, that of tryptophan pyrrolase appears not to be related in any understood fashion to known enzyme inducers. The level of tryptophan oxygenase in rat liver rises following the production of experimental porphyria with allylisopropyl-acetamide (AIA) and was thought to be secondary to the induction of δ-aminolaevulinic acid (ALA) synthetase activity and consequent haeme production. It seems, however, that some other mechanism must be involved (Wetterberg et al., 1969) because allylisopropylacetamide induces a rise in the oxygenase in the absence of the adrenals, whereas injection of δ-aminolaevulinate does not.

The trans-sulphuration pathway

In the adult mammal, including man, cystine or cysteine is formed from dietary methionine via the trans-sulphuration pathway. Two enzymes in this pathway (Fig. 2.1), cystathionine synthase (EC 4.2.1.21) which catalyses the conversion of L-homocysteine to L-cystathionine, and cystathionase (or homoserine dehydratase, EC 4.2.1.15) which catalyses the conversion of L-cystathionine to L-cysteine, have been studied in developing human and rat liver. In human liver, the activity of cystathionine synthase was found to increase little during gestation and at term had about 25% of the activity found in liver from individuals over $2\frac{1}{2}$ years of age (Räihä, 1971). No effect of birth on the activity of this enzyme was detected. This was in marked contrast to the developmental patterns of cystathionase, which was not detected until after birth when it rose sharply to about 75% of the adult value.

In the rat the activity of both these enzymes is low in foetal liver until the 20th day of gestation, when the activity rises rapidly up to 2 days post partum. After the second postnatal day, however, the activity of cystathionine synthase continues to rise gradually to the adult value over the first 3 postnatal weeks (Volpe and Laster, 1972), whereas that of homoserine dehydratase has already reached the adult level (Snell, 1972/73). An important consequence of the different pattern of development of cystathionase in the human and the rat is that cysteine is probably an essential amino acid in the human foetus and neonate up to 2 days after birth (Gaull et al., 1972), whereas it is not essential for the rat.

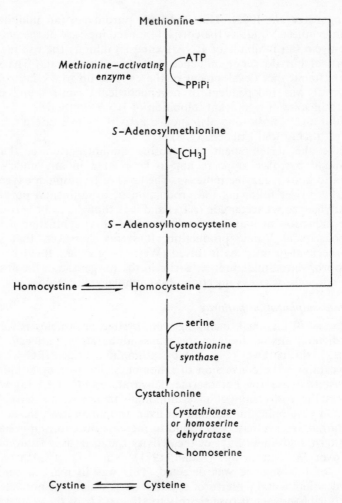

Fig. 2.1. Conversion of methionine to cysteine

The physiological stimulus for the rapid rise in homoserine dehy-dratase activity after the 20th day of gestation is not known. Snell (1972/73) found that cortisol and glucagon were without effect. However, glucagon did induce a 34% rise in activity of the enzyme in the 30-day-old rat, and it has also been reported to cause a rise in adult rats fed for 3 days on a low-protein diet and then starved overnight (Finkelstein, 1967; Pestaña, 1969). Thyroxine causes a fall in the activity of homo-serine dehydratase after the 20th day of gestation (Snell, 1972/73) and in the adult (Finkelstein, 1967).

The hormonal induction of other enzymes in the trans-sulphuration pathway has also been studied. Thus the methionine activating enzyme (ATP:1-methionine 5-adenosyltransferase; EC 2.5.1.6) is induced in neonatal rat liver by cortisol, whereas this hormone is without effect in the adult or foetal rat (Chase *et al.*, 1968). Growth hormone, progesterone, triodothyronine and testosterone are also ineffective pre-natally (Chase *et al.*, 1968). Cystathionine synthase, which develops in a very similar manner to homoserine dehydratase in foetal rat liver, is not induced pre-natally by glucagon, thyroxine or conjugated oestrogens, and not significantly by cortisol (Volpe and Laster; 1972). In the adult, the activity of cystathionine synthase is affected by the availability of methionine (Chatagner and Trautmann, 1962) and by the protein content of the diet (Pestaña, 1969). It is unlikely, however, that the foetus is subjected to changes of this kind that are likely to be responsible for the induction of the enzymes.

Arginine and ornithine metabolism

Arginine and ornithine are involved in a number of important metabolic processes in mammals. Those that have been most studied from the developmental viewpoint are the biosynthesis of urea, glutamate and polyamines (Fig. 2.2).

(a) *Urea biosynthesis* The arginine synthetase system (argininosuccinate synthetase, EC 6.3.4.5; argininosuccinase, EC 4.3.2.1) catalyses the rate-limiting step in the synthesis of urea, and first becomes measurable in the liver of the rat foetus at the 18th day of gestation (Kennan and Cohen, 1959; Illnerova, 1968; Räihä and Suihkonen, 1968a). The capacity of the liver to synthesize urea begins at about this time, and according to the data of Kretchmer *et al.* (1966) attains an adult level by about the 2nd or 3rd postnatal day. Arginine synthetase activity rises from about 2% of the adult value at birth to 45% at 1 day after birth, and from 55% at 10 days of age to 100% at 200 days. Both the increase in the capacity to synthesize urea, and the rise in the activity of arginine synthetase are inhibited by puromycin, and it may therefore be concluded that the synthesis of new protein is involved (Kretchmer *et al.*, 1966; Räihä and Suihkonen, 1968a).

The rate of development of the enzymes involved in urea biosynthesis in different species is possibly related to the degree of kidney development (Kennan and Cohen, 1959). Thus in the pig, where there is early development of a mesonephric kidney which, with the progress of gestation, excretes an increasing amount of urea into the allantoic vesicle (McCance and Dickerson, 1957), all the urea cycle enzymes have been found in the liver from the 28th day of gestation (Kennan and Cohen, 1959). In the human, there is also an early development of the arginine synthetase system (Räihä and Suihkonen, 1968b), and although in this species there is no allantoic vesicle, urea diffuses

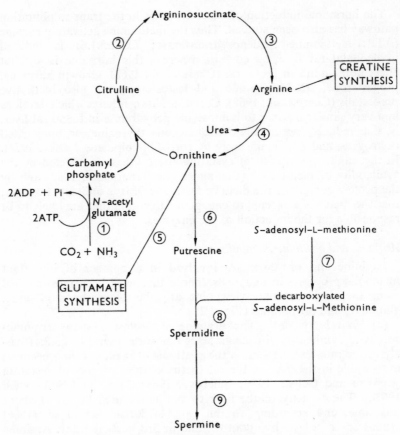

Fig. 2.2. The urea cycle and the metabolism of ornithine in mammals
Enzymes: 1 Carbamyl phosphate synthetase
         2 Argininosuccinate synthetase
         3 Argininosuccinase
         4 Arginase
         5 Ornithine ketoacid aminotransferase
         6 Ornithine decarboxylase
         7 S-adenosyl L-methionine decarboxylase
         8 Spermidine synthetase
         9 Spermine synthetase

across the placental membrane very efficiently for excretion by the mother.

In the rat, birth itself does not affect the rate of development of arginine synthetase, but adrenalectomy causes an almost complete inhibition of the increase in activity at 24 hours after birth, and this is completely reversed by administration of the synthetic glucocorticoid, triamcinolone (Räihä and Suihkonen, 1968a; Räihä and Schwartz, 1973). Between 1 and 6 days after birth the rise in arginine synthetase activity is further increased by glucocorticoids. Steroids have no effect, however, when given to the rat foetus before the natural rise in enzyme activity has begun.

It is possible that the substantial increase in the activity of arginine synthetase during the late suckling period, when animals are beginning to wean themselves, is related to the increase in the protein content of the diet (Illnerova, 1968). Such an explanation would be in agreement with observations in adult rats in which it has been shown that the activity of the urea cycle enzymes is directly proportional to the daily consumption of protein (Schimke, 1962). However, starvation of newborn rats has no effect on the activity of the enzymes, and limiting the milk intake during the suckling period has no effect on the incorporation of ^{14}C-labelled sodium bicarbonate into urea by liver slices (Räihä, 1971). It could be that under these conditions additional protein is utilized for more rapid growth, and there is evidence that in the neonate, growth does indeed have an homeostatic function (McCance and Widdowson, 1956).

(b) *Glutamate biosynthesis* The activity of ornithine ketoacid aminotransferase (EC 2.6.1.13) is low in foetal rat liver. There may be a small rise at the time that the pups are born, but the major rise in enzyme activity starts at about 14 days of age and adult values are reached by 20 to 25 days (Räihä and Kekomaki, 1968). Other workers have reported that the principal increase takes place after weaning, with the enzyme appearing earlier in the liver than in the kidney and reaching a higher activity in the female kidney than in the male (Herzfeld and Knox, 1968). Oestrogens were found to increase significantly the activity of the enzyme in the kidney, but not that in the liver, of adult rats. Oestrogens did not induce precocious development of the enzyme when given to foetal rats, but in 20-day-old animals they increased the activity in the kidney and, to a smaller extent, also increased the activity in the liver.

As with arginine synthetase, the administration of triamcinolone to the foetus had no effect on the activity of ornithine-ketoacid aminotransferase. During the first two weeks of postnatal life, however, the glucocorticoid induced normal adult activity and this induction was prevented by puromycin (Räihä, 1971). A further similarity with arginine synthetase is that the main rise in enzyme activity occurs during the time when the animals are beginning to wean themselves. The relative importance of substrate-induction and hormone-induction in producing the normal developmental pattern has apparently yet to be elucidated.

The behaviour of ornithine-ketoacid aminotransferase in circumstances in which protein retention is stimulated is of particular interest, for the enzyme is primarily concerned with the catabolism of arginine and ornithine. Two such conditions have been studied, hepatectomy and the administration of growth hormone, and in both of these the activity of the enzyme is repressed (Räihä et al., 1967). Both these conditions favour the incorporation of arginine into protein and the utilization of ornithine for polyamine synthesis (Jänne, 1967).

(c) *Polyamine biosynthesis* The polyamines, spermidine and spermine, are found in high concentrations in rapidly growing tissues (Snyder et al., 1970), and in such systems are probably essential for RNA and protein synthesis. They appear to affect RNA synthesis in the rat liver by stimulating DNA-dependent RNA polymerase in nuclei (MacGregor and Mahler, 1967) and nucleoli (Russell et al., 1971), although this effect may be attributable to stabilization of newly formed RNA (Heby and Agrell, 1971) or inhibition of ribonuclease (Brewer, 1972). The effects on protein synthesis are possibly secondary to the effects on RNA synthesis, as polyamines can bind to, and associate, ribosomal subunits (Siekevitz and Palade, 1962; Raina and Telaranta, 1967; Datta et al., 1969).

There are four enzymes involved in polyamine synthesis (Fig. 2.2), ornithine and S-adenosyl-L-methionine decarboxylase, and spermidine and spermine synthetases. Ornithine decarboxylase (EC 4.1.1.17) catalyses the formation of the diamine, putrescine, from ornithine (Jänne and Raina, 1968), and S-adenosyl-L-methionine decarboxylase (EC 4.1.1.50) catalyses the formation of S-adenosyl (5′)-3-methyl-thiopropylamine from S-adenosyl-L-methionine (Pegg and Williams-Ashman, 1968). The "decarboxylated S-adenosyl-L-methionine" provides a propylamine moiety in the reaction with putrescine to form spermidine, and this is catalysed by spermidine synthetase. Spermidine combines with a further propylamine moiety from "decarboxylated S-adenosyl-L-methionine", catalysed by spermine synthetase, to form spermine (Hannonen et al., 1972). Ornithine decarboxylase has the shortest half-life reported for an enzyme (as little as 10 minutes), and appears to be the rate-limiting enzyme in the whole process of polyamine synthesis (Russell and Snyder, 1968; 1969).

There are no direct studies on changes in ornithine decarboxylase activity during postnatal development, but activity of the enzyme declines in the liver during foetal development of the rat, and pre-natal levels are higher than postnatal (Russell and McVicker, 1972).

Administration of growth hormone to young rats causes a marked elevation in liver ornithine decarboxylase activity, with a peak at 4 hours after injection (Russell et al., 1970b; Richman et al., 1971), and the hormone causes a greater elevation in this enzyme in weanling rats than in adults (Russell et al., 1970b). Correspondingly, hypo-physectomy results in a reduction in ornithine decarboxylase activity in young rats, but the deficiency can be rectified by administering growth hormone (Richman *et al.*, 1971). The enhancement of this enzyme following administration of growth hormone can be prevented by actinomycin D or cycloheximide (Jänne *et al.*, 1968; Jänne and Raina, 1968). The effect of actinomycin D, however, decreases with time after administration of growth hormone to weanling rats, and when cycloheximide is given 3 hours after administration of growth hormone there is a rapid decrease of the enzyme activity with a half-life of about 15 min (Russell *et al.*, 1970b). The normal half-life of the growth-hormone induced activity was calculated to be 24 min.

Parallel increases in ornithine decarboxylase and RNA polymerase occur during the regeneration following hepatectomy (Raina *et al.*, 1966) and can be markedly reduced by hypophysectomy, whereas thyroidectomy, adrenalectomy, ovariectomy and castration are without effect (Russell and Snyder, 1969).

Cortisol also stimulates ornithine decarboxylase in the liver of the young rat, but not to the same extent as does growth hormone. However, adrenalectomy has no effect on this enzyme activity. If growth hormone and cortisol are given together they act synergistically, and cause a greater elevation of liver ornithine decarboxylase activity than growth hormone alone (Richman *et al.*, 1971). The hormones insulin, thyroxine and glucagon also cause slight increases in liver ornithine decarboxylase activity in the young rat (Richman *et al.*, 1971), but growth hormone and corticosteroids appear to be the most important regulators of induction of this enzyme in the young rat.

Little information is available on developmental profiles and induction of the other polyamine-synthesizing enzymes in the liver, but growth hormone causes enhanced *S*-adenosyl-L-methionine decarboxylase activity in the liver of the weanling rat (Russell and Lombardini, 1971).

The relationship between the activity of the enzyme histidine decarboxylase (EC 4.1.1.22), which catalyses the formation of histamine from histidine, and that of ornithine decarboxylase in foetal rat liver near term, and in some rat tumours (Russell and Snyder, 1968) suggests that histamine may also play a role in the initiation of the growth process. In rat foetal liver, the activity of histidine decarboxylase decreases continuously from the 16th day of gestation to reach an adult level 2 days post partum (Kunze *et al.*, 1972).

Another enzyme concerned with histidine metabolism, histidase (L-histidine ammonia-lyase, EC 4.3.1.3) shows an interesting sex difference in its developmental pattern. In the female rat there is a steep rise in its activity at puberty (Feigelson, 1968), and ovariectomy results in the same pattern as for the male, but the enzyme can be induced by giving 17β-oestradiol to the ovariectomized animals. The normal developmental pattern of the enzyme is suppressed by substances that block protein synthesis (e.g. ethionine) and RNA synthesis (e.g. actinomycin D).

The activity of ornithine decarboxylase in the rat ovary varies during the oestrous cycle (Kobayashi et al., 1971) and is dependent on endogenous luteinizing hormone (LH) (Kaye et al., 1973) but not on follicle-stimulating hormone (FSH) (Kobayashi et al., 1971). In young rats both LH and prostaglandin E_2 cause a 15-fold increase in ornithine decarboxylase activity in the ovary (Lamprecht et al., 1973); the effect of LH is age-dependent, with no effect before 9 days of age, an increasing capacity to respond up to 20 days of age, and then a gradual decline in response with increasing maturity (Kaye et al., 1973). The activity of ornithine decarboxylase in the rat uterus can be elevated by 17β-oestradiol, but testosterone has no effect (Kaye et al., 1972); the response is not age-dependent, as similar increase in activity is obtained at various times between 2 and 30 days of age (Kaye et al., 1972; Kaye et al., 1973). 17β-Oestradiol also causes a slight increase in uterine S-adenosyl-L-methionine decarboxylase activity in young rats (Kaye et al., 1972).

Drug Metabolism

A variety of drugs are metabolized by enzyme systems localized predominantly in liver microsomes (Brodie et al., 1958), and reactions catalysed by these enzymes include N-dealkylation, deamination, aromatic hydroxylation, alkyl chain oxidation, nitro group reduction, azo link cleavage and glucuronide formation. These reactions occur in the body in two phases (Williams, 1967) (Fig. 2.3). In the first phase

Drug or foreign compound → (Phase I asynthetic reactions) → oxidation reduction and/or hydrolysis → (Phase II synthetic reactions) → conjugation products

Fig. 2.3. The metabolism of drugs and "foreign" compounds

there occur all the reactions which can be classified as oxidations, reductions and hydrolyses, and which result in the introduction or unmasking of functional groups rendering a non-polar compound polar, or a polar compound to be more polar and hence more rapidly excreted than the original compound. These functional groups often act as a centre for the second phase of metabolism in which a synthetic step occurs so that the compound is conjugated with an endogenous molecule. The products of the second phase, so-called conjugated products, are usually water-soluble compounds that are readily excreted. Most drugs are metabolized by both phases of reactions. However, there are some compounds, particularly those that already possess functional groups, such as hydroxyl, carboxyl and amino groups, that often undergo only second phase reactions.

The enzymes responsible for oxidation phase I reactions are NADPH and O_2 dependent, and occur primarily, but not exclusively, in the hepatic endoplasmic reticulum (microsomes) (Gillette, 1967). Cytochrome P-450, a microsomal haemoprotein (Omura and Sato, 1964), is believed to be involved in the activation of oxygen and its transfer to the substrate in the hydroxylation reactions (Omura et al., 1965). This cytochrome P-450-linked mono-oxygenase system of liver microsomes also catalyses the hydroxylation of a variety of endogenous compounds such as steroid hormones (Conney and Klutch, 1963) and fatty acids (ω-oxidation) (Das et al., 1968).

The activity of the drug-metabolizing enzymes, like that of many other enzymes, is subject to substrate or hormonal induction.

Developmental pattern of drug-metabolizing enzymes

Newborn and foetal animals metabolize many xenobiotic substances more slowly than mature animals of the same species due to deficiencies of the hepatic microsomal enzymes. This was first reported in newborn mice and guinea-pigs by Jondorf and his colleagues (Jondorf et al., 1959) and included studies of the enzyme system that N-demethylates monomethyl-4-aminoantipyrine and aminopyrine, O-dealkylates phenacetin, oxidizes hexobarbital and conjugates phenolphthalein as the glucuronide. Fouts and Adamson (1959) studied the oxidation of hexobarbital, the N-dealkylation of aminopyrine, the deamination of amphetamine, the hydroxylation of acetanilide, the oxidation of ring sulphur of chlorpromazine, and the reduction of the nitro group of p-nitrobenzoic acid, and found that newborn rabbits did not have any activity of these enzymes, but that some activity appeared two weeks after birth, and at four weeks after birth the activity was almost the same as that found in the adult. Kato et al., (1964) studied the *in vitro* metabolism of hexobarbital, pentobarbital, meprobamate, carisoprodol and strychnine, and the *in vivo* metabolism of pentobarbital, meprobamate and carisoprodol in female rats. Newborn rats showed a very low metabolic activity, but this increased progressively up to 30 days of age after which it gradually decreased. The hippuric acid synthesizing system in the rat develops similarly (Brandt, 1964). Other workers have studied the development of microsomal glucuronyl transferase (Flint et al., 1964) and of aspirin hydrolase (Eyring et al., 1973) in the liver of rabbit, the glutathione conjugation of sulphobromophthalein in mouse liver (Krasner, 1968) and a number of oxidative, reductive, hydrolytic and synthetic drug-metabolizing enzymes in the liver of the pig (Short and Davis, 1970). All these studies have shown a characteristic development of hepatic microsomal drug-metabolizing enzymes during the first 4 to 6 weeks after birth. A systematic study in rats (Basu et al., 1971a) of the pattern of development of cytochrome P-450 (Fig. 2.4), biphenyl hydroxylation (Fig. 2.5),

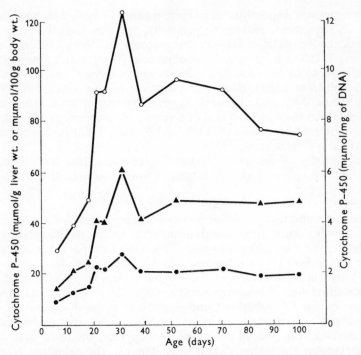

Fig. 2.4. Effect of age on the concentration of cytochrome P-450 in rat liver
Results are expressed per 100 g body wt (○), per g liver wt (●) and per
mg of DNA (▲).

p-nitrobenzoate reduction (Fig. 2.6) and 4-methylumbelliferone
conjugation with glucuronide (Fig. 2.7), showed that the activity
of the enzymes catalysing these reactions and the hepatic content
of cytochrome P-450 at 6 to 70 days postnatum fell into three distinct
phases. The first phase, as measured on the 6th day after birth, was one
of very little activity; this was followed by a rapid increase in activity
for 3 to 5 weeks after birth; and finally, the activity fell to the adult
value at about 50 to 70 days of age.

The enzyme biphenyl-2-hydroxylase is an exception to this general
pattern of development, for this enzyme has a higher activity in the
liver of the newborn rat than in that of the adult (Basu *et al.*, 1971a)
(Fig. 2.8).

A number of components of the hepatic drug-metabolizing enzyme
system—various cytochromes and the enzymes concerned with the
metabolism of benzphetamine and benzo[a]pyrene—have a similar
age-related development in the rabbit (Fouts and Devereux, 1972).
Extra-hepatic drug-metabolizing systems, such as those in the lung,

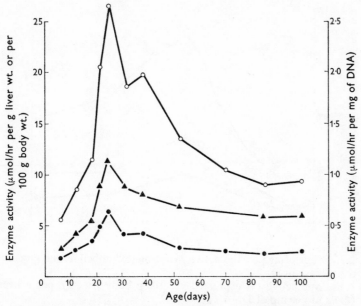

Fig. 2.5. Effect of age on the activity of biphenyl 4-hydroxylase in rat liver.
Enzyme activity is expressed per 100 g body wt (○), per g liver wt (●)
and per mg of DNA (▲).

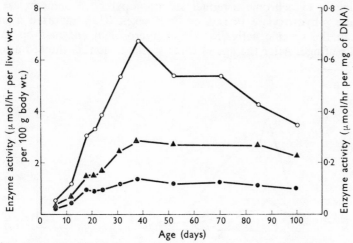

Fig. 2.6. Effect of age on the activity of p-nitrobenzoate reductase in rat
liver.
Enzyme activity is expressed per 100 g body wt (○), per g liver wt. (●)
and per mg of DNA (▲).

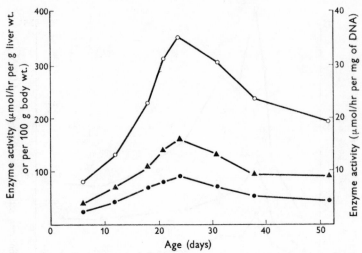

Fig. 2.7. Effect of age on the activity of 4-methylumbelliferone glucuronyltransferase in rat liver.
    Enzyme activity is expressed per 100 g body wt ($\bigcirc$), per g liver wt
($\bullet$) and per mg of DNA ($\blacktriangle$).

appear to develop according to a somewhat different pattern (Fouts
and Devereux, 1972).

    MacLeod *et al.* (1972) studied the development of cytochrome P-450
reductase, cytochrome c reductase, aminopyrene N-demethylase and
aniline p-hydroxylase in rats of both sexes. They reported a steady
rise in the specific activities of the microsomal enzymes up to five
weeks of age. After the age of three weeks male rats showed a higher

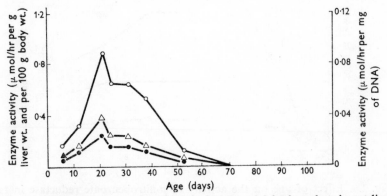

Fig. 2.8. Effect of age on the activity of biphenyl 2-hydroxylase in rat liver.
    Enzyme activity is expressed per 100 g body wt ($\bigcirc$), per g liver wt ($\bullet$)
and per mg of DNA ($\blacktriangle$).

activity for aminopyrine N-demethylation than did females. For aniline p-hydroxylation there was no difference between the sexes. These workers found that in males the pattern of maturation of drug oxidation followed most closely the increasing mean specific activity of NADPH cytochrome P-450 reductase, whereas in females there was a closer relationship with the increasing mean specific activity of NADPH cytochrome c reductase. Other workers have also demonstrated a correlation between the haemoprotein and the activity of some microsomal drug oxidative enzymes during maturation in the rat (Basu et al., 1971a) and also in swine (Short and Davis, 1970).

From the above evidence it seems that the rate of development of the hepatic drug-metabolizing enzyme systems in various mammals, such as the mouse, rat, rabbit, guinea-pig and swine, is generally similar. The phase of most rapid development occurs between birth and the third to fifth week post partum, regardless of such factors as the size of the species, the length of its gestation period, the degree of maturation at birth and the normal life span of the animal.

The hepatic microsomal drug-metabolizing enzyme systems of both immature and mature animals can be stimulated by pretreating the animals with drugs which act as enzyme inducers (Conney, 1967). Fouts and Hart (1965) first reported that treatment of newborn rabbits with phenobarbital enhanced the activity of liver enzymes that metabolize drugs, such as hexobarbital, aminopyrine and p-nitrobenzoic acid. Treatment of rabbits with phenobarbital during the final week of pregnancy also increased the activity of these enzymes in the newborn animal, but the drug-metabolizing enzymes could not be stimulated before the last 4 to 8 days of foetal life (Hart et al., 1962; Rane et al., 1973). These workers suggested that there was a defective enzyme-forming system or a lack of stimulus to develop these systems before this time.

The liver of young rapidly growing animals or liver undergoing regeneration after partial hepatectomy is more sensitive to inducers than the adult organ (Cramer et al., 1960; Chiesara et al., 1967; Kato and Takanaka, 1968; Müller et al., 1971; Basu et al., 1971a). Thus there is a fall in the percentage induction by phenobarbital of four enzymes in rat liver between 12 and 52 days of age (Table 2.1). This fall in inductive response continues after the rats have reached maturity (Table 2.2), and for some enzymes, e.g. aminopyrine N-demethylase and hexobarbital hydroxylase, was considerable during the second year of life. The liver enlargement accompanying this induction is due to cell multiplication in the actively growing organ and to cell enlargement in the mature animal (Paulini et al., 1971).

The inductive effects of carcinogenic polycyclic hydrocarbons on hepatic and extrahepatic microsomal enzymes in developing animals have also been studied. Bresnick and Stevenson (1968) found that

3

TABLE 2.1

Effect of Age on the Induction of Liver Microsomal Enzymes in the Rat (Basu *et al.*, 1971a)
Rats treated with phenobarbital (10 mg/kg body weight) for 3 days. Increase in activity expressed as a percentage of the basal value.

	Age (days)		
	12	21	52
Biphenyl 4-hydroxylase	161	63	50
Biphenyl 2-hydroxylase	260	30	25
p-Nitrobenzoate reductase	60	18	8
Cytochrome P-450	84	65	30

3-methylcholanthrene did not induce microsomal *N*-demethylase activity in foetal liver when administered to the mother or put directly into the amniotic sac, despite the fact that the compound was shown to be able to cross the placenta. However, induction of *N*-demethylase by 3-methylcholanthrene did occur in the neonate, preweanling and weanling rat, and these workers suggested that some aspect of the maternal environment prevented elaboration of the foetal drug-metabolizing enzymes after administration of 3-methylcholanthrene. It is perhaps of interest that the activity of biphenyl 2-hydroxylase, which normally disappears by 50 days of age in rat liver, is induced preferentially by 3-methylcholanthrene (Creaven and Parke, 1966).

Administration of benzo[α]pyrene to pregnant rats induces a large increase in the activity of benzo[α]pyrene hydroxylase activity in

TABLE 2.2

Age Difference in Phenobarbital Induction of the Activities of Electron Transport System and Drug Oxidation and Reduction in Rat Liver Microsomes (Kato and Takanaka, 1968)
The results indicate the increase by treatment with phenobarbital (60 mg per kg body weight for three days) expressed as a percentage of the original activity.

Age (days)	40	100	300	600
NADPH mxidase	181	84	46	20
NADPH-cyt.c reductase	249	116	52	22
Cytochrome P-450	259	131	60	37
Aminopyrine *N*-demethylation	713	395	173	36
Hexobarbital hydroxylation	492	293	134	34
Aniline hydroxylation	165	94	40	31
p-Nitrobenzoate reduction	256	104	45	32

maternal liver (Schlede and Merker, 1972). In the placenta the degree of induction of benzo[α]pyrene hydroxylase activity was significant on the 13th day of gestation and rose to maximum level by the 15th day. In the foetus, induction of this enzyme occurred in 22 % of animals on the 13th day and 14th day, but in all foetuses from the 15th day; the degree of induction increased progressively with age. No activity of this enzyme could be detected in the foetus and placenta of control animals. There is some evidence that cigarette smoking induces the activity of a number of enzymes in the human placenta (Welch *et al.*, 1969) and benzo[α]pyrene hydroxylase activity can be detected in the livers of human foetuses from smokers at 11 to 13 weeks gestation (Juchau, 1971). Pelkonen (1973) found that livers from foetuses weighing only 9 to 10 g (8 weeks gestation) had measurable drug-metabolizing activity and that this increased until the foetuses weighed 100 g (13 weeks gestation). Subsequently, the activity levelled off and there was no further change up to a body weight of 300 to 500 g (21 weeks gestation). Pelkonen and his colleagues (Pelkonen *et al.*, 1973) have obtained tentative evidence that maternal ingestion of pheno-barbital for 7 to 25 days induces the activity of a number of drug-metabolizing enzymes in human foetal liver. The drug had no effect on the enzymes in the placenta during the first half of pregnancy.

The structural framework of the hepatic endoplasmic reticulum of rats and rabbits suggests that the smooth-surfaced endoplasmic elements (enzyme-rich component of the liver) appear after birth (Dallner *et al.*, 1966; Rane *et al.*, 1973) when the animals attain the capacity to metabolize drugs. These findings are, however, in contrast to results obtained with human foetal liver, which has been shown to contain an enzymically-active cytochrome P-450 mono-oxygenase system early in gestation (Yaffe *et al.*, 1970). Ultrastructural studies of the human foetal liver have further shown that the endoplasmic reticulum is already present during the first half of the gestation period (Zamboni, 1965). In the 7th to 9th weeks of gestation, it has the appearance of a tubular system with ribosome-studded membrane, and around the 12th week there seems to be an increase in agranular endoplasmic reticulum membranes concomitant with a deposition of glycogen and iron in the hepatocytes. These changes correlate well with the development of drug-metabolizing enzymes in human foetal liver (Pelkonen, 1973).

The regulation of the development of drug-metabolizing enzymes

A regulatory role of cytochrome P-450 in the development of drug-metabolizing enzymes can be adduced from the parallel changes in its concentration (Kato *et al.*, 1964; Short and Davis, 1970; Basu *et al.*, 1971a). This parallelism does not always occur, however, for Gram *et al.*, (1969) reported that microsomal aniline hydroxylase reaches its highest activity at 2 weeks of age and then gradually declines, whereas ethylmorphine demethylation remains low for 3 weeks, increases abruptly to a peak at about $4\frac{1}{2}$ weeks of age and declines only slightly thereafter. Cytochrome P-450, though tending to increase, changes only slightly.

Traces of "foreign substances" in food may also act as inducers of enzymes in the suckling rat. Some drugs, such as phenobarbitone pass into milk and have been shown to induce an increase in the activity of cytochrome P-450, and in the rates of metabolism of other drugs in the suckling pups (Darby, 1971); the administration of chlorpromazine to the mother had a smaller effect. The developmental pattern of some drug-metabolizing enzymes is unaffected by giving the mother a wholly synthetic diet during gestation and lactation (Basu et al., 1971a).

In the semi-mature male rat corticosterone induces a considerable growth of the liver and an even greater increase in the amount of biphenyl 4-hydroxylase (Table 2.3). The role of corticosterone in the normal development of this enzyme has not been elucidated, but it is known that the circulating levels are high after birth and fall with age (Holt and Oliver, 1968b).

There is a certain parallelism between the changes in liver glycogen and the activities of certain drug-metabolizing enzymes (Fouts, 1963). However, studies of the relationship of hepatic glycogen concentration and hexobarbital sleeping time (Wools and McPhillips, 1966) do not support the concept of a cause and effect relationship. In rats whose growth is retarded by feeding a protein-deficient diet there is evidence that the activity of some drug-metabolizing enzymes may be maintained (Dickerson et al., 1971) as the result of the elevated plasma corticosterone levels caused by the malnutrition (Basu et al., 1971b). The relationship of nutrition to drug metabolism has been reviewed elsewhere (Basu and Dickerson, 1974).

TABLE 2.3

Effect of Corticosterone on Liver Weights and Activity of Biphenyl 4-Hydroxylase (Basu, 1971)
Rats were treated with two successive doses of corticosterone (10 mg/animal subcutaneously) at an interval of 6 hours.

	Control	Corticosterone
Liver weight (g)	8.0	14.0**
Biphenyl 4-hydroxylase (μmole biphenyl/mg microsomal protein/h)	0.15	0.19*

Significantly different from control value ** where $P < 0.001$ and * where $P < 0.05$.

The prominent changes in drug-metabolizing enzyme activity do not appear to be related in time to changes in testicular or seminal vesicle weight (Gram *et al.*, 1969) as might be expected if the changes were associated with puberty.

Several studies have suggested that high blood levels of growth hormone repress the capacity of the liver of the neonate to metabolize drugs. Thus, implantation of a growth hormone secreting pituitary tumour into young rats prevents the normal postnatal development of the hepatic microsomal drug-metabolizing enzymes responsible for hexobarbital metabolism (Wilson, 1968). Furthermore, the adminis-tration of the hormone to immature rats also impairs drug metabolism and prevents its age-related development (Wilson, 1970), and higher serum levels of immunoreactive growth hormone are associated with low hepatic drug-metabolizing enzyme activity in 1 to 10 day old rats as compared to 30-day-old animals (Wilson, 1972).

There is evidence that the hormonal metabolism of the mother may exert a control on the development of the drug-metabolizing enzymes of the neonate (Feuer and Liscio, 1969; 1970). In the female rat, hydroxylation of progesterone decreases with age, and the reductive pathway of metabolism increases (Kardish and Feuer, 1972). The balance is further shifted to the reductive pathway during pregnancy, and reduced progesterone metabolites have been shown to inhibit drug and progesterone hydroxylation in the newborn. These changes in the offspring are reversed by removing the pups from the mother at birth.

Differences in the patterns of development of steroid hormone metabolism may account for the fact that drug-metabolizing activity has been found in the early human foetus, but only in the near-term rat foetus. The distinctive pattern in the human is probably confined to the foetal zone of the foetal adrenal cortex which develops during gestation and involutes after birth. Although unable to synthesize progesterone, the early foetal gland can metabolize it to a variety of hydroxylated steroids (Villee, 1973). There is a preponderance of 16α-hydroxylated $\Delta 5\text{-}3\beta$-hydroxysteroids, most of which circulate in the sulphurylated form, and the ratio of corticosterone to cortisol is much higher in the foetus than in the child or adult. There is some evidence that a sharp rise in blood cortisol may occur just prior to parturition in several species (Murphy and Diez d'Aux, 1972) and that this may play an important role in parturition itself and in inducing enzymes necessary for extra-uterine existence.

Bilirubin metabolism

Studies of the sources of bile pigment have shown that bilirubin from non-erythropoietic sources accounts for about 30% of the total bilirubin load in the newborn, compared with about half this amount in the adult (Schwartz *et al.*, 1964). The majority of these non-erythro-poietic haem compounds are produced predominantly by hepatic amino-levulinic acid synthetase. This enzyme, in contrast to most of those mentioned previously, has a specific activity in the newborn rabbit that is eight times higher at birth than in the adult (Woods and Dixon, 1970). However, the administration of 3,5-dicarbethoxy-1,4-dihydro-collidine (DDC) produces a much smaller increase in enzyme activity in the newborn than in the adult. It therefore appears that haem production in the neonate may be less strictly regulated than in the adult and this could have some bearing on the occurrence of physio-logical jaundice in the newborn. Song *et al.*, (1971) obtained a similar result in neonatal rats treated with allylisopropylacetamide (AIA). These workers were able to show that the induction of cytochrome P-450, another haemoprotein, occurred independently of that of porphyrin. The production of hyperbilirubinaemia by the administration of large amounts of δ-aminolevulinic acid interfered with the induction of cytochrome P-450 by phenobarbital in the young rat, but not in the adult.

Wilson (1972) has reviewed the problem of hyperbilirubinaemia in children. A number of studies have suggested that this condition is often associated with low activity of hepatic glucuronyl transferase(s), since the metabolism of various drugs such as salicylic acid, paracetamol, menthol and steroids that are known to form glucuronides, is also reduced. Furthermore, in normal 5-day-old infants with the usual neonatal deficiency of the glucuronide-conjugating system there is a statistically significant inverse relationship between the formation of salicylamide glucuronide and the concentration of unconjugated bilirubin in the serum (Stern *et al.*, 1970). It has therefore been presumed that the decrease in unconjugated bilirubin levels due to therapy with phenobarbital is due to the induction of glucuronyl transferase(s). There is the possibility, however, that the hyperbilirubinaemia in these children actually inhibits glucuronide conjugation of drugs. In two sisters with congenital unconjugated hyperbilirubinaemia, Levy and Ertel (1971) were able to show a normal response to pheno-barbital of the glucuronide conjugation of drugs, whereas the hyper-bilirubinaemia was unresponsive. They concluded that in the case of these children there was probably a deficiency in hepatic anion-binding protein rather than glucuronyl transferase. However, there is some evidence (Bakken, 1970) that bilirubin UDP-glucuronyl transferase is activated by the presence of high concentrations of bilirubin in foetuses with erythroblastosis foetalis. The administration of immunosuppressive drugs to mothers after the 20th week of gestation significantly reduces the concentration of conjugated bilirubin in the serum and probably suppresses foetal activation of the conjugating enzyme (Rubaltelli, 1972).

Nucleic Acid Synthesis

The same hormone may have a different effect upon immature and mature liver tissues. This phenomenon is well illustrated by the effect of corticosteroids on nuclear RNA synthesis (Sereni and Barnabei, 1967). In adult rats, corticosteroids enhance the rate of RNA synthesis in the liver, and decrease the rate of hydrolysis of the nuclear and microsomal RNA fractions, whereas in the newborn rat, corticosteroids have no effect on the activity of RNA polymerase (EC 2.7.7.6) or on the rate of incorporation of pyrimidine precursors into nuclear RNA.

During the development of the liver there is an increase in the proportion of the tissue attributed to hepatocytes and a corresponding decrease in the amount attributed to haemopoietic tissue (Greengard et al., 1972). The enzyme, thymidine kinase (TK) (EC 2.7.1.21) is necessary for DNA synthesis and therefore must be present in hepatocytes and haemopoietic tissue. However, when foetal rats are given an intraperitoneal injection of 0.12 mg of cortisol acetate on the 18th day of gestation there is a prompt decrease in thymidine kinase activity per gram of liver, and the maximal rate of decrease of activity occurs before there is any sign of loss of haemopoietic cells (Greengard and Machovich, 1972). This may indicate that the two processes are independent, or that the loss of thymidine kinase activity contributes to the subsequent disappearance of the haemopoietic cells. This suggestion is supported by the finding that corticosteroid-induced involution of thymus and spleen is also associated with a fall in thymidine kinase activity (Greengard and Machovitch, 1972).

2.4. THE BRAIN

There is now a considerable literature describing changes in the activities of enzymes during the development of the brain, and particularly in rat brain. There are peculiar difficulties, however, in the interpretation of some of this data, for the brain is an extremely complex and heterogeneous organ and certain enzymes may be located in high concentrations in small clusters of cells, or in certain nuclei (Balázs and Richter, 1973). The regulation of enzyme development in the brain has received comparatively little attention until recently.

Hormones play an important role in the growth and function of the organ (Campbell and Eayrs, 1965) and a considerable amount of work has been done in the rat on the role of thyroxine. In this species the major growth spurt of the brain occurs after birth (Davison and Dobbing, 1966) and is thus easily accessible for study. Thyroidectomy at birth retards practically all aspects of the growth and development of the organ (Balázs and Richter, 1973), whereas the administration of thyroxine cause precocious development (Grave

et al., 1973). Normal thyroid function in the neonate is necessary
for the normal development of the activity of a number of enzymes
including succinic dehydrogenase (EC 1.3.99.1), acetylcholinesterase
and cholinesterase, particulate aminotransferase, glutamic decarboxylase (EC 4.1.1.1.15), and aminobutyric acid transaminase (EC 2.6.1),
and adenosine triphosphatase (EC 3.6.1.4) (Argiz *et al.*, 1967;
Rappoport and Fritz, 1972), and in neonatally thyroidectomized
rats the administration of thyroxine up to the 10th day restores the
normal activities of the enzymes. Hexokinase (EC 2.7.1.1), phosphofructokinase (EC 2.7.1.11) and pyruvate kinase (EC 2.7.1.40) are
inhibited by neonatal thyroidectomy induced with [131]I (Schwark
et al., 1972). Thyroxine also plays a role in the regulation of fatty
acid synthetase (Volpe and Kishimoto, 1972). The activity of this
enzyme in the brain, unlike that in the liver, is only affected by the
long-term ingestion of a high-fat diet. The enzymes whose activity is
depressed by thyroidectomy are located in subcellular organelles.
Studies of two of these, succinic dehydrogenase and aspartate aminotransferase (EC 2.6.1.1), have shown that the specific activity per mg
protein is not affected by thyroidectomy, which suggests (Szijan *et al.*,
1970) that thyroidectomy leads to a reduced formation of subcellular
membranes and organelles without changes in their composition.

The mode of mediation of the effects of thyroxine on brain development is not clear. It appears to be independent of the adenosine
3′, 5′-monophosphate system in the brain, because neonatal thyroidectomy has been found not to adversely affect the development of
the system in the organ (Schmidt and Robison, 1972).

Treatment of mice (Howard, 1965) and rats (Bálazs and Cotterrell,
1972) early in postnatal life with corticosteroids causes a marked
inhibition of cell division. These hormones are, however, powerful
enzyme inducers in the liver, and there is some evidence for similar
effects in the brain and nervous tissue. Perhaps the first clear evidence
for a role in enzyme induction in nervous tissue was obtained in
cultures of embryonic chick retina. In this tissue the activity of glutamine synthetase is low until the 16th to 17th day when it begins to
rise sharply in close correspondence with the terminal differentiation
and maturation of the retina. This change can be made to take place
precociously several days before the normal time by exposing the
culture to cortisol or by the administration of cortisol into the embryonic
yolk sac (Moscona and Piddington, 1966; Reif and Amos, 1966).
Further work showed that the precocious induction of glutamine
synthetase activity requires protein synthesis and is initially dependent
on the formation of new mRNA (Piddington and Moscona, 1967;
Reif-Lehrer and Amos, 1968).

In adult rats, hypophysectomy or adrenalectomy causes an exponential fall in the activity of glycerophosphate dehydrogenase (EC 1.1.1.8)

in the cerebral hemispheres and a greater fall in the brain stem (De Vellis and Inglish, 1968). In young rats the developmental rise is inhibited by hypophysectomy at 20 days of age, whereas the injection of cortisol during the period 7 to 15 days of age accelerates the rise in enzyme activity.

The activity of $Na^+ - K^+ - ATPase$ in the brain of chick embryos normally rises between the 13th day of incubation and hatching. A premature rise in activity of this enzyme, but not in that of Mg^{2+} ATPase, is elicited by injecting cortisol into the eggs (Šťastný, 1971).

The rate-limiting enzyme in the synthesis of the neurotransmitters 5-hydroxytryptamine (5-HT) and the catecholamines are thought to be the hydroxylating enzymes. Adrenalectomy causes a decrease in tryptophan hydroxylase activity in the midbrain of the rat, and the level is restored by the administration of corticosterone (Azmitia and McEwen, 1969). Further work (Azmitia et al., 1970) showed that the conversion of ^3H-L-tryptophan to serotonin was greatly reduced in the brain stem by adrenalectomy, whereas in the telediencephalon this was unaffected. Drug-induced sympatho-adrenal excitation for long periods produces an increase in the amounts of tyrosine hydroxylase in adrenergic areas of the central and peripheral nervous systems (Müeller et al., 1969), and exposure of rats to cold has also been reported to induce tyrosine hydroxylase activity in peripheral and central adrenergic neurones (Thoenen, 1970).

The enzyme phenyl N-methyltransferase converts noradrenaline to adrenaline in the mammalian adrenal medulla. It requires 5-adenosyl-methionine as the methyl donor and methylates β-hydroxylated phenylethylamine derivatives as well as noradrenaline. The normal levels of this enzyme in the adrenal gland are maintained by ACTH and by glucocorticoids, whereas tyrosine hydroxylase is maintained by ACTH (Axelrod et al., 1970). The activities of both enzymes are increased above normal levels by increased prolonged stimulation of sympathetic nerves. The action of glucocorticoids on phenyl N-methyl-transferase is prevented by actinomycin D and puromycin and therefore involves the synthesis of new protein (Wurtman and Axelrod, 1966). ACTH and corticosterone are also involved in the rapid development of phenyl N-methyltransferase in the adrenals of the newborn rat (Margolies et al., 1966).

Whilst glucose is usually considered to be the normal fuel for the brain, it is now clear that during early development, and in the adult under nutritional stress, ketones may also be used (Hawkins et al., 1971). The pathways involved in ketone body (acetoacetate and 3-hydroxybutyrate) utilization in rat brain are shown in Fig. 2.9 (Williamson and Buckley, 1973). The carbon present in acetoacetate can be made available for lipid synthesis by the action of acetoacetyl-CoA synthetase (EC 6.2.1.3) and acetoacetyl-CoA thiolase (EC 2.3.1.9).

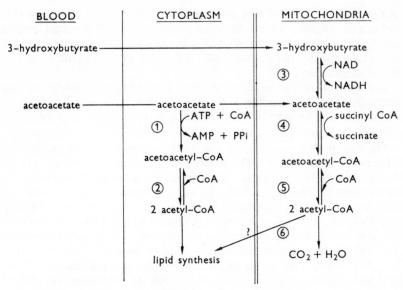

Fig. 2.9. Pathways for the utilization of ketone bodies in the brain (adapted from Williamson and Buckley, 1973)

Enzymes involved: Cytoplasm (1) Acetoacetyl-CoA synthetase
(2) Acetoacetyl-CoA thiolase
Mitochondria (3) 3-Hydroxybutyrate dehydrogenase
(4) 3-Oxoacid CoA-transferase
(5) Acetoacetyl-CoA thiolase
(6) Tricarboxylic acid cycle

The activity of each of these enzymes falls during postnatal life to a value in the adult which is about 30% of that at birth. The activities of two of the mitochondrial enzymes, β-3-hydroxybutyrate dehydrogenase (EC 1.1.1.30) and 3-oxoacid CoA-transferase (EC 2.8.3.5) rise during the suckling period in the rat and fall to a lower level in the adult. Acetoacetyl-CoA thiolase shows a rather different pattern of development, remaining at the neonatal level during suckling and subsequently falling to a lower level in the adult.

High levels of ketones in rat blood during the suckling period have been reported by a number of workers (Williamson and Buckley, 1973) and an increase in enzyme activity in the brain of the suckling rat could therefore be the result of substrate availability. Induction of ketone-body enzymes has been shown to result from changes in the nutritional status of the mother. Thus, starvation of pregnant rats from the 16th day of gestation results in a doubling of the activity of 3-hydroxybutyrate dehydrogenase in the foetal brain at term (Thaler, 1972). Hyperthyroidism, induced from birth with exogenous thyroxine, accelerates the development of D-3-hydroxybutyrate dehydrogenase by approximately 2 days (Grave et al., 1973). Similarly, when pregnant rats were changed from the usual high-carbohydrate diet to one containing 45% fat, the activities of acetoacetyl-CoA thiolase and 3-oxoacid-CoA-transferase were raised in the brains of the pregnant animals and of 20-day-old foetuses (Dierks-Ventling, 1971). No change was noted in the plasma cholesterol of the mothers, and the changes in the enzyme levels were attributed to the ketosis.

In contrast to the enzyme changes reported by Dierks-Ventling (1971) in the brains of pregnant female rats on a high-fat diet, Williamson et al. (1971) reported that enzyme levels were not increased by giving male rats a high-fat diet. The latter workers, in agreement with Daniel et al. (1971) concluded that the utilization of ketone bodies was proportional to the blood concentrations. The absence of an effect of the diet on the enzymes involved in ketone-body utilization could, however, be due to a difference in sex, or to the shorter time (3 days compared with 10 days), that the animals were on the diet.

Feeding diets containing 10% monosodium glutamate to rats through four generations does not interfere with the normal developmental pattern of enzymes (glutamic pyruvic transaminases, EC 2.6.1.2; glutamic-oxaloacetic transaminase, EC 2.6.1.1; and glutamic dehydrogenase, EC 1.4.1.2) concerned with the metabolism of glutamic acid in the brain (Prosky and O'Dell, 1972).

Polyamine biosynthesis has been studied in developing rat brain, although no information seems to be available on the factors involved in its regulation. The activity of ornithine decarboxylase plateaus for the first 11 days of postnatal life and, subsequently, declines to negligible levels (Anderson and Schanberg, 1972). The activity of 5-adenosyl-L-methionine decarboxylase, on the other hand, is fairly constant at a low level during the first 10 days of postnatal life, and then increases rapidly (Schmidt and Cantoni, 1973; Snyder et al., 1973). This increase continues into adulthood in the crebellum and cerebral cortex, but declines in the rest of the brain after 22 days of age (Snyder et al., 1973). Schmidt and Cantoni (1973) suggest that the increase is necessary to maintain the concentration of spermidine, which has a half-life of only 5 days in the brain (Russell et al., 1970a).

2.5. CONCLUSIONS

Realization of genetic potential in animals involves increase in size, and changes in structure and function. Changes in the activities of the multitude of enzymes in the body's tissues play a key role in bringing about these developmental changes. The regulation of the changes of a few enzymes in the small intestine, liver and brain has been

64 J. W. T. DICKERSON AND T. K. BASU

reviewed. Induction of enzyme activity in the foetus is normally effected by hormones, although in some cases the raising of the concentration of a substrate can also induce activity of the appropriate enzyme. An example of this, which is used therapeutically, is the administration of phenobarbital to induce the activity of UDP-glucuronyl transferase in the liver of the neonate and so reduce the likelihood of neonatal hyperbilirubinaemia (Trolle, 1968). Inductive competence is, however,

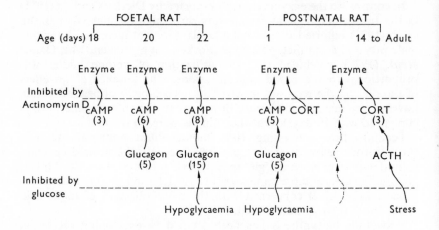

Fig. 2.10. Scheme of sequential development of competence for regulating tyrosine aminotransferase in rat liver (Greengard, 1969)
 cAMP, adenosine cyclic 3′,5′-monophosphate
 CORT, glucocorticoids; ACTH, adrenocorticotrophic hormone
 Numbers shown in parenthesis indicate the ratio:

$$\frac{\text{activity in liver of treated animals}}{\text{activity in liver of untreated animals}}$$

very closely related to the age of the foetus. After birth, substrate induction gradually predominates, although the glucocorticoids continue to play an important role. It seems likely that adenosine 3′, 5′-(cyclic)-monophosphate is an important mediator of hormone action prenatally and in the neonate, but its importance diminishes with age (Fig. 2.10) (Greengard, 1969).

Further elucidation of the factors that regulate the activity of enzymes may well be the key to understanding critical periods of growth such as that described for the brain (Davison and Dobbing, 1966). A better understanding of the role of enzyme induction in the whole process of development might also lead to new approaches to the treatment of inborn errors of metabolism.

ACKNOWLEDGEMENTS

We greatly appreciate the help given to us by our colleagues Dr. K. Snell and Mr. P. A. McAnulty in the preparation of this chapter. We are grateful to the departmental secretaries for typing the manuscript. Professor D. V. Parke has given us constant encouragement throughout its preparation.

REFERENCES

Alpers, D. H. (1969). Separation and isolation of rat and human intestinal β-galactosidases. *J. biol. Chem.*, **244**, 1238

Anderson, T. R. and Schanberg, S. M. (1972). Ornithine decarboxylase activity in developing rat brain. *J. Neurochem.*, **19**, 1471

Argiz, C. A. G., Pasquini, J. M., Kaplún, B. and Gomez, C. J. (1967). Hormonal regulation of brain development II. Effect of neonatal thyroidectomy on succinate dehydrogenase and other enzymes in developing cerebral cortex and cerebellum of the rat. *Brain Res.*, **6**, 635

Assan, R. and Boillot, J. (1971). Pancreatic glucagon, and glucagon-like material in tissues and plasmas from human foetuses 6–26 weeks old. In *Metabolic Processes in the Foetus and Newborn Infant*, p. 210 (Eds J. H. P. Jonxis, H. K. A. Visser and J. A. Troelstra) Kroese; Leiden

Auerbach, V. H. and Weisman, H. A. (1959). Tryptophan peroxidase-oxidase, histidase, and transaminase activity in the liver of the developing rat. *J. biol. Chem.*, **234**, 304

Auricchio, S., Rubino, A. and Mürset, G. (1965). Intestinal glycosidase activities in the human embryo, foetus and newborn. *Pediat.*, **35**, 944

Axelrod, J., Müeller, R. A. and Thoenen, H. (1970). Neuronal and hormonal control of tyrosine hydroxylase and phenylethanolamine *N*-methyltransferase activity. In *New Aspects of Storage and Release Mechanisms of Catecholamines*, p. 212 (Eds. H. J. Schümann and G. Kroneberg) Springer-Verlag; Berlin

Azmitia, E. C., Jr., Algeri, S. and Costa, E. (1970). *In vivo* conversion of ³H-L-tryptophan into ³H-serotonin in brain areas of adrenalectomised rats. *Science*, **169**, 201

Azmitia, E. C. and McEwen, B. S. (1969). Corticosterone regulation of tryptophan hydrolase in the midbrain of the rat. *Science*, **166**, 1274

Baird, J. D. and Farquhar, J. W. (1962). Insulin-secreting capacity in newborn infants of normal and diabetic women. *Lancet*, **1**, 71

Balázs, R. and Cotterrell, M. (1972). Effect of hormonal state on cell number and functional maturation of the brain. *Nature (Lond.)*, **236**, 348

Balázs, R. and Richter, D. (1973). Effects of hormones on the biochemical maturation of the brain. In *Biochemistry of the Developing Brain*, p. 253 (Ed. W. A. Himwich) Marcel Dekker; New York

Bakken, A. F. (1970). Bilirubin excretion in newborn human infants. I. Unconjugated bilirubin as a possible trigger for bilirubin conjugation. *Acta Paediat. Scand.*, **59**, 148

Ballard, F. J. (1971). The development of gluconeogenesis in rat liver. *Biochem. J.*, **124**, 265

Ballard, F. J. and Hanson, R. W. (1967). Phosphoenolpyruvate carboxykinase and pyruvate carboxylase in developing rat liver. *Biochem. J.*, **104**, 866

Ballard, F. J. and Oliver, I. T. (1965). Carbohydrate metabolism in liver from foetal and neonatal sheep. *Biochem. J.*, **95**, 191

66   J. W. T. DICKERSON AND T. K. BASU

Barnabei, O. and Sereni, F. (1964). Cortisol-induced increase of tyrosine-α-keto-glutarate transaminase in the isolated perfused rat liver and its relation to ribonucleic acid synthesis. *Biochim. Biophys. Acta*, **91**, 239

Basu, T. K. (1971). Effect of nutrition on hepatic microsomal drug metabolizing enzymes in growing rats. *Ph.D. Thesis*, University of Surrey

Basu, T. K. and Dickerson, J. W. T. (1974). Inter-relationships of nutrition and the metabolism of drugs. *Chemico-Biological Interactions*, **8**, 193

Basu, T. K., Dickerson, J. W. T. and Parke, D. V. W. (1971a). Effect of development on the activity of microsomal drug metabolising enzymes in rat liver. *Biochem. J.*, **124**, 19

Basu, T. K., Dickerson, J. W. T. and Parke, D. V. (1971b). The effect of diet on rat plasma corticosteroids and liver aromatic hydroxylase activity. *Biochem. J.*, **125**, 16p

Brandt, I. K. (1964). The development of the hippuric acid-synthesizing system in the rat. *Developmental Biology*, **10**, 202

Bresnick, E. and Stevenson, J. G. (1968). Microsomal N-demethylase activity in developing rat liver after administration of 3-methylcholanthrene. *Biochem. Pharmacol.*, **17**, 1815

Brewer, E. N. (1972). Inhibition of alkaline ribonuclease activity by polyamines. *Exp. Cell Res.*, **72**, 586

Brodie, B. B., Gillette, J. R. and LaDu, B. N. (1958). Enzymatic metabolism of drugs and other foreign compounds. *Ann. Rev. Biochem.*, **27**, 427

Burch, H. B., Lowry, O. H., Kuhlman, A. M., Skerjance, J., Diamont, E. J., Lowry, S. R. and Van Dippe, P. (1963). Changes in patterns of enzymes of carbohydrate metabolism in the developing rat liver. *J. biol. Chem.*, **238**, 2267

Burch, H. B., Kuhlman, A. M., Skerjance, J. and Lowry, O. H. (1971). Changes in patterns of enzymes of carbohydrate metabolism in the developing rat kidney. *Pediatrics*, **47**, 199

Cake, M. H. and Oliver, I. T. (1969). The activation of phosphorylase in neonatal rat liver. *Eur. J. Biochem.*, **11**, 576

Campbell, H. J. and Eayrs, J. T. (1965). Influence of hormones on the central nervous system. *Brit. med. Bull.*, **21**, 81

Capková, A. and Jirásek, J. E. (1968). Glycogen reserves in organs of human foetuses in the first half of pregnancy. *Biol. Neonat.*, **13**, 129

Chase, H. P., Volpe, J. J. and Laster, L. (1968). Transsulphuration in mammals: foetal and early development of methionine-activating enzyme and its relation to hormonal influences. *J. clin. Invest.* **47**, 2099

Chatagner, F. and Trautmann, O. (1962). 'Soluble' cysteine desulphurase of rat liver as an adaptive enzyme. *Nature, Lond.*, **194**, 1281

Chiesara, E., Clementi, F., Conti, F. and Meldolesi, J. (1967). The induction of drug-metabolizing enzymes in rat liver during growth and regeneration. *Lab. Invest.*, **16**, 254

Conney, A. H. (1967). Pharmacological implications of microsomal enzyme induction. *Pharmacol. Rev.*, **19**, 317

Conney, A. H. and Klutch, A. (1963). Increased activity and androgen hydroxylases in liver microsomes of rats pretreated with phenobarbital and other drugs. *J. biol. Chem.*, **238**, 1611

Cornblath, M. (1968). Hypoglycaemia. In *Carbohydrate Metabolism and its Disorders*, Vol. 2, p. 51 (Eds. F. Dickens, P. Randle and W. J. Whelan), Academic Press; London

Cramer, J. W., Miller, J. A. and Miller, E. C. (1960). The hydroxylation of the carcinogen 2-acetylamine-fluorene by rat liver: stimulation by pretreatment *in vivo* with 3-methylcholanthrene. *J. biol. Chem.*, **235**, 250

Creaven, P. J. and Parke, D. V. (1966). The stimulation of hydroxylation by carcinogenic and non-carcinogenic compounds. *Biochem. Pharmac.*, **15**, 7

Cuatrecases, P., and Segal, S. (1965). Mammalian galactokinases. Developmental and adaptive characteristics in the rat liver. *J. biol. Chem.*, **240**, 2382

Dallner, G., Siekevitz, P. and Palade, G. E. (1966). Biogenesis of endoplasmic reticulum membranes: structural and chemical differentiation in developing rat hepatocyte. *J. Cell Biol.*, **30**, 73

Daniel, P. M., Love, E. R., Moorehouse, S. R., Pratt, O. E. and Wilson, P. (1971). Factors influencing utilisation of ketone-bodies by brain in normal rats and rats with ketoacidosis. *Lancet*, **ii**, 637

Danovitch, S. H. and Laster, L. (1969). The development of arylsulphatase in the small intestine of the rat. *Biochem. J.*, **114**, 343

Darby, F. J. (1971). Changes in drug-metabolizing activities in the livers of suckling rats as a result of treatment of the lactating mother with phenobarbitone and chlorpromazine. *Biochem. J.*, **122**, 41

Das, M. L., Orrenius, S. and Ernster, L. (1968). On the fatty acid and hydrocarbon hydroxylation in rat liver microsomes. *Europ. J. Biochem.*, **4**, 519

Datta, R. K., Sen, S. and Ghosh, J. J. (1969). Effect of polyamines on the stability of brain-cortex ribosomes. *Biochem. J.*, **114**, 847

Davison, A. N. and Dobbing, J. (1966). Myelination as a vulnerable period in brain development. *Brit. med. Bull.*, **22**, 40

Dawes, G. S. and Shelley, H. J. (1968). Physiological aspects of carbohydrate metabolism in the foetus and newborn. In *Carbohydrate Metabolism and its Disorders*, Vol. 2, p. 87 (Eds. F. Dickens, P. J. Randle and W. J. Whelan), Academic Press; London

Dawkins, M. J. R. (1963). Glycogen synthesis and breakdown in rat liver at birth. *Quart. J. exp. Physiol.*, **48**, 265

De Vellis, J. and Inglish, D. (1968). Hormonal control of glycerol phosphate dehydrogenase in the rat brain. *J. Neurochem.*, **15**, 1061

Dickerson, J. W. T., Basu, T. K. and Parke, D. V. (1971). Protein nutrition and drug-metabolising enzymes in the liver of the growing rat. *Proc. Nutr. Soc.*, **30**, 5A

Dierks-Ventling, C. (1971). Prenatal induction of ketone-body enzymes in the rat. *Biol. Neonate*, **19**, 426

Doell, R. G. and Kretchmer, N. (1962). Studies of the small intestine during development. 1. Distribution and activity of β-galactosidase. *Biochim. Biophys. Acta*, **62**, 353

Doell, R. G. and Kretchmer, N. (1964). Intestinal invertase: precocious development of activity after injection of hydrocortisone. *Science*, **143**, 42

Drahota, Z., Hahn, P., Kleinzeller, A. and Kostolanska, A. (1964). Acetoacetate formation by liver slices from adult and infant rats. *Biochem. J.*, **93**, 61

Dymsza, H. A., Czajka, D. M. and Miller, S. A. (1964). Influence of artificial diet on weight gain and body composition of the neonatal rat. *J. Nutr.*, **84**, 100

Exton, J. H., Mallette, L. E., Jefferson, L. S., Wong, E. H. A., Friedmann, 72, N. and Park, C. R. (1970). Role of adenosine 3',5'-monophosphate in the control of gluconeogenesis. *Amer. J. Clin. Nutr.*, **23**, 993

Eyring, E. J., Crosfeld, J. L. and Connelly, P. A. (1973). Development of aspirin hydrolase in foetal, newborn and adult tissues. *Ann. Surg.*, **177**, 307

Feigelson, M. (1968). Estrogenic regulation of hepatic histidase during postnatal development and adulthood. *J. biol. Chem.*, **243**, 5088

Feuer, G. and Liscio, A. (1969). Origin of delayed development of drug metabolism in the newborn rat. *Nature*, **223**, 68

Feuer, G. and Liscio, A. (1970). Effect of drugs on hepatic drug metabolism in the fetus and newborn. *Int. J. clin. Pharmacol.*, **3**, 30

Finkelstein, J. D. (1967). Methionine metabolism in mammals. Effects of age, diet and hormones on three enzymes of the pathway in rat tissues. *Arch. Biochem. Biophys.*, **122**, 583

Flint, M., Lathe, G. H., Ricketts, T. R. and Silman, G. (1964). Development of glucuronyl transferase and other enzyme systems in the newborn rabbit. *Quart. J. exp. Physiol.*, **49**, 66

Fouts, F. R. (1963). Physiological impairment of drug metabolism. In *Metabolic Factors Controlling Duration of Drug Action*, Vol. 6 (Eds. B. B. Brodie and E. G. Erdos). Proceedings of the First International Pharmacological Meeting, Stockholm. Pergamon; New York

Fouts, J. R. and Adamson, R. H. (1959). Drug metabolism in the newborn rabbit. *Science (N.Y.)*, **129**, 897

Fouts, J. R. and Devereux, T. R. (1972). Development aspects of hepatic and extra-hepatic drug-metabolising enzyme systems: microsomal enzymes and components in rabbit liver and lung during the first month of life. *J. Pharmacol. exp. Therap.*, **183**, 458

Fouts, J. R. and Hart, L. G. (1965). Hepatic drug metabolism during the perinatal period. *Ann. N.Y. Acad. Sci.*, **123**, 245

Friedman, P. A. and Kaufman, S. (1971). A study of the development of phenyl-alanine hydroxylase in fetuses of several mammalian species. *Arch. Biochem. Biophys.*, **146**, 321

Gaull, G., Sturman, J. A., and Räihä, N. C. R. (1972). Development of mammalian sulfur metabolites. Absence of cystathionase in human fetal tissues. *Pediat. Res.*, **6**, 538

Gelehrter, T. D. and Tomkins, G. M. (1967). The role of RNA in the hormonal induction of tyrosine aminotransferase in mammalian cells in tissue culture. *J. molec. Biol.*, **29**, 59

Gentz, J. G., Bengtsson, G., Hakkarainen, J., Hellström, R. and Persson, B. (1970). Metabolic effects of starvation during neonatal period in the piglet. *Am. J. Physiol.*, **218**, 662

Gillette, J. R. (1967). Biochemistry of drug oxidation and reduction by enzymes in hepatic endoplasmic reticulum. *Advanc. Pharmacol.*, **4**, 219

Goldstein, L., Stella, E. J. and Knox, W. E. (1962). The effect of hydrocortisone on tyrosine α-ketoglutarate transaminase and tryptophan pyrrolase activities in the isolated, perfused rat liver. *J. biol. Chem.*, **237**, 1723

Gram, T. E., Guarino, A. M., Schroeder, D. H. and Gillette, G. R. (1969). Changes in certain kinetic properties of hepatic microsomal aniline hydroxylase and ethylmorphine demethylase associated with postnatal development and maturation in male rats. *Biochem. J.*, **113**, 681

Granner, D. K., Hayashi, S., Thompson, E. B. and Tomkins, G. M. (1968). Stimulation of tyrosine aminotransferase synthesis by dexamethasone phosphate in cell culture. *J. Molec. Biol.*, **35**, 291

Grasso, S., Saporito, N., Messina, A. and Reitano, G. (1968). Plasma insulin, glucose and free fatty acid (FFA) response to various stimuli in the premature infant. *Diabetes*, **17**, 306

Grave, G. D., Satterthwaite, S., Kennedy, C. and Sokoloff, L. (1973). Accelerated postnatal development of D(-)-β-hydroxybutyrate activity in the brain in hyperthyroidism. *J. Neurochem.*, **20**, 495

Greenfield, P. C. and Boell, E. J. (1968). Succinic dehydrogenase and cytochrome oxidase of mitochondria of chick liver, heart and skeletal muscle during embryonic development. *J. Exptl. Zool.*, **168**, 491

Greengard, O. (1969). The hormonal regulation of enzymes in prenatal and postnatal rat liver. Effects of adenosine-3′,5′-(cyclic)-monophosphate. *Biochem. J.*, **115**, 19

Greengard, O. (1973). Effects of hormones on development of fetal enzymes. *Clin. Pharmacol. Therap.*, **14**, 721

Greengard, O. and Dewey, H. K. (1967). Initiation by glucagon of the premature development of tyrosine aminotransferase, serine dehydratase, and glucose 6-phosphatase in fetal rat liver. *J. biol. Chem.*, **242**, 2986

Greengard, O. and Dewey, H. K. (1970). The premature deposition or lysis of glycogen in liver of fetal rats injected with hydrocortisone or glucagon. *Dev. Biol.*, **1**, 452

Greengard, O. and Machovich, R. (1972). Hydrocortisone regulation of thymidine kinase in thymus involution and haemopoietic tissues. *Biochim. Biophys. Acta*, **286**, 382

Greengard, O., Federman, M. and Knox, W. E. (1972). Cytomorphometry of developing rat liver and its application to enzymic differentiation. *J. Cell Biol.*, **52**, 261

Greengard, O., Smith, M. A. and Acs, G. (1963). Relation of cortisone and synthesis of ribonucleic acid to induced and developmental enzyme formation. *J. biol. Chem.*, **238**, 1548

Hahn, P. and Skala, J. (1970). Some enzymes of glucose metabolism in the human fetus. *Biol. Neonate*, **16**, 362

Hahn, P. and Skala, J. (1971). The development of some enzyme activities in the gut of the rat. *Biol. Neonate*, **18**, 433

Hahn, P., Vavrouskova, E., Jirasek, J. and Uher, J. (1964). Acetoacetate formation by livers from human fetuses aged 8–17 weeks. *Biol. Neonat.*, **7**, 348

Hannonen, P., Jänne, J. and Raina, A. (1972). Separation and partial purification of S-adenosylmethionine decarboxylase and spermidine and spermine synthases from rat liver. *Biochem. Biophys. Res. Commun.*, **46**, 341

Hart, L. G., Adamson, R. H., Dixon, R. L. and Fouts, J. R. (1962). Stimulation of hepatic microsomal drug metabolism in the newborn and fetal rabbit. *J. Pharmac. exp. Ther.*, **137**, 103

Hawkins, R. A., Williamson, D. H. and Krebs, H. A. (1971). Ketone body utilization by adult and suckling rat brain in vivo. *Biochem. J.*, **122**, 13

Heby, O. and Agrell, I. (1971). Observations on the affinity between polyamines and nucleic acids. *Hoppe-Seyl. Z.*, **352**, 29

Heim, T. and Hull, D. (1966). The effect of propanolol on the calorigenic response in brown adipose tissue of newborn rabbits to catecholamines, glucagon, corticotrophin and cold exposure. *J. Physiol.*, **187**, 271

Herbst, J. J. and Sunshine, P. (1969). Postnatal development of the small intestine of the rat. *Pediat. Res.*, **3**, 27

Hers, H. G., de Wulf, H. and Stalmans, W. (1970). The control of glycogen metabolism in the liver. *FEBS Letters*, **12**, 73

Herzfeld, A. and Knox, W. E. (1968). The properties, developmental formation and estrogen induction of ornithine aminotransferase in rat tissues. *J. biol. Chem.*, **243**, 3327

Holt, P. G. and Oliver, I. T. (1968a). Factors affecting the premature induction of tyrosine aminotransferase in foetal rat liver. *Biochem. J.*, **108**, 333

Holt, P. G. and Oliver, I. T. (1968b). Plasma corticosterone concentrations in the perinatal rat. *Biochem. J.*, **108**, 339

Holt, P. G. and Oliver, I. T. (1969). Multiple forms of tyrosine aminotransferase in rat liver and their hormonal induction in the neonate. *FEBS Letters*, **5**, 89

Holten, D. D. and Kenney, F. T. (1967). Regulation of tyrosine α-ketogluturate transaminase in rat liver. VI. Induction by pancreatic hormones. *J. biol. Chem.*, **242**, 4372

Hommes, F. A. and Beere, A. (1971). The development of adenyl cyclase in rat liver, kidney, brain and skeletal muscle. *Biochim. Biophys. Acta*, **237**, 296

Hommes, F. A. and Wilmink, C. W. (1968). Developmental changes of glycolytic enzymes in rat brain, liver and skeletal muscle. *Biol. Neonat.*, **12**, 181

Howard, E. (1965). Effects of corticosterone and food restriction on growth and on DNA, RNA and cholesterol contents of the brain and liver in infant mice. *J. Neurochem.*, **12**, 181

Hull, D. (1966). The structure and function of brown adipose tissue. *Brit. med. Bull.*, **22**, 92

Illnerova, H. (1968). Activity of urea cycle enzymes in the liver and urea excretion in the rat during development. *Physiol. bohemoslov.*, **17**, 70

Jacquot, R. and Kretchmer, N. (1964). Effect of fetal decapitation on enzymes of glycogen metabolism. *J. biol. Chem.*, **239**, 1301

Jänne, J. (1967). Effect of growth hormone on putrescine and polyamine synthesis in rat liver. *Scand. J. clin. Lab. Invest. Suppl.*, **95**, 46

Jänne, J. and Raina, A. (1968). Stimulation of spermidine synthesis in the regenerating rat liver: relation to increased ornithine decarboxylase activity. *Acta Chem. Scand.*, **22**, 1349

Jänne, J., Raina, A. and Siimes, M. (1968). Mechanism of stimulation of polyamine synthesis by growth hormone in rat liver. *Biochim. Biophys. Acta*, **166**, 419

Johnston, D. I., Bloom, S. R., Greene, K. R. and Beard, R. W. (1972). Failure of the human placenta to transfer pancreatic glucagon. *Biol. Neonate*, **21**, 375

Jondorf, W. R., Maickel, R. P. and Brodie, B. B. (1959). Inability of newborn mice and guinea pigs to metabolize drugs. *Biochem. Pharmacol.*, **1**, 352

Jost, A. (1961). The role of fetal hormones in pre-natal development, *Harvey Lect.*, **55**, 201

Juchau, M. R. (1971). Human placental hydroxylation of 3,4-benzpyrene during early gestation and at term. *Toxicol. appl. Pharmacol.*, **18**, 665

Jumawan, J., Koldovsky, O. and Palmieri, M. (1972). Postnatal changes in α-galactosidase activity in the small intestine of the rat. *Biol. Neonate*, **20**, 380

Kardish, R. and Feuer, G. (1972). Relationship between maternal progesterones and the delayed drug metabolism in the neonate. *Biol. Neonate*, **20**, 58

Kato, R. and Takanaka, A. (1968). Effect of phenobarbital on electron transport system, oxidation and reduction of drugs in liver microsomes of rats of different age. *J. Biochem.*, **63**, 406

Kato, R., Vassanelli, P., Frontino, G. and Chiesara, E. (1964). Variations of the activity of liver microsomal drug metabolizing enzymes in relation to age. *Biochem. Pharmacol.*, **13**, 1037

Kaye, A. M., Icekson, I. and Lindner, H. R. (1972). Stimulation by oestrogens of ornithine and S-adenosylmethionine decarboxylases in the immature rat uterus. *Biochim. Biophys. Acta*, **252**, 150

Kaye, A. M., Icekson, I., Lamprecht, S. A., Gruss, R., Tsafriri, A. and Lindner, H. R. (1973). Stimulation of ornithine decarboxylase activity by luteinizing hormone in immature and adult rat ovaries. *Biochemistry, U.S.A.*, **12**, 3072

Kennan, A. L. and Cohen, P. P. (1959). Biochemical studies of the developing mammalian fetus. *Develop. Biol.*, **1**, 511

Kenney, F. T. (1967). Regulation of tyrosine α-ketoglutarate transaminase in rat liver. V. Repression in growth hormone-treated rats. *J. biol. Chem.*, **242**, 4367

Kenney, F. T. and Flora, R. M. (1961). Induction of tyrosine-α-ketoglutarate transaminase in rat liver. I. Hormonal nature. *J. biol. Chem.* **236**, 2699

Kirby, L. and Hahn, P. (1973). Enzyme induction in human fetal liver. *Pediat. Res.*, **7**, 75

Knox, W. E. and Auerbach, V. H. (1955). The hormonal control of tryptophan peroxidase in the rat. *J. biol. Chem.*, **214**, 307

Kobayashi, Y., Kupelian, J. and Maudsley, D. V. (1971). Ornithine decarboxylase stimulation in rat ovary by luteinizing hormone. *Science, N. Y.*, **172**, 379

Koldovský, O. (1972). Hormonal and dietary factors in the development of digestion and absorption. In *Nutrition and Development*, p. 135 (Ed. M. Winick), Wiley; London and N. Y.

Koldovský, O. and Sunshine, P. (1970). Effect of cortisone on the developmental pattern of the neutral and the acid β-galactosidase of the small intestine of the rat. *Biochem. J.*, **117**, 467

Krasner, J. (1968). Sulfobromophthalein-glutathione conjugating enzyme during mammalian development. *Amer. J. Dis. Child.*, **115**, 267

Kretchmer, N., Hurwitz, R. and Räihä, N. C. R (1966) Some aspects of urea and pyrimidine metabolism during development. *Biol. neonat.* (*Basel*), **9**, 187

Kunze, E., Schauer, A., Aures, D. and Gärtner, H. (1972). Influence of cortisol-21-hemisuccinate on the specific histidine decarboxylase of the foetal rat liver. *Biol. Neonate*, **21**, 463

Lamprecht, S. A., Zor, U., Tsafriri, A. and Lindner, H. R. (1973). Action of prostaglandin E. and of luteinizing hormone on ovarian adenylate cyclase, protein kinase and ornithine decarboxylase activity during postnatal development and maturity in the rat. *J. Endocr.*, **57**, 217

Lea, M. A. and Walker, D. G. (1964). The metabolism of glucose 6-phosphate in developing mammalian tissues. *Biochem. J.*, **91**, 417

Levine, S., Glick, D. and Nakane, P. K. (1967). Adrenal and plasma corticosterone and vitamin A in rat adrenal glands during post-natal development. *Endocrinology*, **80**, 910

Levy, G. and Ertel, I. J. (1971). Effect of bilirubin on drug conjugation in children. *Pediatrics*, **47**, 811

Light, I. J., Berry, H. K. and Sutherland, J. M. (1966). Aminoacidemia of prematurity. *Amer. J. Dis. Child.*, **112**, 229

Lin, E. C. C. and Knox, W. E. (1957). Adaptation of the rat liver. Tyrosine-α-ketoglutarate transaminase. *Biochim. Biophys. Acta*, **26**, 85

Lockwood, E. A., Bailey, E. and Taylor, C. B. (1970). Factors involved in changes in hepatic lipogenesis during development of the rat. *Biochem. J.*, **118**, 155

McCance, R. A. and Dickerson, J. W. T. (1957). The composition and origin of the fetal fluids of the pig. *J. embryol. exp. Morphol.*, **5**, 43

McCance, R. A. and Widdowson, E. M. (1956). Metabolism, growth and renal function of piglets in the first days of life. *J. Physiol.*, **133**, 373

McCance, R. A. and Widdowson, E. M. (1959). The effect of lowering the ambient temperature on the metabolism of the newborn pig. *J. Physiol.*, **147**, 124

Macgregor, R. K. and Mahler, H. R. (1967). RNA synthesis in intact rat liver nuclei. *Arch. Biochem. Biophys*, **120**, 136

McLain, C. R. (1963). Amniography studies of the gastrointestinal motility of the human fetus. *Amer. J. Obstet. Gynecol.*, **86**, 1079

Macleod, S. M., Renton, K. W. and Eade, N. R. (1972). Development of hepatic microsomal drug-oxidizing enzymes in immature male and female rats. *J. Pharmacol. exp. Therap.*, **183**, 489

Margolies, F., Roffi, J. and Jost, A. (1966). Norepinephrine in fetal rats. *Science*, **154**, 275

Masters, C. J. and Holmes, R. S. (1972). Isoenzymes and ontogeny. *Biol. Rev.*, **47**, 309

Menkes, J. H., Welcher, D. W., Levi, H. S., Dallas, J., and Gretsky, N. E. (1972). Relationship of evaluated blood tyrosine to the ultimate intellectual performance of premature babies. *Pediatrics*, **49**, 218

Mersmann, H. J. (1971). Glycolytic and gluconeogenic enzyme levels in pre- and postnatal pigs. *Amer. J. Physiol.*, **220**, 1297

Miller, T. B., Exton, J. H. and Park, C. R. (1971). A block in epinephrine-induced glycogenolysis in hearts from adrenalectomised rats. *J. biol. Chem.*, **246**, 3672

Milner, R. D. G. (1971). The development of insulin secretion in man. In *Metabolic Processes in the Foetus and Newborn Infant*, p. 193 (Eds. J. H. P. Jonxis, H. K. A. Visser and J. A. Troelstra), Kroese; Leiden.

Milner, R. D. G. and Wright, A. D. (1967). Plasma glucose, non-esterified fatty acid, insulin and growth hormone response to glucagon in the newborn. *Clin. Sci.*, **32**, 249

Moog, F. (1953). The functional differentiation of the small intestine. III. The influence of the pituitary-adrenal system on the differentiation of phosphatase in the duodenum of the suckling mouse. *J. Exp. Zool.*, **124**, 329

Moog, F. and Thomas, E. T. (1955). The influence of various adrenal and gonadal steroids on the accumulation of alkaline phosphatase in the duodenum of the suckling mouse. *Endocrinology*, **56**, 187

Moscona, A. A. and Piddington, R. (1966). Stimulation by hydrocortisone of premature changes in the developmental pattern of glutamic synthetase in embryonic retina. *Biochim. Biophys. Acta*, **121**, 409

Müeller, R. A., Thoenen, H. and Axelrod, J. (1969). Increase in tyrosine hydroxylase activity after reserpine administration. *J. Pharmacol. Exp. Ther.*, **169**, 74

Müller, D., Reichenbach, F. and Klinger, W. (1971). Die Aktivität der Nitroreduktase und deren Induzierbarkeit durch Barbital in der Leber von Ratten verschiedenen Alters. *Acta biol. med. germ.*, **27**, 605

Murphy, B. E. P. and Diez d'Aux, R. C. (1972). Steroid levels in the human fetus: cortisol and cortisone. *J. Clin. Endocrinol.*, **35**, 678

Nemeth, A. M. (1954). Glucose-6-phosphatase in the liver of the fetal guinea pig. *J. biol. Chem.*, **208**, 773

Nemeth, A. M. (1959). Mechanisms controlling changes in tryptophan peroxidase activity in developing mammalian liver. *J. biol. Chem.*, **234**, 2921

Nemeth, A. M. and Nachmias, V. T. (1958). Changes in tryptophan peroxidase activity in developing liver. *Science*, **128**, 1085

Newey, H. (1967). Absorption of carbohydrates. *Brit. med. Bull.*, **23**, 236

Omura, T. and Sato, R. (1964). The carbon monoxide binding pigment of liver microsomes. *J. biol. Chem.*, **239**, 2370

Omura, T., Sato, R., Cooper, D. Y., Rosenthal, O. and Estabrook, R. W. (1965). Function of cytochrome P-450 of microsomes. *Fed. Proc.*, **24**, 1181

Orci, L., Lambert, E., Rouiller, C. L., Renold, A. E. and Samols, E. (1969). Evidence for the presence of A-cells in the endocrine fetal pancreas of the rat. *Hormone Metab. Res.*, **1**, 108

Paulini, K., Beneke, G. and Kulka, R. (1971). Zytophotometrische und autoradiographische Untersuchungen zur Lebervergrösserung nach Phenobarbital in Abhängigkeit vom Lebensalter. *Beitr. Path.*, **144**, 319

Pegg, A. E. and Williams-Ashman, H. G. (1968). Stimulation of the decarboxylation of S-adenosylmethionine by putrescine in mammalian tissues. *Biochem. Biophys. Res. Commun.*, **30**, 76

Pelichová, A., Koldovský, O., Heringová, A., Jirsová, V. and Kraml, J. (1967). Postnatal changes of activity and electrophoretic pattern of jejunal and ileal nonspecific esterase and alkaline phosphatase of the rat. Effect of adrenalectomy. *Canad. J. Biochem.*, **45**, 1375

Pelkonen, O. (1973). Drug metabolism in the human fetal liver relationship to fetal age. *Arch. Inter. Pharmacodyn. et Ther.*, **202**, 281

Pelkonen, O., Jouppila, P. and Kärki, N. T. (1973). Attempts to induce drug metabolism in human fetal liver and placenta by the administration of phenobarbital to mothers. *Arch. Inter. Pharmacodyn. et Ther.*, **202**, 288

Pestaña, A. (1969). Dietary and hormonal control of enzymes of amino acid catabolism in liver. *Europ. J. Biochem.*, **11**, 400

Philippidis, H. and Ballard, F. J. (1969). The development of gluconeogenesis in rat liver. Experiments *in vivo*. *Biochem. J.*, **113**, 651

Piddington, R. and Moscona, A. A. (1967). Precocious induction of retinal glutamine synthetase by hydrocortisone in the embryo and in culture. Age dependent differences in tissue response. *Biochim. Biophys. Acta*, **141**, 429

Pitot, H. C. and Yatvin, M. B. (1973). Interrelationships of mammalian hormones and enzyme levels *in vivo*. *Physiol. Revs.*, **53**, 229

Prosky, L. and O'Dell, R. G. (1972). Lack of an effect of dietary monosodium-L-glutamate on some glutamate-metabolizing enzymes in developing rat brain. *J. Neurochem.*, **19**, 1405

Räihä, N. C. R. (1971). The development of some enzymes of amino acid metabolism in human liver. In *Metabolic Processes in the Foetus and Newborn Infant*,

p. 26 (Eds. J. H. P. Jonxis, H. K. A. Visser and J. A. Troelstra), Stenfert Kroese; Leiden

Räihä, N. C. R. (1973). Phenylalanine hydroxylase in human liver during development. *Pediat. Res.*, **7**, 1

Räihä, N. C. R., Jänne, J. and Suihkonen, J. (1967). Effects of partial hepatectomy on enzymes concerned with urea synthesis in rat liver. *Scand. J. clin. Lab. Invest. Suppl.*, **95**, 45

Räihä, N. C. R. and Kekomaki, M. P. (1968). Studies on the development of ornithine-ketoacid aminotransferase activity in rat liver. *Biochem. J.*, **108**, 521

Räihä, N. C. R. and Schwartz, A. L. (1973). Development of urea biosynthesis and factors influencing the activity of the arginine synthetase system in perinatal mammalian liver. In *Inborn Errors of Metabolism*, p. 221 (Eds. F. A. Hommes and C. J. Van Den Berg). Academic Press; London

Räihä, N. C. R., Schwartz, A. L. and Lindroos, M. C. (1971). Induction of tyrosine-α-ketoglutarate transamines in fetal rat and fetal human liver in organ culture. *Pediat. Res.*, **5**, 70

Räihä, N. C. R. and Suihkonen, J. (1968a). Factors influencing the development of urea-synthesizing enzymes in rat liver. *Biochem. J.*, **107**, 793

Räihä, N. C. R. and Suihkonen, J. (1968b). Development of urea-synthesizing enzymes in human liver. *Acta paediat. Scand.*, **57**, 121

Raina, A., Jänne, J. and Siimes, M. (1966). Stimulation of polyamine synthesis in relation to nucleic acids in regenerating rat liver. *Biochim. Biophys. Acta*, **123**, 197

Raina, A. and Telaranta, T. (1967). Association of polyamines and RNA in isolated subcellular particles from rat liver. *Biochim. Biophys. Acta*, **138**, 200

Rane, A., Berggren, M., Yaffe, S. and Ericsson, L. E. (1973). Oxidative drug metabolism in the perinatal rabbit liver and placenta: a biochemical and morphologic study. *Xenobiotica*, **3**, 37

Rappoport, D. A. and Fritz, R. R. (1972). Molecular biology of developing brain. In *The Structure and Function of Nervous Tissue*, Vol. VI, p. 273 (Ed. G. H. Bourne), Academic Press; London

Raychaudhuri, C. and Desai, I. D. (1972). Regulation of lysosomal enzymes—III. Dietary induction and repression of intestinal acid hydrolases during development. *Comp. Biochem. Physiol.*, **41B**, 343

Reif, L. and Amos, H. (1966). A dialyzable inducer for the glutamotransferase of chick embryo retina. *Biochem. biophys. Res. Commun.*, **23**, 39

Reif-Lehrer, L. and Amos, H. (1968). Hydrocorticone requirement for the induction of glutamine synthetase in chick-embryo retinas. *Biochem. J.*, **106**, 425

Reynolds, R. D and Potter, V. R. (1971). Neonatal rise of rat liver tyrosine aminotransferase and serine dehydratase activity associated with lack of food. *Life Sci.*, **10**, Pt. II, 5

Richman, R. A., Underwood, L. E., Van Wyk, J. J. and Voina, S. J. (1971). Synergistic effect of cortisol and growth hormone on hepatic ornithine decarboxylase activity. *Proc. Soc. Exp. Biol. Med.*, **138**, 880

Rizzardini, M. and Abeliuk, P. (1971). Tyrosinemia and tyrosinuria in low birth weight infants. *Amer. J. Dis. Child.*, **121**, 182

Robertson, A. F. and Sprecher, H. (1968). A review of human placental lipid metabolism and transport. *Acta. Ped. Scand. Suppl.*, **183**, 3

Rosensweig, N. S., Herman, R. H. and Stifel, F. B. (1971). Dietary regulations of small intestinal enzyme activity in man. *Amer. J. Clin. Nutr.*, **24**, 65

Ross, L. and Goldsmith, E. B. (1955). Histochemical studies of effects of cortisone on fetal and newborn rats. *Proc. Soc. exp. Biol. Med.*, **90**, 50

Rubaltelli, F. F. (1972). Serum conjugated bilirubin in newborns of mothers treated with immunosuppressive drugs. *Acta. Ped. Scand.*, **61**, 606

Russell, A. and Snyder, S. H. (1968). Amine synthesis in rapidly growing tissues. *Proc. Nat. Acad. Sci.* (Wash.), **60**, 1420

Russell, D. H. and Lombardini, J. B. (1971). Polyamines: (1) Enhanced S-adenosyl-L-methionine decarboxylase in rapid growth systems, and (2) the relationship between polyamine concentrations and RNA accumulation. *Biochim. Biophys. Acta*, **240**, 273

Russell, D. H. and McVicker, T. A. (1972). Polyamines in the developing rat and in supportive tissues. *Biochim. Biophys. Acta*, **259**, 247

Russell, D. H., Levy, C. C., Taylor, R. L., Gfeller, E. E. and Sterns, D. N. (1971). Nucleolar RNA polymerase activity in response to polyamines. *Fed. Proc.*, **30**, 1093

Russell, D. H., Medina, V. J. and Snyder, S. H. (1970a). The dynamics of synthesis and degradation of polyamines in normal and regenerating rat liver and brain. *J. biol. Chem.*, **245**, 6732

Russell, D. H. and Snyder, S. H. (1969). Amine synthesis in regenerating rat liver. *Endocrinology*, **84**, 223

Russell, D. H., Snyder, S. H. and Medina, V. J. (1970b). Growth hormone induction of ornithine decarboxylase in rat liver. *Endocrinology*, **86**, 1414

Ryder, E. (1970). Effect of development on chicken liver acetyl-coenzyme A carboxylase. *Biochem. J.*, **119**, 929

Schaeffer, L. D., Chenoweth, M. and Dunn, A. (1969a). Adrenal corticosteroid involvement in the control of liver glycogen phosphorylase activity. *Biochim. Biophys. Acta*, **192**, 292

Schaeffer, L. D., Chenoweth, M. and Dunn, A. (1969b). Adrenal corticosteroid involvement in the control of phosphorylase in muscle. *Biochim. Biophys. Acta*, **192**, 304

Schapiro, S., Geller, E. and Eiduson, S. (1962). Corticoid response to stress in the steroid-inhibited rat. *Proc. Soc. exp. Biol. Med.*, **109**, 935

Schaub, J., Gutmann, I. and Lippert, H. (1972). Developmental changes of glycolytic and gluconeogenic enzymes in fetal and neonatal rat liver. *Horm. Metab. Res.*, **4**, 110

Schimke, R. T. (1962). Adaptive characteristics of urea cycle enzymes in the rat. *J. biol. Chem.*, **237**, 459

Schimke, R. T. (1966). Studies on the roles of synthesis and degradation in the control of enzyme levels in animal tissues. *Bull. Soc. Chim. Biol.*, **48**, 1009

Schlede, E. and Merker, H.-J. (1972). Effect of benzo(α)pyrene treatment on the benzo(α)pyrene hydroxylase activity in maternal liver, placenta, and fetus of the rat during day 13 to day 18 of gestation. *Naunyn-Schmiederberg's Arch. Pharmak.*, **272**, 89

Schmidt, G. L. and Cantoni, G. L. (1973). Adenosylmethionine decarboxylase in developing rat brain. *J. Neurochem.*, **20**, 1373

Schmidt, M. J. and Robison, G. A. (1972). The effect of neonatal thyroidectomy on the development of the adenosine 3',5'-monophosphate system in the rat brain. *J. Neurochem.*, **19**, 937

Schor, J. M. and Frieden, E. (1958). Induction of tryptophan peroxidase of rat liver by insulin and alloxan. *J. biol. Chem.*, **233**, 612

Schwark, W. S., Singhal, R. L. and Ling, G. M. (1972). Metabolic control mechanisms in mammalian systems—Regulation of key glycolytic enzymes in developing brain during experimental cretinism. *J. Neurochem.*, **19**, 1171

Schwartz, S., Ibrahim, G. and Watson, C. J. (1964). The contribution of non-haemoglobin hemes to the early labelling of bile bilirubin. *J. Lab. clin. Med.*, **64**, 1003

Sereni, F. and Barnabei, O. (1967). Nuclear ribonucleic acid polymerase activity, rate of ribonucleic acid synthesis and hydrolysis during development. The role of glucocorticoids. *Advanc. enzyme Regulat.*, **5**, 165

Serini, F., Kenney, F. T. and Kretchmer, N. (1959). Factors influencing the development of tyrosine α-ketoglutarate transaminase activity in rat liver. *J. biol. Chem.*, **234**, 609

Shelley, H. J. (1961). Glycogen reserves and their changes at birth and in anoxia. *Brit. med. Bull.*, **17**, 137

Shelley, H. J. and Nelligan, G. A. (1966). Neonatal hypoglycaemia. *Brit. med. Bull.*, **22**, 34

Short, C. R. and Davis, L. E. (1970). Perinatal development of drug metabolizing enzyme activity in swine. *J. Pharmacol. Exptl. Ther.*, **174**, 185

Siekevitz, P. and Palade, G. E. (1962). Cytochemical study on the pancreas of the guinea pig. VII. Effects of spermine on ribosomes. *J. cell Biol.*, **13**, 217

Smith, S. and Abraham, S. (1970). Fatty acid synthesis in developing mouse liver. *Arch. Biochem. Biophys.*, **136**, 112

Snell, K. (1972/73). The neonatal development of rat liver L-homoserine dehydratase activity. *Enzyme*, **14**, 193

Snell, K. and Walker, D. G. (1972). The adaptive behaviour of isoenzyme forms of rat liver alanine aminotransferases during development. *Biochem. J.*, **128**, 403

Snell, K. and Walker, D. G. (1973a). Glucose metabolism in the newborn rat. Temporal studies *in vivo*. *Biochem. J.*, **132**, 739

Snell, K. and Walker, D. G. (1973b). Glucose metabolism in the newborn rat. Hormonal effects *in vivo*. *Biochem. J.*, **134**, 899

Snyder, S. H., Kreuz, D. S., Medina, V. J. and Russell, D. H. (1970). Polyamine synthesis and turnover in rapidly growing tissues. *Ann. N.Y. Acad. Sci.*, **171**, 749

Snyder, S. H., Shaskan, E. G. and Harik, S. I. (1973). Polyamine disposition in the central nervous system. In *Polyamines in Normal and Neo-plastic Growth*, p. 199 (Ed. D. H. Russell), Raven Press; New York

Snyderman, S. E. (1971). The protein and amino acid requirements of the premature infant. In *Metabolic Processes in the Foetus and Newborn Infant*, p. 128 (Eds. J. H. P. Jonxis, H. K. A. Visser and J. A. Troelstra) Stenfert Kroese; Leiden

Song, C. S., Moses, H. L., Rosenthal, A. S., Gelb, N. A. and Kappas, A. (1971). The influence of postnatal development on drug-induced hepatic porphyria and the synthesis of cytochrome P-450. *J. exper. Med.*, **134**, 1349

Sriratanaban, A., Smynkywicz, L. A. and Thayer, W. R. (1971). Effect of physiological concentration of lactose on prevention of postweaning decline of intestinal lactase. *Amer. J. Dig. Dis.*, **16**, 839

Šťastný, F. (1971). Hydrocortisone as a possible inductor of Na$^+$-K$^+$-ATPase in the chick embryo cerebral hemispheres. *Brain Res.*, **25**, 397

Stern, L., Khanna, N., Levy, G. and Yaffe, S. J. (1970). Effect of phenobarbital on hyperbilirubinemia and glucuronide formation in newborns. *Amer. J. Dis. Child.*, **120**, 26

Sutherland, E. W. and Rall, T. W. (1960). The relation of adenosine-3′,5′-phosphate and phosphorylase to the actions of catecholamines and other hormones. *Pharmacol. Rev.*, **12**, 265

Swiatek, K. R., Kipnis, D. M., Mason, G., Chao, K. L. and Cornblath, M. (1968). Starvation hypoglycemia in newborn pigs. *Am. J. Physiol.*, **214**, 400

Szijan, I., Chepelinsky, A. B. and Piras, M. M. (1970). Effects of neonatal thyroidectomy on enzymes in sub-cellular fraction of rat brain. *Brain Res.*, **20**, 313

Tanaka, T., Harano, Y., Sue, F. and Morimura, H. (1967). Crystallization, characterization and metabolic regulation of two types of pyruvate kinase isolated from rat tissues. *J. Biochem.*, **62**, 71

Taylor, C. B., Bailey, E. and Bartley, W. (1967). Changes in hepatic lipogenesis during development of the rat. *Biochem. J.*, **105**, 717

Tepper, T. and Hommes, F. A. (1970). Changes in activity and isoenzyme patterns of glycolytic enzymes in the developing rat liver. In *Enzymes and Isoenzymes, Structure, Properties and Function*, p. 209 (Ed. D. Shuyar), Academic Press; London

Thaler, M. M. (1972). Effects of starvation on normal development of β-hydroxy-butyrate dehydrogenase activity in foetal and newborn rat brain. *Nature New Biol.*, **236**, 140

Thoenen, H. (1970). Induction of tyrosine hydroxylase in peripheral and central adrenergic neurones by cold-exposure in rats. *Nature*, **228**, 861

Trolle, D. (1968). Decrease of total serum-bilirubin concentration in newborn infants after phenobarbitone treatment. *Lancet*, **ii**, 705

Vernon, R. G. and Walker, D. G. (1968a). Changes in activity of some enzymes involved in glucose utilization and formation in developing rat liver. *Biochem. J.*, **106**, 321

Vernon, R. G. and Walker, D. G. (1968b). Adaptive behaviour of some enzymes involved in glucose utilization and formation in rat liver during the weaning period. *Biochem. J.*, **106**, 331

Villee, D. B. (1973). Changes in fetal steroid metabolism with age. *Clin. Pharmacol. Therap.*, **14**, 705

Volpe, J. J. and Kishimoto, Y. (1972). Fatty acid synthetase of brain: development, influence of nutritional and hormonal factors and comparison with liver enzyme. *J. Neurochem.*, **19**, 737

Volpe, J. J. and Laster, L. (1972). Transsulphuration in fetal and postnatal mammalian liver and brain. Cystathionine synthase, its relation to hormonal influences, and cystathimine. *Biol. Neonat.*, **20**, 385

Volpe, J. J., Lyles, T. O., Roncari, D. A. K. and Vagelos, P. R. (1973). Fatty acid synthetase of developing brain and liver. *J. biol. Chem.*, **248**, 2502

Walker, D. G. (1963). The postnatal development of hepatic fructokinase. *Biochem. J.*, **87**, 576

Walker, D. G. (1971). Development of enzymes for carbohydrate metabolism. In *The Biochemistry of Development*, p. 77 (Ed. P. F. Benson), Heinemann Medical; London

Walker, D. G. and Eaton, S. W. (1967). Regulation of development of hepatic glucokinase in the neonatal rat by the diet. *Biochem. J.*, **105**, 771

Walker, D. G. and Holland, G. (1965). The development of hepatic glucokinase in the neonatal rat. *Biochem. J.*, **97**, 845

Walker, D. G. and Khan, H. H. (1968). Some properties of galactokinase in the developing rat liver. *Biochem. J.*, **108**, 169

Walker, D. G., Khan, H. H. and Eaton, S. W. (1965/66). Enzymes catalyzing the phosphorylation of hexoses in neonatal animals. *Biol. neonat.* (Basel), **9**, 224

Welch, R. M., Harrison, Y. E., Gommi, B. W., Poppers, P. J., Finster, M. and Conney, A. H. (1969). Stimulatory effect of cigarette smoking on the hydroxylation of 3,4-benzpyrene and the *N*-dimethylation of 3-methyl-4-monomethyl aminoazobenzene by enzymes in human placenta. *Clin. Pharmacol. Ther.*, **10**, 100

Wetterberg, L., Yuwiler, A. and Geller, E. (1969). Tryptophan oxygenase changes following δ-aminolevulinic acid administration in the rat. *Life Sci.*, **8**, 1047

Wicks, W. D. (1968a). Induction of tyrosine-α-ketoglutarate transaminase in fetal rat liver. *J. biol. Chem.*, **243**, 900

Wicks, W. D. (1968b). Tyrosine-α-ketoglutarate transaminase: induction by epinephrine and adenosine-3′,5′-cyclic phosphate. *Science*, **160**, 997

Widdowson, E. M. (1950). Chemical composition of newly born mammals. *Nature*, **166**, 626

Williams, R. T. (1967). Comparative patterns of drug metabolism. *Fed. Proc.*, **26**, 1029

Williamson, D. H., Bates, M. W., Page, M. A. and Krebs, H. A. (1971). Activities of enzymes involved in acetoacetate utilization in adult mammalian tissues. *Biochem. J.*, **121**, 41

Williamson, D. H. and Buckley, B. M. (1973). The role of ketones bodies in brain development. In *Inborn Errors of Metabolism*, p. 81 (Eds. F. A. Hommes and C. J. Van Den Berg), Academic Press; London

Wilson, J. T. (1968). Prevention of the normal postnatal increase in drug-metabolizing enzyme activity in rat liver by a pituitary tumour. *Pediat. Res.*, **2**, 514

Wilson, J. T. (1970). Alteration of normal development of drug metabolism by injection of growth hormone. *Nature*, **225**, 861

Wilson, J. T. (1972). Developmental pharmacology: a review of its application to clinical and basic science. *Ann. Rev. Pharmacol.*, **12**, 423

Wittels, B. and Bressler, R. (1965). Lipid metabolism in the newborn heart. *J. clin. Invest.*, **44**, 1639

Woods, J. S. and Dixon, R. L. (1970). Neonatal differences in the induction of hepatic aminolevulinic acid synthetase. *Biochem. Pharmacol.*, **19**, 1951

Wools, W. R. and McPhillips, J. J. (1966). Hepatic glycogen concentration and hexobarbital sleeping time. *Proc. Soc. exp. Biol. Med.*, **121**, 399

Wurtman, R. J. and Axelrod, J. (1966). Control of enzymatic synthesis of adrenaline medulla by adrenal cortical steroid. *J. biol. Chem.*, **241**, 2301

Yaffe, S. J., Rane, A., Sjöqvist, F. Boréus, L. O., and Orrenius, S. (1970). The presence of a monooxygenase system in human foetal liver microsomes. *Life Sci.*, **9**, 1189

Yeung, D. and Oliver, I. T. (1967). Development of gluconeogenesis in neonatal rat liver. Effect of premature delivery. *Biochem. J.* **105**, 1229

Yeung, D. and Oliver, I. T. (1968a). Induction of phosphopyruvate carboxylase in neonatal rat liver by adenosine 3′,5 cyclic monophosphate. *Biochemistry*, **7**, 3231

Yeung, D. and Oliver, I. T. (1968b). Factors affecting the premature induction of phosphopyruvate carboxylase in neonatal rat liver. *Biochem. J.* **108**, 325

Zamboni, L. (1965). Electron microscopic studies of blood embryogenesis in humans. I. The ultrastructure of the foetal liver. *J. Ultrastruct. Res.*, **12**, 509

Chapter 3

Mechanism of Steroid Hormone Action at the Cellular Level

Mels Sluyser
Antoni van Leeuwenhoek-Huis,
The Netherlands Cancer Institute,
Amsterdam, The Netherlands

3.1. INTRODUCTION

Investigators have, in general, approached the study of steroid hormone action in two ways. In the first approach, the hormone is labelled with a radioactive isotope, and its interaction with cellular components is studied. In the second approach, the physiological changes due to the action of the hormone are investigated. These changes are very diverse (Table 3.1), and at first sight it seems an unsurmountable task to find an underlying unitary mechanism. Although the molecular basis of these hormone-induced phenomena is at the present time only poorly understood, it is possible to make some simplifying assumptions. Since the pioneering studies of Williams-Ashman (1965) it is clear that one of the most important ways a hormone influences cell activity is by activating specific genes, However, that is not the complete story. There is evidence that hormones can have several effects at the level of nucleolar and chromosomal transcription, and it has been claimed that they not only control the rate of RNA synthesis and the transport of ribosomes, but also certain translational processes taking place in the cytoplasm. These studies have been reviewed elsewhere (Hamilton, 1971; Tata, 1968). The present paper will confine itself to tracing the whereabouts of the hormone molecules in target tissues, and discussing the interactions which take place between the hormone and specific macromolecular components of the target cell. No attempt is made to give a complete survey of the field. Only those data are

TABLE 3.1

Some of the More Common Physiological Effects of Steroid Hormones
(Grant, 1969)

Hormone	Effects
Oestradiol	Stimulates growth of female reproductive tract and mammary glands; controls oestrus and menstrual cycles; inhibits gonadotrophin secretion; increases plasma lipids, lipoproteins and calcium. Increases protein synthesis in avian liver.
Testosterone	Stimulates (a) differentiation of male reproductive organs *in utero* and adult development, (b) development of male secondary sexual characteristics in mammals, birds, fish, amphibia and reptiles, (c) spermatogenesis, fructose and citrate formation in vesicular glands, (d) nitrogen retention (anabolic effects).
Progesterone	Acts synergistically with oestradiol in maintaining uterine growth; inhibits gonadotrophin production by the pituitary; stimulates galactose metabolism in rabbit liver; may affect central nervous system.
Corticosteroids	Stimulate glycogen deposition in liver and increase tryptophan and amino acid: α-oxoglutarate transaminases in liver; generally anabolic in liver, but catabolic in muscle and lymphoid tissues; anti-inflammatory; stimulate methylation of nor-adrenaline; increase renal tubular sodium transport and maintenance of Na^+/K^+ ratio in muscle and brain, and control secretion of NaCl in salt gland of aquatic birds; inhibit ACTH secretion (cortisol) and angiotensin production (aldosterone).
Ecdysone	Stimulates moulting of epidermis, producing pupal and imaginal ecdysis. A moth or flying insect finally emerges, its cuticle hardened ("sclerotized") to provide rigidity, required for flight. DOPA decarboxylase involved is "increased" by ecdysone.

included which seem pertinent to elucidating how steroid hormones are
able to induce the synthesis of enzymes and other proteins in animal
tissues.

3.2. THE ACTIVE FORM OF THE HORMONE

The first question that comes to mind is: does the steroid molecule
which is injected act as such, or does it first have to be metabolized
in order to exert the hormonal effect? This problem was investigated
by Jensen and his co-workers (Jensen and Jacobson, 1962) with regard

to oestradiol. They found that up to 6 hours after administration of physiological amounts of tritiated oestradiol to rats, the only radioactive substance present in the uterus was free oestradiol, although the amount recovered was small (only about 0.1–0.2% of the amount administered). A striking characteristic of oestradiol-responsive tissues such as uterus and vagina was their ability to incorporate and retain oestradiol for a prolonged period of time. These findings indicate that oestradiol stimulates uterine growth without undergoing metabolic transformation.

In contrast, there is evidence that in some cases testosterone is metabolized to 5α-dihydrotestosterone, and that this metabolite exerts the hormonal effect. When radioactive testosterone is administered to castrated rats, the cell nuclei of seminal vesicle and ventral prostate incorporate and retain 5α-dihydrotestosterone (Anderson and Liao, 1968; Bruchovsky and Wilson, 1968). The metabolite can be found in the prostate cell nuclei within a few minutes after the injection of the radioactive testosterone and can be detected in the nuclei even 6 to 16 hours later (Bruchovsky and Wilson, 1968). The selective retention of 5α-dihydrotestosterone by prostate cell nuclei can be reproduced *in vitro* by incubating minced prostate glands with radioactive testosterone, androstenedione or 5α-dihydrotestosterone. The 5α-dihydrotestosterone appears to be bound to chromatin of the prostate nuclei (Mainwaring, 1971). The androgenic potency of 5α-dihydrotestosterone equals that of testosterone (Jeffcoate and Short, 1970) Radioactive 5α-dihydrotestosterone has also been detected in the human prostate after infusion of radioactive testosterone via the cephalic vein (Pike, *et al.*, 1970). A 5α-reductase enzyme which converts testosterone to 5α-dihydrotestosterone is present not only in the accessory sex organs of the male rat (prostate, seminal vesicles, etc.) but also in the hypothalamus and pituitary gland (Jaffe, 1969; Pérez-Palacios *et al.*, 1970).

Although administered progesterone undergoes extensive metabolic transformation in the guinea-pig (Falk and Bardin, 1970) and is reduced to 5α-pregnane-3, 2-dione by isolated rat uterine nuclei (Armstrong and King, 1970), the steroid taken up and specifically bound by target tissues is mainly progesterone itself. After injection of radioactive progesterone into guinea-pigs, most of the radioactivity in uterus represents unchanged progesterone (Falk and Bardin, 1970). Similar types of studies have revealed that it is cortisol itself which is active in rat liver (Beato, *et al.*, 1971). Apparently, aldosterone also does not have to be metabolized in order to exert its hormonal effect in toad bladder (Sharp and Alberti, 1971). Some steroid hormones therefore act as such in target tissues, while others first have to be enzymically converted to an active form. It is possible that whether or not conversion takes place depends on the tissue studied. A certain steroid may have to be metabolized in one tissue in order to be hormonally active, while in another tissue the unchanged molecule is the active form.

3.3. HOW DO STEROID HORMONES ENTER CELLS?

It is still undecided whether or not steroid hormones enter cells associated with a specific plasma protein, by a specific transport mechanism, or by simple diffusion. In the plasma, steroids are bound to various types of high affinity proteins with a rather narrow specificity. For instance in man, testosterone and oestradiol are bound to the sex steroid-binding plasma protein (SBP), which does not bind progesterone or corticosteroids. Other steroid-binding proteins include the corticosteroid-binding globulin (CBG) and the progesterone-binding plasmaprotein (PBP) (Baulieu et al., 1971). The steroid-binding plasma proteins may serve a role in transporting the hormone and regulating their concentration in the body fluids. In any case, they do not seem to impair the ability of the steroid to enter the target cell. Cortisol bound to CBG is able to induce hepatic tyrosine aminotransferase activity, and cause a fall in lymphocyte concentration in peripheral blood. The presence of CBG does not affect either the time-course or the amplitude of the response of these two cortisol-responsive variables (Rosier and Hochsberg, 1972).

3.4. STEROID RECEPTORS IN CYTOSOL

Tissues responsive to steroid hormones contain characteristic steroid-binding proteins. Evidence for the presence in oestrogen-responsive tissues of oestrogen "receptors" was first conclusively obtained by Jensen and co-workers who found that these tissues have a striking affinity for the hormone. When radioactive oestradiol or hexestrol is given to experimental animals, the uterus, vagina and anterior pituitary take up and retain radioactive hormone against a marked concentration gradient in the blood (Jensen and DeSombre, 1972). Oestradiol incorporation as a function of dose indicates that the interaction with target tissues is a biphasic phenomenon; i.e. steroid uptake is not saturable at levels of injected hormone which saturate hormone retention (Jensen et al., 1967). The binding of oestradiol by the uterus is dependent on the age and hormonal status of the animal. In newborn rats the uterine receptor content is low; it increases until the tenth day of life, and then diminishes with age (Clark and Gorski, 1970; Lee and Jacobson, 1971). The content of the uterine receptor also varies with the ovarian cycle, indicating that there is a hormonal influence on the receptor concentration (Jensen and DeSombre, 1972).

Various chemicals which block the growth response of uterine tissue to oestrogens, do so by inhibiting the binding of the oestrogens to their receptors. The most studied chemicals in this respect are nafoxidine (Upjohn 11,100A), ethamoxy-triphetol (MER-25), clomiphene and Parke-Davis Cl-268. When, for instance, nafoxidine is given to rats, the reduction in oestrogen uptake parallels the inhibition of uterine growth. However, the binding of oestradiol to uterine receptors is not inhibited by the antibiotics puromycin or actinomycin D. This indicates that the steroid-receptor interaction is an early step in hormone action, and that the hormone-induced increases in protein and RNA synthesis (which are blocked by the antibiotics) take place in a later phase (Jensen and DeSombre, 1972; Jensen, 1965).

The hormone-responsive cell usually contains both cytoplasmic and nuclear hormone receptors. For example, when rats are treated with radioactive oestradiol, about 20–30% of the steroid is recovered in the cytosol, and about 70–80% in the nuclear fraction; this distribution can be confirmed by autoradiography. In the immature rat there are about 100,000 cytosol receptor sites per uterine cell (Jensen and DeSombre, 1972). The cytosol receptor of the immature rat uterus sediments with a coefficient of about 8S in a sucrose gradient (Erdos, 1968). In the presence of 0.2 M or higher concentrations of salt, this 8S complex is transformed into material which sediments at about 4S in salt-containing sucrose (Jensen et al., 1969). In the uterus of the adult rat, a protein of the cytosol receptor sediments in the 4S form in the absence of salt and this is also observed with the uterine cytosols from women and from adult monkeys (Wyss et al., 1968).

Various detailed schemes have been proposed for the way cytosol receptors dissociate in vitro. Muller (1971) has proposed that when the 8S oestradiol receptor of rat uterus dissociates in a medium containing 0.3M salt, not only is the 4S oestradiol-binding protein released, but also a 4–5S subunit which does not bind oestradiol. Puca et al. (1972) report that the oestrogen-binding protein of calf uterus cytosol is a protein with a molecular weight of 240,000 dalton, which sediments at 8.6S in buffers of low ionic strength. When the ionic strength is raised to higher values, the protein dissociates into protein units with a molecular weight of 118,000 dalton which sediment at 5.3S. This appears to be a reversible transition which consists of longitudinal dissociation of the 8.6S form into two halves. The 5.3S protein is then converted into a smaller 4.5S protein with a molecular weight of 61,000 dalton by the action of a calcium-activated molecular factor which is present in the cytosol (Fig. 3.1).

Other steroid hormones also bind to receptors in their target tissues. In target tissues of the oestrogen-prestimulated animal, progesterone is specifically bound in the cytosol and nucleus. Progesterone localization

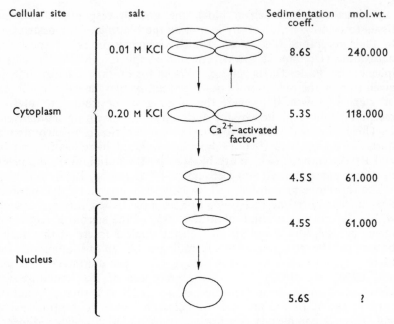

Fig. 3.1. Transitions of the oestradiol receptor in calf uterus (as proposed by Puca *et al.*, 1972)

is predominantly in the nucleus in some tissues (guinea-pig uterus and pregnant rat myometrium), while in other tissues cytosol and nuclear binding are about equal (ovariectomized rabbit uterus). In chick oviduct more radioactive progesterone appears in the cytosol than in the nucleus (Jensen and DeSombre, 1972). After injection of the hormone, the progesterone receptor of chick oviduct sediments at about 4S in sucrose gradients which contain potassium chloride (O'Malley *et al.*, 1970; 1971). When radioactive progesterone is added to cytosol, the receptor sediments at 3.8S in salt-containing gradients, but as a mixture of 5S and 8S proteins in salt-free gradients. Oestrogen pretreatment increases the total amount of cytosol progesterone-receptor, and also affects the relative amounts of 4S and 8S components detected in salt-free gradients (Jensen and DeSombre, 1972). Cytosol receptors for 5α-dihydrotestosterone have been found in rat prostate (Fang *et al.*, 1969), rat epididymis (Ritzén *et al.*, 1971) and rat seminal vesicle (Liao *et al.*, 1971). Receptors for cortisol exist in rat liver (Beato *et al.*, 1970) and rat thymus (Wira and Munck, 1970). A receptor for aldosterone has been found in rat kidney (Swanek *et al.*, 1970; Edelman, 1971). In general, the content of hormone receptors in the cytosol correlates fairly well with the hormone-response of the tissue.

From this it has been inferred that it might be possible to determine the hormone-sensitivity of tumours by assaying the receptor content of the cytosol. Rat mammary carcinomas which did not regress after ovariectomy were found to have a reduced level of cytoplasmic oestradiol-binding protein (McGuire *et al.*, 1972), and a parallelism has been observed between the oestrogen-binding capacity and the hormone responsiveness of mammary tumours in mice (Terenius, 1972).

On the other hand, Shyamala (1972) has reported that some oestradiol-independent mouse mammary tumours have cytoplasmic 8S oestradiol receptors with similar characteristics to the uterine receptors. However, these tumours have no intra-nuclear localization of the hormone. This suggests that some tissues may contain cytoplasmic receptors which can bind hormone, but which cannot subsequently exert a physiological effect in the nucleus. In these cases, assay of the hormone-binding capacity of the cytosol does not suffice as a measure of hormone responsiveness. The report by Mester and Baulieu (1972) that chicken liver nuclei contain oestradiol receptor, but that no receptor is detectable in the cytosol, emphasizes this point.

3.5. THE 5S NUCLEAR RECEPTOR

The oestradiol that is bound in the nucleus of uterine cells can be effectively extracted with 0.4 M KCl at pH 8.5. The oestradiol-receptor complex that is then isolated sediments at 5S, and can be easily distinguished from the 4S receptor of the cytosol (Puca and Bresciani, 1968). No detectable 5S binding protein exists in nuclei from uteri which have not been exposed to oestrogen (Jensen *et al.*, 1968), and no 5S complex is formed by adding oestradiol to isolated uterine nuclei unless cytosol is also present. The production of the 5S complex in isolated nuclei is temperature dependent; after incubation at 37°C the 5S complex is detected but at 2°C no 5S complex is produced, and heating to 45°C destroys the receptor protein and the ability of cytosol to form a 5S complex when incubated with nuclei (Jensen *et al.*, 1968). The transformation of the 4S receptor of uterine cytosol into what appears to be the 5S complex, otherwise only observed in nuclear extracts, can also be effected by incubating the cytosol with oestradiol in the absence of nuclei (Brecher *et al.*, 1970). The rate of this conversion is dependent on temperature, pH and association with oestradiol; warming the cytosol in the absence of added hormone does not yield 5S material (Jensen *et al.*, 1971).

On the basis of these results, a two-step mechanism has been proposed by Jensen and DeSombre (1972), in which the hormone first binds to the cytosol receptor protein and the receptor-hormone complex then moves to the nucleus. In the case of oestrogen, nuclear transfer is accompanied by a conversion of the 4S to a 5S form. With progesterone,

4

no alteration of the receptor takes place during transfer to the nucleus. This two-step mechanism rests principally on four pieces of experimental evidence (Jensen and DeSombre, 1972):

1. The nuclear 5S component is not formed in isolated nuclei unless cytosol is added.
2. The oestrogen-induced transformation of the cytosol 4S complex into a 5S complex which binds to nuclei.
3. The temperature-dependent shift of cytosol 8S to nuclear 5S in uterine tissue *in vitro*.
4. The depletion of cytosol receptor which takes place as oestradiol becomes bound in rat uterine nuclei.

3.6. ACCEPTOR SITES IN THE NUCLEUS

When uterine nuclei are incubated with 5S oestradiol receptor, RNA synthesis in these nuclei increases (Arnaud *et al.*, 1971; Mohla *et al.*, 1972). This suggests that formation of the 5S oestradiol-receptor complex is an intermediate step in the mechanism by which steroid hormones activate nuclear genes. In order to unravel the actual mechanism taking place in the nucleus, it is important to know what the acceptor sites are for 5S receptor molecules in the nucleus.

These studies have yielded different results. Some data suggest that the receptors bind to acidic proteins which are constituents of nuclear chromatin (Steggles *et al.*, 1971; Schrader *et al.*, 1972; Spellsberg *et al.*, 1972), while other reports indicate a binding to DNA (Yamamoto and Alberts, 1972; Clemens and Kleinsmith, 1972; Baxter *et al.*, 1972; Musliner and Chader, 1972; King and Gordon, 1972). However, these results are not necessarily mutually exclusive, since it is possible that both proteins and nucleic acids of the chromatin are involved in the binding of hormone receptors.

The interaction of receptors with DNA has been studied using columns of DNA-cellulose (Yamamoto and Alberts, 1972; Clemens and Kleinsmith, 1972), or by adding DNA to uterine cytosol (Mosliner and Chader, 1972). In our laboratory a simple assay for receptor-DNA binding has been devised which is carried out as follows. Calf uterus cytosol is incubated with tritiated oestradiol and the mixture is fractionated on a column of Sephadex G200. The 5S receptor-oestradiol complex is not retarded on the column (Fraction A, Fig. 3.2). Portions of this fraction are incubated with DNA and the DNA-receptor complex isolated by Sepharose 2B chromatography (Fig. 3.3). The radioactivity of the DNA peak is taken as a measure of receptor binding to DNA. Incubation of DNA with tritiated oestradiol in the absence of 5S receptor does not result in binding of radioactivity to the DNA peak (Sluyser *et al.*, 1974).

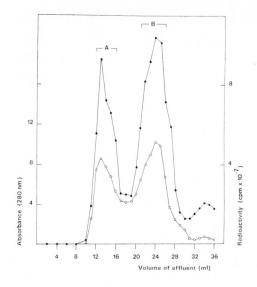

Fig. 3.2. Sephadex G200 chromatography (at 4°C) of calf uterus cytosol after incubation (30 min at 20°C) with 2n molar tritiated oestradiol. Incubation and chromatography were carried out in 10 mM Tris-HCl (pH 7.4), 1.5 mM EDTA, 1 mM dithiothreitol (TED buffer). Column size 1 cm × 78 cm. Fractions of 1 ml were collected. Absorbance at 280 nm is shown by ○—○; and radioactivity by ●—●.

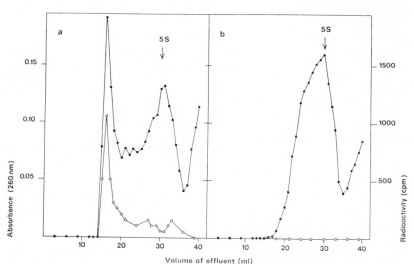

Fig. 3.3. Binding of 5S oestradiol receptor to DNA. Sepharose 2B chromatography of (a) Fraction A + 21 μg calf thymus DNA; (b) Fraction A. Each sample had a total volume 0.7 ml in TED buffer and contained 0.7 mg Fraction A (65,600 cpm/mg protein). The mixtures were incubated at 4°C for 5 min and were then applied to a column of Sepharose 2B (column size 1 cm × 78 cm). The column was eluted with TED buffer at 4°C, and fractions of 1 ml were collected. Absorbance at 260 nm is shown by ○—○; and radioactivity by ●—● (Sluyser et al., 1974).

3.7. INVOLVEMENT OF HISTONES

There is evidence that histones serve a structural role in chromatin. The arginine-rich histones F2a2 and F3 (and possibly also the F2a1 histone) force DNA into assuming a superhelical structure (Bradbury and Crane-Robinson, 1971). The lysine-rich histones form bridges between DNAs (Sluyser and Snellen-Jurgens, 1970) and between chromatin strands (Littau *et al.*, 1965), but do not constrain the DNA to form a supercoil (Bradbury *et al.*, 1972). It is therefore of interest to know whether histones are involved in the mechanism of steroid hormone action. Conceivably, hormone-bearing receptors might in some way interact with histones, thereby diminishing their constraint on nucleoprotein structure. In this way unwinding of the supercoil would occur, and specific structural genes could be released for transcription.

Evidence that histones may play a role in hormone action, was first obtained when, after injection of ^{3}H-testosterone to the duck, the isotopic label was found to be localized within the nuclei of the preen gland (Wilson and Loeb, 1965). More radioactivity was found to be attached to the euchromatin than to the heterochromatin, which was consistent with the finding that the euchromatin was also the major site of RNA synthesis in the nucleus of the preen gland (Wilson and Loeb, 1965). In order to further localize the binding site, euchromatin which had been prelabelled with ^{3}H-testosterone was centrifuged on a CsCl$_2$ gradient, and preliminary experiments suggested that part of the hormone was attached to the histone component of euchromatin (Wilson, 1965). More extensive studies on this interaction were carried out by Sekeris and Lang (Sekeris and Lang, 1965) using rat liver. They injected ^{3}H-cortisol into rats, and then prepared the liver nuclei in 0.25 M sucrose-containing buffer. Total histone was extracted from these nuclei in 0.25 M-H$_2$SO$_4$ and precipitated with ammonia and methanol; it was found to contain radioactivity.

A study on the distribution of ^{3}H-cortisol in rat liver histone fractions was made in our laboratory (Sluyser, 1966), using a differential extraction procedure for the separation of the histone fractions. The F3 histone (arginine-rich) was found to have the highest specific radioactivity. In order to establish whether histones could also bind steroid hormones *in vitro*, various histone fractions were isolated from rat liver. Each fraction was incubated with ^{3}H-cortisol and then submitted to gel filtration on a column of Sephadex G25. The run-through protein peak, which represented the histone, contained radioactivity, indicating that rat liver histones can bind cortisol *in vitro* and the F3 histone was found to have the highest binding capacity (Sluyser, 1966). Detailed studies which have been presented elsewhere, indicated that the *in vivo* interactions between histones and steroids have a low specificity and non-saturability (Sluyser, 1969; 1971). If specific *in vivo* binding occurred, other factors would therefore have to confer specificity to the reaction (Sluyser, 1972).

In order to establish whether the receptor which carries the hormone to the chromatin, confers this specificity, we have carried out investigations to determine whether receptors bind to histones. Rat liver nuclei were incubated with receptor-[³H]cortisol complex, and the nuclei were then spun down in the ultracentrifuge, and histone fractions extracted. Figure 3.4 shows that radioactivity was bound to these

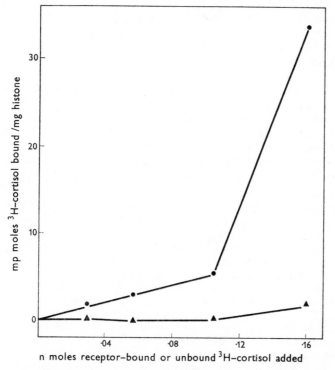

Fig. 3.4. Radioactivity of arginine-rich histones extracted from rat liver nuclei incubated with [³H]cortisol-labelled cytosol proteins or with [³H]cortisol. The experiment with [³H]cortisol-labelled cytosol proteins is shown as ●——●, and that with [³H]cortisol as ▲——▲.

Preparation of [³H]cortisol-labelled cytosol proteins: Rat liver (10 g) was homogenized in 10 ml of buffer A (0.05 M-Tris-HCl, pH 7.55; 0.25 M-sucrose; 0.025 M-KCl; 0.01 M-MgCl₂). The homogenate was centrifuged for 30 min at 105,000 × g in the cold. The supernatant (2 ml) was added to 1 mCi dry ³H-cortisol (44 Ci/mmole; New England Nuclear, Boston, Mass.). After mixing thoroughly, the solution was incubated for 5 min at 37°C and submitted to gel filtration on a column histones, and that the radioactivity increased with increasing amounts of receptor-[³H]cortisol added. In contrast, when nuclei were incubated with ³H-cortisol alone, no radioactivity was bound to the histones.

This result is compatible with the involvement of the receptor in the transfer of cortisol to nuclear histones. However, since rat liver cytosol contains various cortisol-binding proteins, some of which might bind to isolated nuclei unspecifically, further evidence is required before definite conclusions can be drawn, Figure 3.5 shows the result of an experiment in which high doses of ³H-cortisol were injected into rats, liver nuclei were prepared in 2.2 M sucrose and histone fractions extracted. These histones showed increasing specific radioactivities with increasing dose of ³H-cortisol.

The following experiment indicated that the radioactivity observed in the histone preparations was not due to cytoplasmic contamination. ³H-Cortisol was injected into rats, the radioactive cytosol was heated at 95°C, to destroy physiological activity, was then cooled and added to non-radioactive nuclei. After homogenization of the mixture, the nuclei were spun down in the ultracentrifuge and purified. The histones extracted from these nuclei were not radioactive (Fig. 3.5). In order to study the nature of *in vivo* binding, ³H-cortisol was injected into rats and liver nuclei isolated 30 min after injection. Histones were extracted from these nuclei by acid extraction and samples of the histone were

---

of Sephadex G25 (coarse; column size 1 cm × 39 cm). Elution was carried out at 4°C with buffer A. The protein peak that was not retarded on the column was collected.

Preparation of rat liver nuclei: Liver tissue was homogenized in 2.2 M-sucrose (10 ml/g tissue). The homogenate was strained through one layer, and then through 6 layers, of bandage gauze in order to remove gross material. The filtrate was centrifuged for 15 min at 50,000 × g. The pellet was homogenized briefly in 0.25 M-sucrose–10 mM-MgCl₂ solution and centrifuged for 10 min at 900 × g. The second pellet was homogenized and centrifuged twice in 0.14 M-NaCl–0.01 M-sodium citrate, twice in 0.1 M-Tris-HCl buffer (pH 7.6) and twice in 95% ethanol, yielding a final pellet of liver nuclei.

Incubation procedure: Each incubation mixture contained the nuclei of 4 rat livers suspended in 1.5 ml buffer A. Incubation was carried out for 15 min at 37°C with [³H]cortisol-labelled cytosol protein or with [³H]cortisol. The nuclei were re-obtained by centrifugation and washed as described above.

Extraction of arginine-rich histones: The washed nuclei were extracted 3 times with absolute ethanol–1.25 M-HCl mixture (4:1 v/v). Acetone, (20 vols) was added to the extract and the mixture left overnight at 4°C. The arginine-rich histones (F2a + F3) were collected by centrifugation for 20 min at 35,000 × g. They were dissolved in small volumes of distilled water, dialysed overnight and lyophilized. Radioactivity was assayed per mg dry weight.

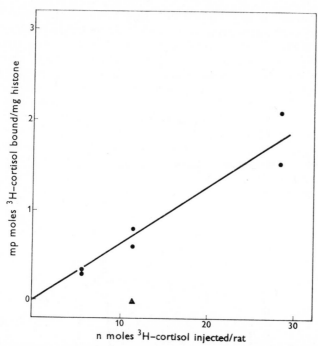

Fig. 3.5. Radioactivity of arginine-rich histones isolated after intraperitoneal injection of [³H]cortisol into rats or after mixing [³H]cortisol-containing cytosol with non-radioactive liver nuclei.

The radioactivity of arginine-rich histones isolated 30 min after i.p. injection of [³H]cortisol is shown as ●——●, and that obtained after mixing [³H]cortisol-containing cytosol (heated for 20 min at 95°C and then cooled) with non-radioactive liver nuclei is shown as ▲.

submitted to gel filtration in phosphate buffer (pH 6.8) which contained either 7% or 40% guanidinium chloride (Fig. 3.6). When gel filtration was carried out in 7% guanidinium chloride, about 6% of the radioactivity was eluted with the histone, but in 40% guanidinium chloride no radioactivity remained bound to histone. Therefore the radioactivity was linked non-covalently to histone, and was dissociated completely in 40% guanidinium chloride solution.

3.8. STRUCTURE OF CHROMATIN

Most theories which have been proposed to explain the mechanism of steroid hormone action, take as a starting point the Jacob–Monod model for gene regulation in bacteria. For instance, in Karlson's model (Karlson, 1961), the hormone is thought to act as an "inducer" which combines with a "repressor", thus allowing the transcription of specific genes. However, such models do not give a clue to the striking effects steroid hormones induce in chromosome structure. How can a small molecule, like for instance ecdysone, cause the upheavals in polytene chromosome structure which are observed as puffs? Perhaps one might best explain this effect by comparing a chromatin strand to a coiled spring. When certain protein molecules, e.g. steroid hormone receptor proteins, interact with the chromatin, the tension in the spring might be released locally, and the DNA might loop out. Since there is some evidence that histones constrain chromosomal DNA into forming a superhelical structure (Bradbury and Crane-Robinson, 1971), it therefore seems pertinent at this point to consider the structural role of histones in some detail.

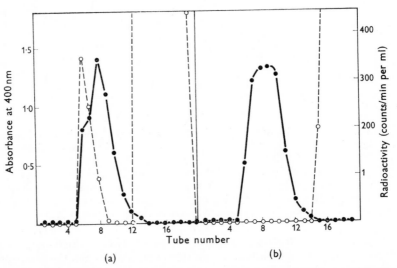

Fig. 3.6. Gel filtration of total histone isolated from rat liver nuclei after injection of ^{3}H-cortisol.

Rats (20) were injected with ^{3}H-cortisol (200 μCi/rat) and after 30 min. the total histone was isolated from rat liver nuclei by acid extraction (Johns, 1971). The histone was subjected to gel filtration on a column (1 × 39 cm) of Sephadex G100; fractions of 1.5 ml were collected. The protein content was determined by turbidimetry (Luck *et al.*, 1958), the absorbance at 400 nm being shown as ●———●; the radioactivity is shown as ○ · · · · · ○.

(a) Elution was effected in 7% guanidinium chloride–0.1 M phosphate (pH 6.8).

(b) Elution was effected in 40% guanidinium chloride–0.1 M phosphate (pH 6.8).

If the nucleate cell required histones only to neutralize negatively charged groups on its DNA, one would expect basic amino acids residues to be distributed in a regular fashion along the histone chains. A histone molecule could then neutralize a certain stretch of the DNA most effectively by placing a positive charge opposite each negatively charged phosphate on DNA. Amino-acid sequence studies have

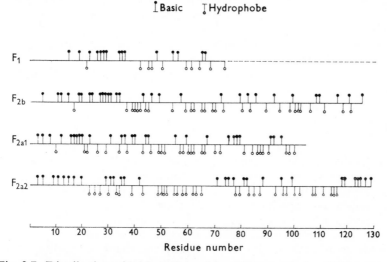

Fig. 3.7. Distribution of basic amino-acid residues and hydrophobic amino-acid residues in calf thymus histones.

The N-terminal regions (CTL-1-N$_2$) of the F1 histone (Rall and Cole, 1971), the F2b histone (Iwai *et al.*, 1970), the F2al histone (Smith *et al.*, 1970), and the F2a2 histone (Yeoman *et al.*, 1972) are shown.

The basic amino-acid residues are lysine, arginine and histidine, the hydrophobic amino-acid residues are valine, methionine, leucine, iso-leucine, tyrosine and phenylalanine

revealed, however, that the lysine, arginine and histidine residues are not evenly distributed but are clustered in certain regions of the histone chains. Other regions have predominantly hydrophobic amino-acid residues (Fig. 3.7).

The histone/DNA ratio (weight) is about 1.3/1 and on average the number of basic groups in histones is about equal to the number of phosphate groups in the associated DNA (Vendrely *et al.*, 1960). However, the irregularity in distribution of the basic groups in histones rules out neutralization of each phosphate on DNA by a basic group in histone. In fact, there is good evidence, from the binding of dyes such as azure A and toluidine blue, that a large number (perhaps

40–50%) of phosphate groups in nucleohistone is free to bind with added cations (Klein and Szirmai, 1963; Miura and Ohba, 1967).

Studies involving an analysis of dissociation of nucleohistone as a function of increasing salt or acid concentration have shown that the strength of attachment of DNA to histones does not correlate with the content of basic residues in the latter. For instance, the lysine-rich histones (F1) which have the highest content of basic residues are more easily detached from DNA than are the arginine-rich histones (Johns, 1971). Therefore besides ionic bonds, other types of interactions occur. Recent work has revealed that hydrophobic bonding takes place, and that this is most marked for the argine-rich histones F3 and F2a1, and least for the lysine-rich histones (F1). In addition, hydrogen bonds may contribute to the specificity of these interactions (Bartley and Chalkley, 1972).

The electrostatic interactions between DNA and histones are modulated by enzyme modifications of the basic regions of the histones. A large number of modified basic amino-acids have deen detected in histones. These include ε-N-acetyl-lysine, ε-N-monomethyl-lysine, ε-N-dimethyl-lysine, ε-N-trimethyl-lysine, 3-methylhistidine, ω-N-mono-methylarginine and α-N-methyl-guanidinomethylated arginine (DeLange and Smith, 1971). The side-chain modifications affect only certain loci, indicating a specific function for these changes. Non-basic amino-acid residues located near to basic regions are also liable to enzymic change and in certain molecular species of rabbit thymus F1 histones, the serine residue at site 40 is readily phosphorylated; this serine residue being located very near to a highly basic region (Langan et al., 1971; Rall and Cole, 1971). It appears, therefore, that amino-acid residues in, or near, basic regions of histones are especially susceptible to enzymic modification and in a number of cases these changes can be correlated with certain physiological activities of the cells, and may be influenced by hormones (Allfrey, 1971).

3.9. ROLE OF STEROID HORMONE RECEPTORS

The question may be raised as to what are the respective roles of the protein and steroid moieties of the receptor-hormone complex. There are three main possibilities:

1. The receptor serves to transport the steroid to the nucleus but does not itself play any further functional role in the nucleus;
2. The steroid induces a conformational change in the receptor molecule which enables the latter to move to the nucleus but the steroid itself plays no further role in the nucleus;
3. The protein and steroid moieties of the receptor-hormone complex

both have specific functions in the nucleus, i.e. both moieties are involved in the mechanism by which the RNA synthesizing capacity of nuclear chromatin is increased.

The evidence for or against each of these alternatives is meagre, but it seems possible to draw some tentative conclusions. Alternative 1 is unlikely to be true in view of the highly specific conformational changes the receptor undergoes before entering the nucleus (Puca *et al.*, 1972). This suggests that the receptor plays a specific role in nuclear events. Furthermore, as already mentioned, receptors bind to macromolecular components of the nuclear chromatin. Alternative 2 is also unlikely to be true for the following reasons. As pointed out by Dannenberg (1963), the area covered by a flat surface of steroid hormones is very similar to that of base pairs of DNA. Huggins and Yang (1962) have found that Courtauld models of progesterone, testosterone and oestradiol fit neatly into a frame constructed around a model of guanine-cytosine. If one takes into account the hydrogen bonds between complementary bases, the molecular geometry of a steroid resembles that of a base pair (Sluyser, 1971). These similarities strongly suggest that steroid molecules themselves act in some way at the level of DNA. In conclusion, therefore, alternative 3 is most likely to be true, i.e. both the receptor and the steroid probably play specific roles in the cell nucleus.

3.10. SUPER-HELICAL DNA

A clue to a possible binding mechanism of hormone-bearing receptor molecules with chromatin can possibly be obtained by considering some further structural aspects of chromatin. Vogel (1964) has proposed that most of the DNA of higher organisms does not code for protein synthesis, but is used for control purposes. Britten and Davidson (1969) have put forward the suggestion that multiple control elements may well be adjacent to each particular coding sequence, and Crick (1971) has proposed that the control elements are located in regions of the genome which contain intricately wound DNA, and that these regions correspond to the bands seen in polytene chromosomes. X-ray diffraction studies (Pardon and Wilkins, 1972) indicate that the DNA of nucleohistone is constrained to form a superhelix with pitch of 120 Å and radius 50 Å, and there is evidence that the arginine-rich histones F2a2 and F3 (and perhaps also the F2a1 histone) force the DNA into assuming this super-helical conformation (Bradbury and Crane-Robinson, 1971). The F3 and F2a1 histones are attached to regions with a high content of guanine-cytosine base pairs and cover stretches of 40–50 nucleotides (Clark and Felsenfeld, 1972). The sites of attachment of F3 and F2a1 histones are spaced at intervals of (on the average) about 900 nucleotides (Varshavsky and Georgiev, 1972).

On the basis of these data the following tentative model is proposed for a region of the eukaryotic genome which contains control elements (Fig. 3.8). The region contains a number of address loci for the binding of hormone receptors or other directional protein molecules.

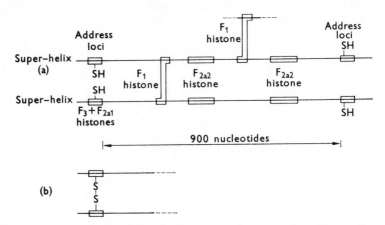

Fig. 3.8. Proposed model for the structure of a control region on the eukaryotic genome.

(a) The region contains address loci which have F3 and F2a1 histones attached. F2a2 histones are bound further along the superhelical DNA chain at discrete points. F1 histones form bridges between super-helices. In euchromatin the F3 histones contain mainly thiol groups.

(b) In heterochromatin these thiol groups are oxidized to form disulphide bridges between address loci. This may prevent the directional control of these loci.

The F1 histones may be attached to the DNA at AT-rich regions (Šponar and Šormova, 1972), while the F2b histones (not illustrated) may be complexed with the F2a1 histones (D'Anna and Isenberg, 1973).

F3 and F2a1 histones are attached to these address loci and together with F2a2 histones which are attached further along the DNA chain, they constrain an adjacent stretch of DNA, on the average 900 nucleotides in length, into forming a super-helical conformation. The F1 histones form extended bridges between the super-helices. With such a three-dimensional framework, torsion of individual super-helices would cause the structure as a whole to form a coil of larger dimension such as seen in many chromosomes with the light microscope, and, as has been pointed by Pardon and Wilkins, such coiling could account for the poor orientation in X-ray diffraction patterns (Pardon and Wilkins, 1972).

Figure 3.9 shows the proposed structure of an address locus in more detail. The locus contains a guanine-cytosine -rich stretch of DNA, 40–50 nucleotides in length, which is covered with F2al and F3 histones. Since the F2al and F3 histones contain about 100 and 130 amino acid residues, respectively, the size of the protected DNA

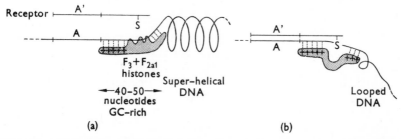

Fig. 3.9. Proposed structure for an address locus. (*a*) The locus is thought to contain a stretch of DNA with a specific nucleotide sequence (A) which is recognized by a structure (A′) of the hormone receptor protein. In addition, the address locus contains a stretch of DNA of length 40–50 nucleotides, having a high content of guanine-cytosine base pairs. A cluster of F3 and F2al histone molecules attaches to this DNA by polar and non-polar bonds. In this way the histones constrain the adjacent DNA to form a superhelical structure.

(*b*) Structure A′ of the hormone receptor binds to site A of the address locus. This aligns the steroid (S) with the non-polar region of the F3 + F2al complex, and allows it to interfere with the non-polar interaction of these histones with DNA. The constraint on the super-helix is lifted and the DNA loops out. This model could explain why histones do not require to be gene-specific, since the gene-specificity of the interaction between the hormone receptor and the genome is determined by nucleotide sequence A.

would be comparable with the degree of extension of single histone molecules if these were wound in one of the grooves of the DNA in a fully extended conformation. However, in that case much less F2al and F3 histones would occur, relative to the amount of DNA, than is actually found. This indicates that the F2al and F3 histones may be compact, or that they partly overlap. If portions are not in contact with DNA, then more than one molecule of histone would be required to cover the DNA segment of 40–50 nucleotides (Clark and Felsenfeld, 1972). There is evidence indicating that histones complexed with DNA have previously unexposed functional groups which can bind to free histone or stabilize the binding of histone to DNA (Rubin and Moudrianakis, 1972). This suggests that the F2al and F3 histones at the address locus may form a complex structure in which certain histone regions interact with each other, while others interact with DNA. It is reasonable to assume that both polar and non-polar interactions could be involved in the binding of the F2a1 + F3 complex to DNA at the address locus. This is illustrated very schematically in Fig. 3.9.

3.11. HORMONE-INDUCED UNWINDING OF SUPER-HELICAL DNA

Let us now consider how an hormone receptor might bind to the address locus. The available data do not make clear whether hormone receptors bind to DNA or to acceptor sites on acidic proteins. However, if acidic proteins are the recognition sites one might expect the cell to contain a very large variety of these proteins, since they would have to be gene-specific. Evidence for such an extreme heterogeneity of acidic proteins of the chromatin is lacking. As has been summarized by Hnilica (1972), the difficulties associated with their isolation and fractionation warn against premature conclusions concerning their tissue specificity, heterogeneity and possible function (Hnilica, 1972). It therefore seems more reasonable to favour a model in which hormone receptor proteins recognize specific nucleotide sequences of DNA. The hormone receptor protein could then bind to a stretch of DNA which contains such a sequence (Fig. 3.9). This would align the steroid with the hydrophobic region of the F2a1 + F3 complex, and interaction of the steroid with these histones might dissociate the latter from DNA. Whereas in this way the non-polar region of the F2a1 + F3 complex might become detached from DNA, the basic regions could remain linked to the phosphate groups of the DNA strand. This type of DNA-histone interaction has been discussed by Lewin (1970). If the constraint exerted by the F2a1 + F3 complex on the adjacent DNA is diminished, the supercoil might loop out, unmasking other address loci for the attachment of hormone receptor proteins or acidic proteins of the chromatin. This might cause further unwinding, and in this way the hormonal effect could be amplified and the looped conformation of the genome stabilized. In polytene chromsomes this might lead to the appearance of puffs.

Pardon and Wilkins (1972) have pointed out that almost 90% of the DNA would require to be coiled in super-helices in order to provide an untwisting mechanism for the remaining 10% of DNA. They add, however, that about 98% of the DNA might be non-informational. This DNA could be assigned the role of untwisting the small amount of DNA which is informational.

The present hypothetical model is in agreement with that of Paul (1972) in postulating the existence of address loci for the binding of directional protein molecules. Paul proposed that acidic proteins bind to these loci and cause unwinding of the supercoil. However, the concentration of acidic proteins in chromatin begins to increase only late after administration of hormone, whereas other hormonal effects such as an increase in RNA polymerase activity becomes apparent very rapidly (Glasser et al., 1972). It therefore seems likely that the binding of hormone receptor molecules to address loci would be the triggering event, and that this unmasks additional loci for the attachment of acidic proteins.

Finally the model suggests a possible explanation for the finding by Stocken (1966) that the thiol/disulphide ratio of F3 histones in diffuse chromatin is higher than in condensed, inactive chromatin. In heterochromatin, disulphide bridges between the F3 histone of different address loci might prevent the binding of hormone receptor proteins to these loci (Fig. 3.7). This might explain why steroid hormones manage to enhance the RNA synthetic capacity of euchromatin to a much larger extent than that of heterochromatin.

3.12. INTRODUCTION OF ENZYME SYNTHESIS BY STEROID HORMONES

Knox, in 1951, showed that cortisol induces the liver enzyme trypto-phan pyrrolase (Knox, 1951). Immunological studies by Feigelson and Greengard (1962) proved that the increase in enzyme activity was due to an increase in net amount of protein molecules. This was also found to apply to the enzymes tyrosine transaminase (Kenney and Flora, 1961), pyruvate-glutamate transaminase (Segal and Kim, 1963), pyruvate carboxylase (Henning et al., 1963), phosphoenol pyruvate-carboxykinase (Schrago et al., 1963), glycogen synthetase (Hilz et al., 1963), glucose-6-phosphatase (Weber et al., 1961) and fructose 1,6-diphosphatase (Kvam and Parks, 1960). Of these enzymes, tyrosine transaminase and tryptophane pyrrolase are induced within 30 min after cortisol administration, reaching a maximum after 6–8 hours. In contrast, glucose-6-phosphatase and fructose-1,6-diphosphatase are induced only after 2–6 hours, and do not reach a maximum activity before 24 hours. (Lang, 1971). A difficulty in studying the significance of hormone effect on the liver, is that even in the absence of the hormone there are already enzyme molecules present. The hormone therefore brings about a quantitative, rather than a qualitative, effect (Lang, 1971).

The mechanism of action of ecdysone on insect development appears to be a simpler model to study. The hormone induces the appearance of the epidermal enzyme DOPA-decarboxylase in the third larval stage of Calliphora erythrocephala. This opens the pathway to an anabolic reaction in tyrosine metabolism (Sekeris and Karlson, 1962), the tyrosine being metabolized to *N*-acetyl-dopamine, which is the precursor of the tanning agent. After injection of ecdysone into the larvae of the midge *Chironomus tentans*, puffs appear at certain loci on the polytene chromosomes of the salivary gland (Clever and Karlson, 1960). Within 1 hour after injection of the hormone, incorporation of precursors into RNA in epidermis nuclei is markedly increased, while the cytoplasmic RNA shows an increase only 3 hours after hormone injection. Evidence has been presented that some of this RNA is messenger RNA which can act as a template for DOPA-decarboxylase (Lang, 1971; Sekeris *et al.*, 1965).

Similarly, one of the earliest actions of oestrogens on target cells is to enhance DNA-dependent synthesis of RNA. The initial stimulation of RNA synthesis by oestradiol occurs within minutes, shortly after the entrance of the hormone into the nucleus (Hamilton, 1968; Teng and Hamilton, 1968) and precedes the earliest rise in specific protein synthesis (Knowles and Smellie, 1971). Workers studying the action of oestrogens or other steroids are divided in their opinions concerning the type of RNA synthesized first. Some postulate that control by steroid hormones is at the level of the new generation of ribosomes produced, or in their specific location within the cytoplasm (Tata, 1970). Others suggest that the primary action of steroid hormones in general, is the induction of messenger RNA synthesis, and that all other events are secondary to this (Sekeris, 1965). The translation of the mRNA molecules would then result in the synthesis of specific proteins (Notides and Gorski, 1966). Recent work by Knowles and Smellie (1971) substantiate the latter view. They showed that high-molecular-weight RNA (probably messenger RNA) is synthesized within 30 minutes after administration of oestradiol and that this event takes place before the formation of ribosomal RNA is stimulated. Results by Glasser, Chytil and Spellsberg (1972) indicate that the stimulation of RNA synthesis in rat uteri by oestrogen is mediated by activating the RNA polymerase enzyme that synthesizes DNA-like RNA (polymerase II). The template capacity of the chromatin and the polymerase enzyme activity that synthesizes ribosomal RNA (polymerase I) appear to be stimulated at a later time. Besides affecting messenger and ribosomal RNA, hormones also influence the synthesis of transfer RNA. Thus, cortisol has been reported to induce a new leucine-accepting tRNA and its synthetase in rat liver (Altman *et al.*, 1972). This all implies a highly sensitive control mechanism in protein synthesis.

The picture that emerges from these studies is that steroid hormones influence the number and types of RNA molecules synthesized and the transportation of these molecules to the cytoplasm. This is brought about by activating certain RNA polymerases and increasing the template activity of chromatin. It is possible that when the mechanism of gene regulation in the eukaryotic cell is more fully understood, the relationship between these processes will become more clear. The major problem is recognizing the primary events in which steroids participate, and distinguishing them from secondary events which may be remote in space and time. The model proposed in this paper for hormone-induced unwinding of super-helical chromatin can perhaps be employed as a useful working hypothesis for the initial action of the hormone at the level of DNA. However, as pointed out by Grant (1969), the situation may be aptly compared with peeling the layers off an onion: "not only may it bring tears to the eyes, but after each successful step one is left with the layer underneath".

REFERENCES

Allfrey, V. G. (1971). In *Histones and Nucleohistones*, p. 241 (D. M. P. Phillips, ed.), Plenum Press; London and New York

Altman, K., Southren, A. L., Uretsky, S. C., Zabos, P. and Acs, G. (1972). *Proc. Nat. Acad. Sci. U.S.* **69**, 3567

Anderson, K. M. and Liao, S. (1968). *Nature*, **219**, 277

Armstrong, D. T. and King, E. R. (1970). *Fed. Proc.*, **29**, 250

Arnaud, M., Beziat, Y., Guilleux, J. C., Hough, A., Hough, D. and Mousseron, M. (1971). *Biochim. Biophys. Acta*, **232**, 117

Bartley, J. A. and Chalkley, R. (1972). *J. Biol. Chem.*, **247**, 3647

Baulieu, E. E., Alberga, A., Jung, I., Lebeau, M. C., Mercier-Bodard, C., Milgrom, E., Raynaud, J. P., Raynaud-Jammet, C., Rochefort, H., Truong, H. and Robel, P. (1971). *Rec. Progress Hormone Res.*, **27**, 351

Baxter, J. D., Rousseau, G. G., Benson, M. C., Garcia, R. L., Itto, J. and Tomkins, G. M. (1972). *Proc. Nat. Acad. Sci. U.S.*, **69**, 1892

Beato, M., Braendle, W., Biesewig, D. and Sekeris, C. E. (1970). *Biochim. Biophys. Acta*, **208**, 125

Beato, M., Schmid, W., Braendle, W., Biesewig, D. and Sekeris, C. E. (1971). In *Adv. in the Biosciences*, Vol. 7, p. 349 (G. Raspé, ed.), Pergamon Press; Vieweg, Braunschweig, Germany

Bradbury, E. M. and Crane-Robinson, C. (1971). In *Histones and Nucleohistones*, p. 85 (D. M. P. Phillips, ed.), Plenum Press; London and New York

Bradbury, E. M., Molgaard, H. V., Stephens, R. M., Bolund, L. A. and Johns, E. W. (1972). *Europ. J. Biochem.*, **31**, 474

Brecher, P. I., Numata, M., DeSombre, E. R. and Jensen, E. V. (1970). *Fed. Proc. Fedn. Am. Socs. exp. Biol.*, **29**, 249

Britten, R. J. and Davidson, E. H. (1969). *Science*, **165**, 349

Bruchovsky, N. and Wilson, J. D. (1968). *J. Biol. Chem.*, **243**, 2012

Clark, J. and Gorski, J. (1970). *Science*, **169**, 76

Clark, R. J. and Felsenfeld, G. (1972). *Nature New Biol.*, **240**, 226

Clemens, L. E. and Kleinsmith, L. J. (1972). *Nature New Biol.*, **237**, 204

Clever, U. and Karlson, P. (1960). *Exp. Cell Res.*, **20**, 623

Crick, F. H. (1971). *Nature*, **234**, 25

D'Anna, J. A. and Isenberg, I. (1973). *Biochemistry*, **12**, 1035

Dannenberg, H. (1963). *Dtch. med. Wschr.*, **88**, 605

DeLange, R. J. and Smith, E. L. (1971). *Ann. Rev. Biochem.*, **40**, 279

Edelman, S. (1971). In *Adv. in the Biosciences*, p. 267 (G. Raspé, ed.), Pergamon Press; Vieweg, Braunschweig, Germany

Erdos, T. (1968). *Biochem. Biophys. Res. Comm.*, **32**, 338
Falk, R. J. and Bardin, C. W. (1970). *Endocrinology*, **86**, 1059
Fang, S., Anderson, K. M. and Liao, S. (1969). *J. Biol. Chem.*, **244**, 6584
Feigelson, P. and Greengard, O. (1962). *J. Biol. Chem.*, **237**, 3714
Glasser, S. R., Chytil, F. and Spelsberg, T. C. (1972). *Biochem. J.*, **130**, 947
Grant, J. K. (1969). In *Essays in Biochemistry*, Vol. 5, p. 1 (P. N. Campbell and
    G. D. Greville, eds.), Academic Press; London and New York
Hamilton, T. H. (1968). *Science*, **161**, 649
Hamilton, T. H. (1971). In *The Biochemistry of Steroid Hormone Action*, p. 49
    (R. M. S. Smellie, ed.). Academic Press; London and New York
Henning, I. V., Seifert, J. and Leubert, W. (1963). *Biochim. Biiphys. Acta*, **77**, 345
Hilz, H., Tarnowski, W. and Arend, P. (1963). *Biochem. Biophys. Res. Comm.*, **10**,
    492
Hnilica, L. S. (1972). *The Structure and Biological Function of Histones*, CRC Press;
    Cleveland, Ohio
Huggins, C. and Yang, N. C. (1962). *Science*, **137**, 257
Iwai, K., Ishikawa, K. and Hayashi, H. (1970). *Nature*, **226**, 1056
Jaffe, R. B. (1969). *Steroids*, **14**, 483
Jeffcoate, W. J. and Short, R. V. (1970). *Endocrinology*, **48**, 199
Jensen, E. V. (1965). *Proc. Can. Cancer Res. Conf.*, **6**, 143
Jensen, E. V. and DeSombre, E. R. (1972). *Ann. Rev. Biochem.*, **41**, 203
Jensen, E. V. and Jacobson, H. I. (1962). *Rec. Progr. in Hormone Res.*, **18**, 387 and
    461
Jensen, E. V., DeSombre, E. R. and Jungblut, P. W. (1967). In *Hormonal Steroids*,
    p. 492 (L. Martini, F. Fraschini and M. Motta, eds.), Exerpta Medica; Am-
    sterdam
Jensen, E. V., Suzuki, T., Kawashima, T., Stumpf, W. E., Jungblut, P. W. and
    DeSombre, E. R. (1968). *Proc. Nat. Acad. Sci. U.S.*, **59**, 632
Jensen, E. V., Suzuki, T., Numata, M., Smith, S. and DeSombre, E. R. (1969).
    Steroids, **13**, 417
Jensen, E. V., Numata, M., Brecher, P. I. and DeSombre, E. R. (1971). In *The
    Biochemistry of Steroid Hormone Action*, p. 133 (R. M. S. Smellie, ed.), Academic
    Press; London and New York
Johns, E. W. (1971). In *Histones and Nucleohistones*, p. 1 (D. M. P. Phillips, ed.),
    Plenum Press; London and New York
Karlson, P. (1961). *Dtch. med. Wschr.*, **86**, 668
Kenney, F. T. and Flora, R. M. (1961). *J. Biol. Chem.*, **236**, 2699
King, R. J. B. and Gordon, J. (1972). *Nature New Biol.*, **240**, 185
Klein, F. and Szirmai, J. A. (1963). *Biochim. Biophys. Acta*, **72**, 48
Knowles, J. T. and Smellie, R. M. S. (1971). *Biochem. J.*, **125**, 605
Knox, W. E. (1951). *Brit. J. exp. Path.*, **32**, 462
Kvam, D. C. and Parks, R. E. (1960). *Am. J. Physiol.*, **198**, 21
Lang, N. (1971). In *The Biochemistry of Steroid Hormone Action*, p. 85 (R. M. S.
    Smellie, ed.), Academic Press; London and New York
Langan, T. A., Rall, S. C. and Cole, R. D. (1971). *J. Biol. Chem.*, **246**, 1942
Lee, C. and Jacobson, H. I. (1971). *Endocrinology*, **88**, 596
Lewin, S. (1970). *Biochem. J.*, **117**, 19P
Liao, S., Tymoczko, J. L., Liang, T., Anderson, K. M. and Fang, S. (1971). In
    Adv. in the Biosciences, p. 155 (G. Raspé, ed.), Pergamon Press; Vieweg, Braun-
    schweig, Germany
Littau, V. C., Burdick, C. J., Allfrey, V. G. and Mirsky, A. E. (1965). *J. Cell Biol.*,
    27, 124A
Luck, J. M., Rasmussen, P. S., Satake, K. and Tsvetikow, A. N. (1958). *J. Biol.
    Chem.*, **233**, 1407
Mainwaring, W. I. P. (1971). *Biochem. J.*, **124**, 42P

McGuire, W. L., Huff, K., Jennings, A. and Chamness, G. C. (1972). *Science*, **175**, 235

Mester, J. and Baulieu, E. E. (1972). *Biochim. Biophys. Acta*, **261**, 236

Miura, A. and Ohba, Y. (1967). *Biochim. Biophys. Acta*, **145**, 436

Mohla, S., DeSombre, E. R. and Jensen, E. V. (1972). *Biochem. Biophys. Res. Comm.*, **46**, 661

Mueller, G. C. (1971). In *The Biochemistry of Steroid Hormone Action*, p. 1 (R. M. S. Smellie, ed.), Academic Press; London and New York

Musliner, T. A. and Chader, G. J. (1972). *Biochim. Biophys. Acta*, **262**, 256

Notides, A. and Gorski, J. (1966). *Proc. Nat. Acad. Sci. U.S.*, **56**, 230

O'Malley, B. W., Sherman, M. R. and Toft, D. O. (1970). *Proc. Nat. Acad. Sci. U.S.*, **67**, 501

O'Malley, B. W., Toft, D. O. and Sherman, M. R. (1971). *J. Biol. Chem.*, **246**, 1117

Pardon, J. F. and Wilkins, M. H. F. (1972). *J. Mol. Biol.*, **68**, 115

Paul, J. (1972). *Nature*, **238**, 444

Pérez-Palacios, G., Castañeda, E., Gómez-Pérez, F., Pérez, A. E. and Gual, C. (1970). *Biol. Reprod.*, **3**, 205

Pike, A., Peeling, W. B., Harper, M. E., Pierrepoint, C. G. and Griffiths, K. (1970). *Biochem. J.*, **120**, 443

Puca, G. A. and Bresciani, F. (1968). *Nature*, **218**, 967

Puca, G. A., Nola, E., Sica, V. and Besciani, F. (1972). *Biochemistry*, **11**, 4157

Rall, S. C. and Cole, R. D. (1971). *J. Biol. Chem.*, **246**, 7175

Ritzén, E. M., Nayfeh, S. N., French, F. S. and Dobbins, M. C. (1971). *Endocrinology*, **89**, 143

Rosier, W. and Hochsberg, R. (1972). *Endocrinology*, **91**, 626

Rubin, R. L. and Moudrianakis, E. N. (1972). *J. Mol. Biol.*, **67**, 361

Schrader, W. T., Toft, D. O. and O'Malley, B. W. (1972). *J. Biol. Chem.*, **247**, 2401

Segal, H. L. and Kim, X. S. (1963). *Proc. Nat. Acad. Sci. U.S.*, **50**, 912

Sekeris, C. E. (1965). In *Mechanism of Hormone Action*, p. 149 (P. Karlson, ed.), G. Thieme Verlag; Stuttgart

Sekeris, C. E. and Karlson, P. (1962). *Biochim. Biophys. Acta*, **63**, 489

Sekeris, C. E. and Lang, N. (1965). *Hoppe-Seyler's Zeitschr. Physiol. Chem.*, **340**, 92

Sekeris, C. E., Lang, N. and Karlson, P. (1965). *Z. physiol. Chem.*, **341**, 36

Sharp, G. W. G. and Alberti, K. G. M. M. (1971). In *Adv. in the Biosciences*, p. 281 (G. Raspé, ed.), Pergamon Press; Vieweg, Braunschweig, Germany

Schrago, E. H., Lardy, A., Nordlie, R. C. and Forster, D. O. (1963). *J. Biol. Chem.*, **238**, 3188 (1963).

Shyamala, G. (1972). *Biochem. Biophys. Res. Comm.*, **46**, 1623

Sluyser, M. (1966). *J. Mol. Biol.*, **19**, 591

Sluyser, M. (1969). *Biochim. Biophys. Acta*, **182**, 235

Sluyser, M. (1971). In *The Biochemistry of Steroid Hormone Action*, p. 31 (R. M. S. Smellie, ed.). Academic Press; London and New York

Sluyser, M. (1972). *Biochem. J.*, **130**, 49P

Sluyser, M., Evers, S. G. and Nijssen, T. (1974). *Biochem. Biophys. Res. Comm.* (in press)

Sluyser, M. and Snellen-Jurgens, N. H. (1970). *Biochim. Biophys. Acta*, **199**, 490

Smith, E. L., DeLange, R. J. and Bonner, J. (1970). *Physiol. Rev.*, **50**, 159

Spelsberg, T. C., Steggles, A. W., Chytil, F. and O'Malley, B. W. (1972). *J. Biol. Chem.*, **247**, 1386

Šponar, J. and Šormová, Z. (1972). *Europ. J. Biochem.*, **29**, 99

Steggles, A. W., Spelsberg, T. S. and O'Malley, B. W. (1971). *Biochem. Biophys. Comm.*, **43**, 20

Stocken, L. A. (1966). In *Histones* (A. V. S. DeReuck and J. Knight, eds.), J. and A. Churchill Ltd.; London

Swaneck, G. E., Chu, L. L. H. and Edelman, I. S. (1970). *J. Biol. Chem.*, **245**, 5382

Tata, J. R. (1968). *Nature*, **219**, 331
Tata, J. R. (1970). In *Biochemical Actions of Hormones*, p. 89 (G. Litwack, ed.),
Academic Press; New York
Teng, C. and Hamilton, T. H. (1968). *Proc. Nat. Acad. Sci. U.S.*, **60**, 1410
Terenius, L. (1972). *Europ. J. Cancer*, **8**, 55
Varshavsky, A. J. and Georgiev, G. P. (1972). *Biochim. Biophys. Acta*, **281**, 669
Vendrely, R., Knobloch-Mazen, A. and Vendrely, C. (1960). In *The Cell Nucleus*,
Butterworth's; London
Vogel, F. (1964). *Nature*, **201**, 847
Weber, G., Banergee, G. and Bronstein, S. B. (1961). *Biochem. Biophys. Res. Comm.*,
4, 331
Williams-Ashman, H. G. (1965). *Cancer Res.*, **25**, 1096
Wilson, J. D. (1965). 19th Ann. Symp. in Fundam. Cancer Res., p. 38
Wilson, J. D. and Loeb, P. M. (1965). In *Developmental and Metabolic Control
Mechanisms and Neoplasia*, p. 375, The Williams and Wilkins Company, Balti-
more
Wira, C. and Munck, A. (1970). *J. Biol. Chem.*, **245**, 3436
Wyss, R. H., Heinrichs, W. L. and Herrmann, W. L. (1968). *J. Clin. Endocrinol.
Metab.*, **28**, 1227
Yamamoto, K. R. and Alberts, B. M. (1972). *Proc. Nat. Acad. Sci. U.S.*, **69**, 2105
Yeoman, L. C., Olson, M. O. J., Sugano, N., Jordan, J. J., Taylor, C. W., Starbuck,
W. C. and Busch, H. (1972). *J. Biol. Chem.*, **247**, 6108

Chapter 4

Action of Aldosterone on Transepithelial Sodium Transport

George A. Porter
Division of Nephrology
Department of Medicine
University of Oregon Medical School
Portland, Oregon, U.S.A.

4.1. INTRODUCTION

This review of the mechanism of action of aldosterone will draw principally from experimental observations derived from isolated anuran (tailless amphibian) skin and urinary bladder. When appropriate, information from mammalian kidney experiments will be cited. Previous reviews of this subject include those of Crabbé (1963a), Sharp and Leaf (1966), Edelman (1968; 1969), Edelman and Fanestil (1970), Fanestil (1969) and Pelletier *et al.* (1972). In addition, Voûte (1972) has edited a recent symposium of aldosterone and active sodium transport across epithelia.

4.2. HISTORICAL BACKGROUND

The evolution of species adapted to terrestrial and freshwater habitats required the development of regulatory mechanisms capable of maintaining the ionic compositions of the organism at dynamic equilibrium. To cope with hourly fluctuations in salt and water intake, land forms have evolved a system for regulating rates of excretion of these chemical species.

The initial suggestion of steroid regulation of ionic transport in mammals was that of Lucas (1926) when he recorded hypochloraemia as a consequence of adrenalectomy in the dog. Baumann and Kurland (1927) extended this observation when they reported that adrenal ablation in the cat, in addition to hypochloraemia, caused hyponatraemia, hyperkalaemia and hypermagnesaemia. Furthermore, supplemental sodium chloride injections prolonged feline survival following adrenalectomy (Marine and Baumann, 1927). Renal sodium wasting, in the presence of hyponatraemia, was identified in the study of adrenalectomized dogs by Loeb and co-workers (1933), and a renal locus of steroid action was soon confirmed by demonstrating that injection of adrenal extract into adrenal insufficient man (Thorn et al., 1936) or adrenalectomized dog (Harrop et al., 1936) led to renal sodium retention and correction of plasma electrolyte abnormalities. It is now well established that mineralocorticoids exert their principal action on sodium–potassium balance by acting on renal tubular transport processes (Ross, 1959; Gaunt and Chart, 1962).

A clue as to the evolution of adrenocortical steroid regulation of ion balance comes from studies in lower vertebrates. Biosynthesis of aldosterone and other adrenocortical steroids by the inter-renal body (the homologue of the adrenal cortex) of bony fishes has been demonstrated in vitro (Phillips et al., 1959) as well as in vivo (Phillips and Mulrow, 1959). Similar observations have been reported for bullfrog adrenal slices by Carstensen et al. (1961), and for adrenal homogenates of the toad Bufo marinus and the frog Rana ridibrinda (Crabbé, 1963a). However, a definition of the physiological mechanism by which mineralocorticoids regulate or modify ion transport in the teleosts has been elusive. Administration of cortisol, 9α-fluorocortisol or aldosterone had no observable effect on the rates of loss of Na^+ or K^+ in either stenohaline or euryhaline marine teleosts immersed in fresh water (Edelman et al., 1960). Injections of adrenocortical extracts failed to increase the saline tolerance of freshwater trout, nor did deoxycorticosterone (DOC) or adrenocorticotropic hormone (ACTH) enable hypophysectomized killifish to survive in fresh water (Smith, 1956; Burden, 1956). The rate of sodium excretion by trout in response to an intraperitoneal injection of saline was increased by injections of deoxycorticosterone or cortisol (Holmes, 1959). In contrast to fish, the involvement of adrenocortical steroids in the physiological control of sodium balance in Amphibia has been better defined.

The initial suggestion of a key role for adrenocortical regulation of salt balance in Amphibia came from the work of Marenzi and Fustinoni (1938) who reported hyponatraemia following adrenalectomy in the toad Bufo arenarum, but were unable to prolong survival in ablated animals by the injection of adrenocortical extract. This observation was extended by the studies of Jørgensen (1947) who found an increased sodium efflux across the ventral skin of hypophysectomized frogs. Using the axolotl, Koefoed-Johnson and Ussing (1949) demonstrated an increased sodium influx after injecting the animals with ACTH but not with deoxycorticosterone. Huf and Wills (1953) reported that net transport of NaCl by isolated frog skin was increased after injection of ACTH into the donor. The importance of aldosterone in the physiological regulation of active sodium transport across the frog skin was established in a series of studies; immersion of frogs in salt solution depressed active transport across the isolated skin and injection of aldosterone reversed this effect (Maetz et al., 1958). Depression of the rate of sodium transport across frog skin induced by adrenalectomy was reversed by injections of aldosterone (Bishop et al., 1961; Scheer et al., 1961; Williams and Angerer, 1959). Crabbé (1963), using the toad Bufo marinus, demonstrated that the rate of endogenous secretion of aldosterone varied with the salinity of the environment. Finally, McAfee and Locke (1961) and Imamura and Sasaki (1962) using frog skin, and Crabbé (1961a; 1963) using the skin and urinary bladder of the toad, obtained significant stimulation of active sodium transport following the in vitro addition of cortisol, deoxycorticosterone or aldosterone. The latter success was made possible by the pioneering work of Ussing and Zerahn (1951) who introduced and confirmed the short-circuit current technique as a convenient method for monitoring active sodium transport across isolated frog skin. Reliability of the short-circuit current (scc) as a measure of active sodium transport by the isolated urinary bladder of Bufo marinus was provided by the work of Leaf and associates (1958).

Armed with the knowledge that active sodium transport activity of amphibian skin and urinary bladder could be modulated by secretions of the adrenal cortex, a number of investigators set out to develop a simple in vitro system suitable for studying the biochemical and biophysical mechanisms involved in the action of aldosterone and related mineralocorticoids. However, before turning to this system and its results, a few words concerning the amphibian epithelia as a model of the mammalian nephron are germane.

4.3. TOAD BLADDER AS A MODEL FOR INVESTIGATING ACTIVE SODIUM TRANSPORT

This section is intended to provide the reader with a superficial survey of the characteristics of the isolated toad urinary bladder as a structure which actively transports sodium. It does not substitute for the detailed review of this subject by Leaf (1965), nor the more recent critique by Dicker (1970) concerning the skin and bladder of amphibians as models of the mammalian nephron.

Morphologically, the toad urinary bladder consists of a thick stroma of connective tissue, smooth muscle bundles and blood vessels which is bounded on the mucosal side by transitional epithelium and on the serosal side by a single layer of mesothelial cells. The functional portion of this membrane seems to reside in the mucosal cell surface, since serosal tissue taken from adjacent portions of the peritoneum lack measurable bioelectric activity (Frazier and Leaf, 1963). The epithelium covering the mucosal surface, which was originally felt to have three cell types according to Keller (1963), i.e. ordinary epithelial cells (83%), mitochondrial rich cells (11%) and goblet cells (6%), is now believed to contain four cell types with identification of a basal cell by Choi (1963). An additional contribution by Choi was the description of a typical desmosome occurring near the mucosal end of the intercellular spaces. The "tightness" of this junction has been the subject of much controversy ever since Pak Poy and Bentley (1960) and Peachey and Rasmussen (1961) reported the existence of larger intercellular lakes extending from the desmosome into the interstitial space of tissues undergoing large net water flow. However, more important to the present discussion is the fact that, although one may view the isolated urinary bladder of the toad as a relatively simple structure, the precise function of each of the four mucosal cell types and their relative contribution to transepithelial sodium transport remains to be defined.

The isolated toad bladder generates a spontaneous potential difference which is always oriented mucosal side negative to serosa, irrespective of the composition of the fluids bathing the two sides of the membrane, provided that a minimal concentration of sodium is present on the mucosal side and likewise a minimal amount of potassium is present on the serosal side. As alluded to previously, Leaf and co-authors (1958) were the first to confirm the reliability of the short-circuit current (scc) technique to monitor active sodium transport in the isolated bladder. In addition, the metabolic dependence of this transport system was verified by numerous investigators (Leaf et al., 1958; Leaf and Dempsey, 1960; Maffly and Edelman, 1963; Maffly and Coggins, 1965), as was its aerobic requirement (Hersey, 1969; McDougal and Sullivan, 1971). The influence of environmental temperature on active sodium transport in the isolated toad bladder preparation has been defined (Porter, 1970a; 1970b), as have the alterations in electrical characteristics and permeability to solute and water following introduction of hypertonicity into either the mucosal or serosal bathing media (Urakabe et al., 1970). Finally, it is noteworthy that, in addition to oxidation of the products of glycolysis, the toad's urinary bladder can also provide energy for sodium transport by oxidation of fatty acids (Ferguson et al., 1968).

Although the principal ionic species transported by the isolated toad bladder is sodium, under specific experimental conditions the bladder has also been reported to secrete potassium ion (Kallus et al., 1971), to secrete hydrogen ion (Ludens and Fanestil, 1972) and to increase the net transport of calcium (Walser, 1971). Speculations concerning the various models proposed for transepithelial transport of sodium and how their activity might be modified by mineralocorticoid will be presented later in this review.

4.4. RESPONSE OF ISOLATED AMPHIBIAN TISSUE TO MINERALOCORTICOIDS

Because of subtle but significant differences in the short-circuit current pattern which results from aldosterone exposure in the urinary bladder preparation as compared to ventral toad skin, the two tissues will be described separately.

Demonstration of a positive effect of aldosterone on active sodium transport in the isolated urinary bladder of the toad was first reported by Crabbé (1961a). In his original studies aldosterone was injected into the live animal and then, following sacrifice and bladder isolation, the short-circuit current was recorded and when compared to untreated controls was found to be significantly higher. He also demonstrated that the environmental salinity of the animals prior to sacrifice was critical in that toads immersed in saline had a plasma aldosterone concentration only half that of animals immersed in distilled water. This initial *in vivo* observation was followed up by Crabbé (1961b) who reported a short-circuit current response to the *in vitro* addition of aldosterone similar to the pattern shown in Fig. 4.1. Interestingly, he noted that the most reproducible results occurred in tissue from toads immersed in distilled water for 5 days prior to sacrifice and that the one hour delay in onset of aldosterone's effect on short-circuit current, in his experience, could not be shortened by increasing the external concentration of aldosterone 100-fold. Two other points of interest were: confirmation of a linearity between short-circuit current and radioisotopic sodium fluxes and the greater depletion of tissue glycogen in the bladders stimulated to transport more sodium under the influence of aldosterone. These observations were soon confirmed and extended by ourselves (Porter and Edelman, 1964) and others (Sharp and Leaf, 1964). However, during our preliminary studies we found that: (*a*) incubating the isolated tissue overnight in the absence of exogenous steroid enhanced the magnitude and reproducibility of the short-circuit current response and (*b*) exogenous glycolytic substrate, in our case glucose, was an absolute requirement for complete manifestation of the bioelectric response (see Fig. 4.1). Additional observations included evidence that aldosterone was effective regardless of whether it was added to the mucosal or serosal surface, a markedly different situation from vasopressin which is only effective when added serosally, and that the aldosterone-induced increase in net sodium transport, as measured by the short-circuit current, was totally the result of an enhanced active flux from mucosa to serosa rather than a reduction in

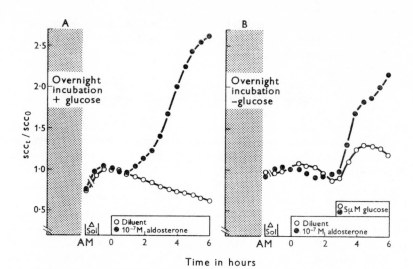

Fig. 4.1. The derived short-circuit current (scc) ratio, scc_t/scc_0 appears on the ordinate, while time in hours relative to steroid addition, t_0, is shown along the abscissa. The sequence of experimental manipulations is depicted immediately above the time axis. Sixteen paired quarter bladder experiments are included in panel A following overnight incubation in the presence of 20 mM glucose. The scc at t_0 for the control quarter bladders (open circles) were 57 ± 9 μA, while for those receiving 10^{-7} M aldosterone (solid circles) the scc at t_0 was 61 ± 11 μA. Panel B represents eight paired quarter bladder experiments in which glucose was deleted from the overnight incubating media. The scc at t_0 for control quarter bladders (open circles) was 42 ± 11 μA at t_0, while for those receiving 10^{-7} M aldosterone the scc at t_0 was 37 ± 6 μA at t_0.

passive back diffusion, i.e. serosal to mucosal flux. Finally, initial exploration of steroid structure–function relationships were studied at equimolar concentrations (7×10^{-7} M) and the order of potency was found to be: aldosterone = 9α-fluorocortisol $\geq$ deoxycorticosterone acetate $\geq$ cortisol $\geq$ corticosterone $\geq$ prednisone $>$ 2-methyl-9α-fluorocortisol $>$ 2-methyl-cortisol $>$ progesterone $\geq$ no steriod (Porter and Edelman, 1964). From these meagre beginnings the field has grown voluminously with many divisions and subdivisions occurring as the quest for a complete description of the mechanism of action of aldosterone was pursued.

A dose response relationship between the short-circuit current response and aldosterone, using the overnight incubated preparation, has been previously reported by our laboratory (Porter, 1968) and the curve is not dissimilar to that reported by Dalton and Snart (1967a) for deoxycorticosterone. Using the double-reciprocal transformation of Lineweaver–Burke we derived a dissociation constant (K_D) for aldosterone of 7.7×10^{-9} M (Porter, 1968) which closely approximates the 2.1×10^8 M^{-1} association constant reported by Snart (1967) for one set of aldosterone binding sites which exist in this membrane preparation.

Another factor which we have shown to alter the basic physiological response of the active Na^+ transport to aldosterone stimulation is environmental temperature. Fig. 4.2 depicts the different short-circuit current response to aldosterone addition at membrane temperatures between 18 and 30°C. As can be seen, there is an inverse relationship between the duration of the latent period (i.e. time between aldosterone addition and increase in short-circuit current) and temperature, and this same relationship holds for the time required to achieve maximum stimulation following aldosterone as influenced by temperature.

In more prolonged incubation studies Mamelak and Maffly (1969) reported that after 40 hours of substrate-steroid depletion, a time at which no recordable short-circuit current was present, re-introduction of aldosterone followed by glucose resulted in complete recovery of the short-circuit current, equivalent to those recorded following initial membrane mounting. They concluded that virtually all transepithelial sodium transport was aldosterone dependent.

Contrary to the results obtained with the isolated urinary bladder the short-circuit current response in amphibian skin to aldosterone has been variable. The earliest report demonstrating a positive response to the *in vitro* application of adrenal steroid came from Taubenhaus *et al.* (1956), following a 16-hour incubation of skin bags obtained from frog hind limbs. Unfortunately, because net ion transfer was measured chemically, thus necessitating a prolonged collection period, a detailed description of the sequential response of sodium transport was not obtained. In 1961, McAfee and Locke reported their results in which short-circuit current was used to measure the sodium transport response to the *in vitro* addition of adrenal steroids. Measurements were made at four-hour intervals for 16 hours following steroid addition. Maximum short-circuit current responses were recorded between the fourth and eighth hour and this effect persisted for the remainder of the observation time. In 1964, Crabbé summarized his early experience concerning the *in vitro* addition of aldosterone to isolated amphibian skin. Whereas three hours of aldosterone exposure were sufficient to stimulate short-circuit current in frog skin, no significant effect in toad skin was evident unless the period of incubation was extended to 12–16 hours. He suggested that this difference might be due to the greater thickness of the toad skin.

Recently, we have reported our detailed observation concerning the short-circuit current response to *in vitro* aldosterone exposure in isolated ventral skin from the toad, *Bufo marinus* (Porter, 1971a). Several

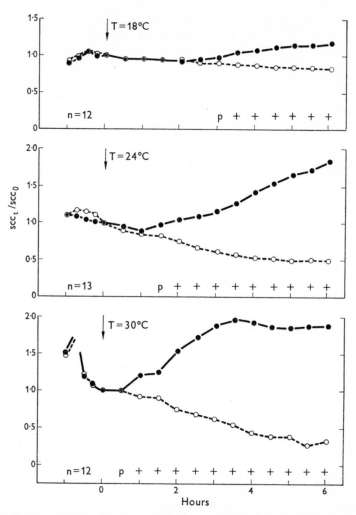

Fig. 4.2. The short-circuit current response, expressed as the ratio scc_t/scc_0, to 10^{-7} M aldosterone (addition point indicated by the arrows) at the 3 temperatures indicated, i.e. 18°, 24° and 30°C, is depicted in paired quarter bladder experiments. The solid circles represent responses in aldosterone-treated membranes, while the open circles are untreated controls. All membranes were incubated overnight in the presence of glucose-fortified Frog Ringers and temperature was adjusted the following morning. Significant differences, assessed by the paired t test are shown immediately above the abscissa with a + signifying $p < 0.05$.

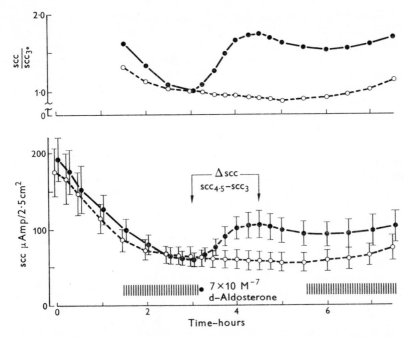

Figure 4.3. Change in short-circuit current (scc) in freshly mounted ventral skin quarters after addition of 7×10^{-7} M aldosterone (solid circles), compared to scc in matched, non-steroid treated control skin quarters (open circles) is plotted in the lower graph. The vertical bars represent ±SEM for the eight pairs summarized. The scc in μA/chamber cross-sectional area is shown on the ordinate, while time elapsed after mounting is shown on the abscissa.

The upper graph is a plot of the change in the scc relative to the value recorded at the time of aldosterone addition, i.e. 3 hours. This provides a visual record of the percentage increase in scc induced by aldosterone [taken from Porter (1971a)].

points bear emphasis concerning both similarities and differences between the response which aldosterone elicits in the skin as compared to the urinary bladder. Figure 4.3 depicts the short-circuit current pattern in freshly mounted skin harvested from toads immersed in 0.6% NaCl solution for 2–3 days prior to sacrifice, a technique reported by Sharp and Leaf (1966) to enhance aldosterone responsiveness in freshly mounted urinary bladder. Although a latent period of approximately 90 min was evident, the maximum short-circuit current response was achieved within 3 hours of aldosterone addition, which differs from results obtained using isolated urinary bladders incubated overnight (see Fig. 4.1). It is noteworthy that an electrical pattern similar to that shown in Fig. 4.3 was recorded by Sharp and Leaf (1966) using freshly mounted bladder preparations. Following over-night incubation of toad skin in steroid-free media we originally failed to demonstrate a significant increase in short-circuit current during five hours of aldosterone exposure (Porter, 1971a); however, in subsequent experiments, which confirm the recent report of Crabbé *et al.* (1971), extending the aldosterone exposure to eight hours did provide evidence of aldosterone-stimulated short-circuit current. Figure 4.4 demonstrates this more prolonged latent period following aldosterone addition to ventral toad skin after overnight incubation and also demonstrated that the presence of exogenous aldosterone during overnight incubation interfers with the magnitude of hormone action. As in the case of iso-lated urinary bladder, glucose is required for complete demonstration of the enhanced short-circuit current following aldosterone (Fig. 4.5). In addition, the dose response range for aldosterone in isolated toad skin is between 10^{-6} M and 10^{-8} M (Porter, 1971a); whereas, that for isolated urinary bladder is between 10^{-7} M and 10^{-9} M (Porter, 1968). Using toad skins incubated overnight with various steroids (aldo-sterone, deoxycorticosterone, estradiol and testosterone), Cirne and Malnic (1972) confirmed and extended the observations of Crabbé

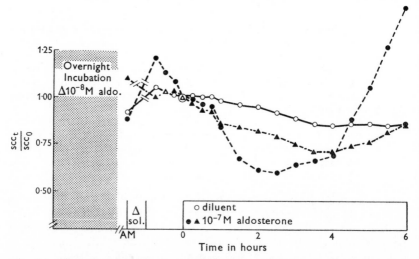

Fig. 4.4. The change in the short-circuit current (scc_t/scc_0) ratio induced by aldosterone in ventral toad skin quarters following overnight incubation. Open circles represent untreated control skin quarters, while solid circles represent skin quarter given 10^{-7} M aldosterone at time zero. Triangles depict skin quarters which were incubated overnight in the presence of 10^{-8} M aldosterone and to which 10^{-7} M aldosterone was added the following morning at time zero.

et al. (1971) by analysing the kinetics of short-circuit current changes associated with progressively decreasing the external sodium concentration. They found that aldosterone increased the maximum velocity of Na^+ transport by the skin, but the affinity between Na^+ transport sites and Na^+ *per se*, as judged by the derived Michaelis–Menten constant (K_m) was unaffected. In addition, these same authors confirmed

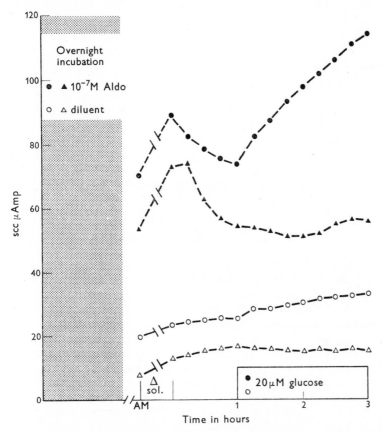

Fig. 4.5. This experiment indicates the glucose-dependency of the aldosterone effect on the short-circuit current (scc) in isolated ventral toad skin. Four skin quarters from the same animal were incubated overnight in the absence of glucose. In addition, two of the four skin quarters had 10^{-7} M aldosterone included in their overnight incubation solution (solid circle and solid triangle). Following the change of solution in the morning and after the scc had stabilized, 20 mM of glucose was provided to one aldosterone-treated (solid circle) and one non-aldosterone treated (open circle) skin quarter and the subsequent scc response recorded. Experimental results from eight separate animals are included.

the steroid-induced fall in electrical resistance which was reported in detail by Civan and Hoffman (1971) using the isolated toad bladder preparation.

Recently, Nielsen (1969) and Voûte et al. (1969) have added a new dimension to the in vitro effect of aldosterone on the isolated skin preparation. These authors have shown that a moult of the skin is associated with the characteristic change in short-circuit current induced by aldosterone. From these observations has arisen a formulation of aldosterone-induced increase in active sodium transport (Huiid–Larsen, 1972; Voûte et al., 1972) which can be summarized as follows: aldosterone increases entry of sodium from the external solution into the cellular compartment; this phase is then followed by a cellular slough formation during which the cornified layer is shed and a shift of the functioning stratification of the epithelium to the deeper cell layer occurs which re-establishes the responsiveness of the skin to aldosterone-stimulated sodium transport.

Finally, aldosterone has been shown by Cofre and Crabbé (1967) to increase sodium transport across toad colon, and a similar effect has also been confirmed for mammalian intestine by work from the laboratory of Edmunds (1972).

4.5. INDUCTION HYPOTHESIS

In 1963, while collaborating with Edelman and Bogoroch, we proposed a hypothesis for the subcellular mechanism of aldosterone action which can be summarized as follows: following exposure of isolated toad urinary bladder to aldosterone, DNA-dependent RNA synthesis is initiated by transcription resulting in de novo synthesis of protein intermediates by translation, which leads to increased active sodium transport. This postulated mechanism was an extension of the earlier suggestion by Crabbé (1961b) regarding a possible explanation for the latent period which characteristically occurs between aldosterone addition and the onset of the physiological response, i.e. increased active sodium transport. After noting that increasing the external concentration of aldosterone failed to influence the latent interval, Crabbé (1963a) inferred the existence of an active intermediate. Elimination of the possibility that delayed aldosterone penetration accounted for the latent period was obtained when the intracellular uptake of ^{3}H-aldosterone was shown to reach steady-state levels after 30 min (Edelman et al., 1963). More recently we have extended these studies with regard to variation in external temperature and, as can be seen in Fig. 4.6, aldosterone uptake plateaus well in advance of the stimulation of the short-circuit current. Other evidence supporting our postulated mechanism of action included: (a) localization of the uptake of

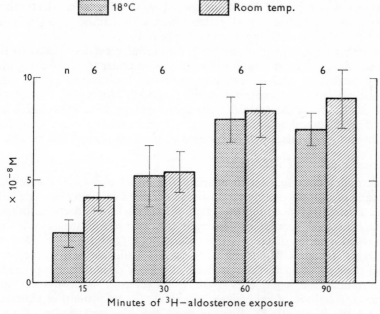

Fig. 4.6. Uptake of ^{3}H-aldosterone by the urinary bladder of the toad as a function of both time and environmental temperature. The ordinate is the concentration of aldosterone in bladder cells expressed in moles per litre cell water. The abscissa indicates the time that elapsed during ^{3}H-aldosterone exposure. Above each column is shown the number of paired hemi-bladders included. Variations in environment temperature are coded in the insert. The vertical bars represent $\pm$SEM. The technique used to measure intracellular aldosterone concentration was identical to that of Edelman *et al.* (1963).

^{3}H-aldosterone in the nuclei of epithelial cells, demonstrated by radio-autography (Edelman *et al.*, 1963); (*b*) failure of the removal of aldo-sterone, after only 45 min exposure, to alter the characteristic short-circuit pattern (Edelman *et al.*, 1963); (*c*) selective sensitivity of the aldosterone-induced increase in the short-circuit current pattern to inhibition of either RNA synthesis by actinomycin D or protein syn-thesis by puromycin (Edelman *et al.*, 1963); and finally (*d*) elimination of the latent period when substrate-depleted tissue was pretreated with aldosterone for a three-hour interval and then given glucose repletion (Edelman *et al.*, 1963). Confirmation and extention of the effect of various metabolic inhibitors soon followed (Crabbé and De Weer, 1964; Fanestil and Edelman, 1966a), as did demonstration that 5 min of aldosterone exposure was sufficient to initiate the subsequent in-crease in the short-circuit current (Sharp and Leaf, 1966). Additional

5

support was reported by Williamson (1963) using adrenalectomized rats who demonstrated that pretreatment with actinomycin D eliminated the sodium-retaining action of aldosterone.

Further exploration of the early intracellular events related to the mechanism of steroid action will be discussed under three subcategories for convenience and do not correspond to the chronology of these investigations. The three areas include: (a) tissue-binding of aldosterone; (b) RNA synthesis; (c) aldosterone-induced protein. Speculation regarding the coupling of this sequence of reactions to the active sodium transport mechanism will be covered in the final part of this review.

Tissue Binding of Aldosterone

A major contribution to defining the initial events in the cellular uptake of aldosterone by target tissue occurred in 1966 when Fanestil and Edelman (1966b) reported that kidney nuclei of adrenalectomized rats contained specific protein receptors of aldosterone. There then followed a series of reports from these authors and their various co-workers further delineating the characteristics of these receptors and their function in the initiation of genetic decoding. Aldosterone-macromolecular complexes were isolated from both nuclear and cytosol fractions of rat kidney and shown to be protein in character (Herman et al., 1968). Furthermore, the binding affinity of these complexes for other steroids paralleled their potency in regulating renal tubular sodium transport, thus confirming their specificity as mineralocorticoid receptors (Herman et al., 1968). In addition to kidney, aldosterone-binding proteins were identified in both nuclei and cytosol of duodenal mucosa, spleen, liver and brain of rats, although the concentration in the latter three tissues was relatively low (Swaneck et al., 1969). The relevance of these aldosterone receptors to the mineralocorticoid activity was supported by the close correlation between the ability of a steroid to displace aldosterone from its nuclear binding sites and its biological potency as either an antagonist or agonist (Herman et al., 1968). As a follow-up to these studies, Swaneck, Chu and Edelman (1970) demonstrated a similar correlation between steroid binding affinities with respect to nuclear chromatin, the site of transcriptional events, and their mineralocorticoid potency.

Aldosterone-binding macromolecules from both the cytosol (Ludens and Fanestil, 1971) and nuclei (Marver et al., 1972) are reported to exist in at least two forms. The significance of the two forms of cytosol-binding protein is unclear since they exhibit multiple similarities with regard to steroid-binding hierarchy and association constant (Ludens and Fanestil, 1971). However, the same is not true for the two nuclear fractions. Based upon reconstitution experiments, in which radiolabelled cytosol is mixed with unlabelled nuclei, a three-step process for the transfer of aldosterone from cytoplasm to nucleus is proposed (Marver *et al.*, 1972). In essence, the quantity of heat-labile, steroid–cytosol complex was shown to decline reciprocally with the appearance of increased quantities of, first, the 0.1 M-Tris-extractable nuclear complex, and later a 0.4 M-KCl-extractable nuclear complex which is bound to chromatin (Marver *et al.*, 1972). A schematic representation of the proposed sequence by which aldosterone is translocated from the cell surface to the nucleus is shown in Fig. 4.7. This proposal is not unique for mineralocorticoids but rather is assumed to be a common feature of all steroid hormones (O'Malley, 1971).

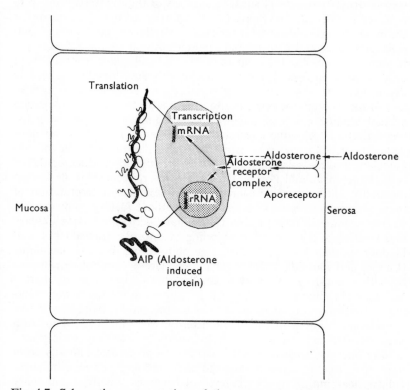

Fig. 4.7. Schematic representation of the proposed model concerning the early events in the action of aldosterone in isolated epithelial cells. The steroid enters by diffusion where it may combine with a cytoplasmic aporeceptor to form an aldosterone-receptor complex which translocates to the nucleus, or it may diffuse freely through the cytoplasm to be taken up by the nucleus. Once in the nucleus the steroid interacts with chromatin to increase both m-RNA and r-RNA synthesis. The various RNAs generated then enter the cytoplasm inducing the translation of protein which is designated AIP (aldosterone-induced protein).

Unfortunately, characterization of the aldosterone-binding protein from toad bladder epithelial cells has not progressed with the same speed noted above for rat kidney cells. Using a Scatchard plot derived from the displacement of ^{3}H-aldosterone from isolated toad urinary bladder, Sharp and co-workers (1966b) defined two sets of binding sites within the physiological concentration range for aldosterone. Since the smaller set of sites was saturated at a concentration far below that which elicits a physiological response in the tissue, they concluded that the larger set of sites was responsible for the hormone's actions. Furthermore, antagonists such as progesterone and spironolactone displaced aldosterone from both sets of sites. Extensions of these observations were reported by Ausiello and Sharp (1968), from which they concluded that the mineralocorticoid-receptor sites resided in the nuclei of the epithelial cells. Attempts at isolation and characterization of the aldosterone-binding macromolecules from toad bladder cells have, to date, been only partially successful. Alberti and Sharp (1969) identified a rapidly dissociable steroid–protein complex which they concluded was derived almost totally from the nucleus and were unable to delineate a separate cytosol binding complex. Likewise, in preliminary studies using the reconstitution technique suggested by Marver *et al.* (1972), we have been unsuccessful in demonstrating a cytoplasmic transfer of ^{3}H-aldosterone into isolated, intact nuclei from toad bladder epithelial cells (Hutchinson and Porter, 1973a). However, because of the known instability of the steroid–macromolecular complex, interpretation of these observations would be hazardous without further confirmation. Recently, Fanestil and Ludens (1973) have performed a preliminary investigation of the aldosterone-binding by cytosol extracted from toad bladder epithelial cells. Utilizing a competitive binding technique (Ludens and Fanestil, 1971) involving displacement of ^{3}H-aldosterone by a thousand-fold excess of unlabelled aldosterone, and comparing the residual radioactivity in the cytosol sample so treated to the counting activity of a non-displaced cytosol sample, no evidence of selective cytoplasmic binding of the ^{3}H-aldosterone was obtained when fresh toad bladder cells were used. Interference by excessive endogenous aldosterone was suspected when a mixture of cytosol aliquots from fresh toad bladder and adrenalectomized rat renal tissue eliminated aldosterone-binding in the rat cytosol fraction. Finally, preliminary evidence for preferential cytosol binding of aldosterone was demonstrated in toad bladder tissue following overnight incubation in steroid-free media. Thus, lacking conclusive proof of a specific mineralocorticoid receptor protein in the cytosol of toad bladder epithelia, Fig. 4.7 includes an alternative entry step in which aldosterone penetrates the nuclear membrane in the unbound form.

Samuels and Tomkins (1970), based upon their examination of the induction of tyrosine aminotransferase in hepatoma tissue culture cells by a wide variety of glucocorticoids, define four categories of inducers: (a) optimal inducers, which give maximal enzyme induction; (b) suboptimal inducers; which cause only submaximal enzyme induction when used alone or, in combination with optimal inducers, cause competitive inhibition; (c) anti-inducers, which are without inducer activity but can competitively inhibit inducers; and (d) inactive steroids, which neither induce nor inhibit.

Recently, Alberti and Sharp (1970) have reported four classes of steroids, classified according to their interaction with mineralocorticoid receptors in the toad bladder, which have many of the properties of the categories enumerated above for glucocorticoids. Using the Samuels and Tomkins classification and based upon the findings of Alberti and Sharp (1970), mineralocorticoids would be categorized as follows: (a) optimal inducers would include aldosterone, deoxycorticosterone and 9α-fluorocortisol; (b) suboptimal inducers–cortexolone; (c) anti-inducers–progesterone and steroidal spironolactones; (d) inactive steroids–17β-estradiol.

Recently, we have reported evidence confirming the competitive inhibition of various mineralocorticoids by steroidal spironolactones (Porter, 1968). Utilizing the double-reciprocal transformation of Lineweaver-Burk, the relative affinity of the toad bladder tissue receptors was evaluated based on the derived dissociation constants. The resulting hierarchy of mineralocorticoid–spironolactone interaction (Porter, 1971b) corresponded to the displacement results for rat kidney nuclear receptor reported by Herman et al. (1968), thus inferring a similar sterospecific mineralocorticoid receptor within toad bladder tissues.

RNA Synthesis

More direct evidence regarding the proposed induction hypothesis for the mechanism of action of aldosterone (Edelman et al., 1963) involves the demonstration of a preferential increase in ^{3}H-uridine incorporation into RNA in aldosterone treated tissue (Porter et al., 1964). Of significance is the fact that a measurable increase in whole bladder RNA synthesis was evident prior to the steroid-induced stimulation of sodium transport. Progesterone, a steroid without effect on active sodium transport failed to enhance ^{3}H-uridine incorporation into RNA. When epithelial cells were scraped free from the submucosal and serosal cellular components, increased ^{3}H-uridine incorporation was likewise evident following aldosterone exposure (Porter et al., 1964).

In addition to confirming the effect of aldosterone on RNA synthesis in the whole bladder, DeWeer and Crabbé (1968) demonstrated that intact sodium transport was not a prerequisite for such an observed effect, thus eliminating changes in intracellular ionic composition as a cause for induction-repression. In a follow-up report from the same laboratory (Rousseau and Crabbé, 1968), again using whole bladder extracts, a rapidly labelled RNA fraction was reported which the authors concluded possessed certain characteristics similar to m-RNA. However, because of certain inconsistencies between various laboratories, regarding the duration of aldosterone exposure necessary to demonstrate enhanced H^3-uridine incorporation into RNA (Fanestil, 1969), and due to aberrant results occasionally encountered in our own laboratory, a modified technique for the estimation of RNA synthesis using an acidified alkaline-extract of perchloric acid-insoluble RNA, rather than a phenol-RNA extract, was developed (Hutchinson and Porter, 1970a). Utilizing differential centrifugation techniques, the significant increase in subcellular RNA specific radioactivity due to aldosterone action was localized in the nuclear fraction of scraped epithelial cells (Hutchinson and Porter, 1970b). Furthermore, the effect of aldosterone on RNA synthesis could be eliminated by pretreatment with the spironolactone (Hutchinson and Porter, 1970b).

Doubt concerning the validity of the uridine incorporation studies was introduced when Vančura et al. (1971) were unable to demonstrate a rapidly labelled RNA fraction from epithelial cell homogenates after aldosterone exposure. Using polyacrylamide gel electrophoresis to analyse labelling in various heterodisperse RNAs, these authors (Vančura et al., 1971) suggested that bacterial contamination might account for previously reported early RNA labelling associated with aldosterone studies in the isolated toad bladder. This question of possible bacterial contamination in the in vitro system employed in our laboratory was examined in detail. The results of these experiments excluded bacterial contamination as an explanation for our results (Hutchinson and Porter, 1972a). However, more recent experiments have suggested a source of potential interpretative error with regard to the use of ^{3}H-uridine as a measure of aldosterone-induced RNA synthesis. In essence we have demonstrated (Hutchinson and Porter, 1972b) that when an excess of exogenous non-isotopic uridine was added to tracer quantities of ^{3}H-uridine, the reproducibility of aldosterone-stimulated isotope incorporation into heterogenous nuclear RNA was improved substantially. From these results it was suggested that internal recycling of endogenous nucleoside might account for this observation (Hutchinson and Porter, 1973). Figure 4.8 schematically depicts this postulated mechanism. Under tracer conditions, i.e. 0.53 μM uridine, the label enters the epithelial cell from the serosal side of the cell, but the amount of dilution by unlabelled uridine in the cells' precursor pool is constantly changing because of nucleoside feedback from unstable, preformed RNA (left-hand side of Fig. 4.8). In contrast, when excess uridine is supplied along with tracer (right-hand side of Fig. 4.8), the dilution effect from nucleoside recycling via

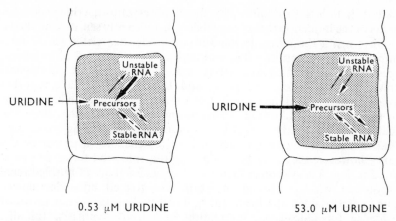

0.53 μM URIDINE   53.0 μM URIDINE

Fig. 4.8. Schematic model of the nucleoside recycling believed to occur within
the toad bladder cell, showing how its direction might be modified by
flooding the system with exogenous uridine. See text for explanation.

an endogenous "salvage" pathway is diminished, thus allowing a more
constant ratio of labelled to unlabelled uridine in the precursor pool.
Under the latter condition any change in uridine incorporation into
newly formed nuclear RNA can be discriminated. As a further check
on this explanation, toad bladder cells were prelabelled with ^{3}H-uridine
for 20 hours—then exposed to either tracer quantities or excess amounts
of unlabelled uridine as a "chaser" and the effect on ^{3}H-uridine distribution between nuclear RNA and the cytosol precursor pool measured
with and without aldosterone being present. When tracer amounts of
unlabelled uridine were employed no difference in the distribution of
^{3}H-uridine between nuclear RNA extract and cytosol fractions from
which cytoplasmic RNA had been removed was evident between control and aldosterone-treated cells. However, when excess unlabelled
uridine was present the distribution ratio of ^{3}H-uridine between nuclear
RNA extract and cytosol was significantly greater in the aldosterone-
treated cells as compared to the control cells, indicating an increased
rate of RNA synthesis by aldosterone (Hutchinson and Porter, 1973b).
In addition, under conditions of excess unlabelled uridine, the specific
activity of the nuclear RNA decreased with aldosterone treatment as
compared to similarly treated control cells. Finally, in a detailed report
on the effects of aldosterone on RNA synthesis in the toad bladder,
Rousseau and Crabbé (1972) confirmed our findings of increased
uridine incorporation into nuclear RNA but could not identify this as
being associated with the m-RNA component. Furthermore, they
(Rousseau and Crabbé, 1972) could not demonstrate an effect of aldosterone on protein synthesis at the level of the polysomes and concluded

that their failure to confirm, by direct observations, critical components of the induction theory may reflect insufficiently sensitive methods.

Once again the adrenalectomized rat model has yielded additional
information concerning the mechanism of action of aldosterone,
specifically supporting the RNA induction hypothesis. Fimognari
et al. (1967a; 1967b) and Forte and Landon (1968) have reported
aldosterone-induced increases in RNA synthesis in subcellular fractions
of the rat kidney following in vivo hormone administration. Recently,
Kalra and Wheldrake (1972) have questioned this finding, suggesting
that aldosterone's principal action involves changes in precursor uptake; however, they may be confirming the enhanced turnover postulated in Fig. 4.8. Although increased template activity of purified renal
chromatin was not found in aldosterone-treated adrenalectomized
rats (Trachewsky and Cheah, 1971), Trachewsky's group has provided
evidence for aldosterone-stimulation of r-RNA synthesis. Initially,
Nandi-Majumdar and Trachewsky (1971) demonstrated that ribosomes of the renal cortex, prepared from aldosterone-treated adrenalectomized rats, incorporated more phenylalanine into polypeptides
than ribosomes derived from non-hormonally treated controls. More
recently, they (Trachewsky et al., 1972) have reported data which
support this effect as being specific for mineralocorticoids. Qualitative
changes in the RNA population from the renal cortical tissue of aldosterone-treated rats are indicated from the DNA hybridization studies
of Congote and Trachewsky (1972), which gain further significance
from the demonstration of a stimulatory effect of aldosterone on
nuclear RNA-polymerase activity (Liew, Liu and Gornall, 1972).
Present evidence suggests that two forms of RNA-polymerase activity
are present within the nucleus; RNA-polymerase I resides in the nucleolus
and synthesizes principally r-RNA, while RNA polymerase II is
present in the nucleoplasm and is involved in the synthesis of heterogenous RNA, including m-RNA (Roeder and Rutter, 1970). Edelman's
group has reported that aldosterone preferentially increased RNA-
polymerase I activity in relation to RNA-polymerase II activity in
adrenalectomized rats (Feldman et al., 1972). A similar finding has
been reported by Mishra et al. (1972). Furthermore, these latter authors
reported a decrease in turnover rate for nuclear, mitochondrial and
ribosomal RNAs following adrenalectomy in the rat, and implied that
adrenocortical steroids modified the synthesis of RNA (Mishra,
et al., 1971). Since gene activation and repression in eukaryotes is
believed to involve chromatin components and histone acetylation.
Trachewsky and Lawrence (1972) have examined the effect of aldosterone on these nuclear events using renal cortical material from
adrenalectomized rats. They (Trachewsky and Lawrence, 1972) reported a sequential increase in the synthesis of non-histone proteins,
RNA, and finally histones of chromatin material, between 15 and 45

min following aldosterone injection. Furthermore, within 15 min after steroid administration, histone acetylation was increased, a necessary prerequisite for the concept of DNA-directed RNA synthesis. Thus, evidence continues to mount affirming the critical role of DNA-directed RNA synthesis as an early event in the biochemical expression of aldosterone's intracellular action.

Aldosterone-induced Protein (AIP)

In contrast to the progress in confirming the effect of aldosterone on RNA synthesis, substantiation of the proposed translational effect on protein synthesis has been disappointing (Fanestil, 1972). That new protein synthesis is an integral part of the proposed mechanism of aldosterone's action comes from studies involving the use of inhibitors of protein synthesis. Both puromycin and cycloheximide, inhibitors of the ribosomal assembly of protein, have been shown to interfere with the characteristic *in vitro* action of aldosterone on the short-circuit current response of the toad bladder (Edelman *et al.*, 1963; Fanestil and Edelman, 1966a). With regard to the possibility that the induced protein might be an enzyme, the only positive evidence concerning this point comes from the observations of Kirsten and co-workers (1968; 1970a; 1970b). They have demonstrated an increase in the activity of certain enzymes of the tricarboxylic acid cycle, particularly citrate synthetase, in both rat kidney and toad bladder. In addition, there is the previously mentioned report by Nandi–Majumdar and Trachewsky (1971) showing an increased rate of incorporation of phenylalanine into polypeptides by ribosomes isolated from aldosterone-treated rat kidney. I share the opinion of Fanestil (1972): "this area of investigation is in need of new technology or an innovative experimental approach." At present it would seem that, although indirect evidence supports the concept that aldosterone-induced protein plays a critical role in the intracellular mechanism of action of aldosterone, to date direct confirmation has eluded experimental demonstrations.

4.6. MECHANISM OF STIMULATION OF ACTIVE Na+ TRANSPORT BY ALDOSTERONE

Sodium transport across toad bladder epithelial cells is postulated to occur through a system of membrane boundaries in series, the so-called Ussing–Skou model of sodium transport (Ussing and Zerahn, 1951; Skou, 1957), in which Na+ passively enters across the apical plasma membrane (Frazier *et al.*, 1962; Frazier and Hammer, 1963; Koefoed–Johnson and Ussing, 1958) and is actively extruded across the basal-lateral cell surface at the expense of ATP breakdown. Recent observations by Ferguson and Smith (1972) demonstrating non-linear mucosal sodium entry in isolated bladder preparations suggests that two steps are involved at this boundary. At lower external Na^+ concentrations, entry involves a saturable mechanism suggesting carrier mediation, while at higher concentrations entry can be shown to be concentration-dependent. Histochemical studies by Keller (1963) show apparent ATPase activity distributed along the basal–lateral surface of the epithelial cells of the urinary bladder. Parallel extracellular pathways capable of net fluid movement have been postulated for amphibian skin (Ussing, 1966) and more recently for the toad bladder (Urakabe et al., 1970; DiBona, 1972). Whether the tight junction (desmosome) of the toad bladder epithelial cells allow direct communication from mucosal surface to lateral intercellular spaces is unresolved. Using an optical system which allowed direct visualization, Grantham et al. concluded that for permeable bladder, water flows across the apical membrane into the cell and hence into the lateral intercellular space by osmosis (Grantham et al., 1971). More recently, Herman and Stiles (1972), using lanthanum as a marker, interpreted electron micro-scopic studies as evidence of a direct communication across the tight junction during increased water permeability.

The correlation between active Na^+ transport and Na^+–K^+ activated ATPase has been demonstrated repeatedly since Skou's (1957) original description of the enzyme. Bonting and Canady (1964) confirmed a relationship between Na^+–K^+ ATPase activity and active Na^+ trans-port in the urinary bladder of Bufo marinus, by demonstrating a nearly identical ouabain concentration curve for inhibition of enzyme and active Na^+ transport. They derived a pI_{50} (half-maximal inhibitor concentration) of 4.6 for ouabain-inhibited short-circuit current, while our results indicate a pI_{50} of 4.2 (Fig. 4.9). From the data of McClane (1965), a pI_{50} of 4.4 can be derived for the ouabain inhibition of the short-circuit current. Similar results have been reported by Asano et al. (1970) for the Na^+–K^+ activated ATPase extracted from the urinary bladder of Bufo bufo japonicus. Based on a kinetic analysis of ouabain-induced changes in rate coefficients for uni-directional Na^+ flux in the in vitro toad bladder preparation, Herrera (1966) concluded that the transport inhibition caused by ouabain involved the active transport system located at the basal barrier of the epithelial cells. This conclusion is supported by the work of Nagel and Dörge (1971) using the frog skin. Cortas and Walser (1971) have suggested a co-operative interaction between enzymic binding sites of Na^+–K^+ activated ATPase, extracted from isolated mucosal cells of toad bladder, and cations. Furthermore, based upon variations in the extraction technique, they (Cortas and Walser, 1971) caution that Na^+–K^+ stimulated ATPase can not be equated with ouabain-inhibited ATPase under all conditions. Additional evidence dissociating ouabain in-hibition of Na^+ transport and Na^+–K^+ activated ATPase comes from Park (1973). Although he found good correspondence between ouabain

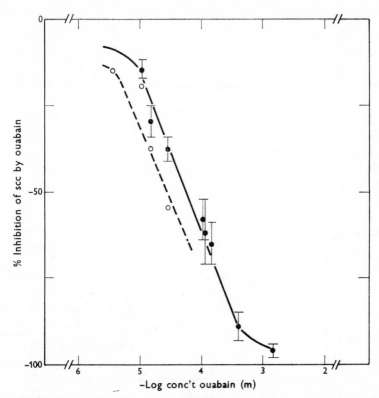

Fig. 4.9. Log concentration curve relating the percentage of short-circuit current (scc) inhibited at various concentrations of ouabain. Solid circles are measurements from bladder quarters exposed to ouabain, following overnight incubation in the presence of glucose only, while open circles represent quarter bladders exposed to ouabain 3 hours after mounting in the presence of glucose but no aldosterone. Vertical bars represent ±SEM. 156 separate studies are summarized on this graph.

inhibition of Na^+–K^+ activated ATPase and the short-circuit current when enzyme activity was assayed at 37°C, when the assay temperature was reduced to 24°C, the temperature at which short-circuit current is typically measured, little evidence of a ouabain-inhibitable component remained in the enzyme.

Based upon the Ussing–Skou model, three theories regarding the mechanism by which aldosterone stimulates active sodium transport have been proposed. They include: (a) the pump theory; (b) the permease theory, and (c) the energy theory. A schematic representation of how these various mechanisms would be coupled to aldosterone-induced protein (AIP) is shown in Fig. 4.10.

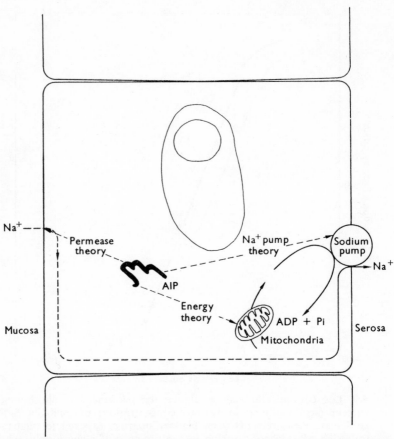

Fig. 4.10. The possible site of action of **AIP** (aldosterone-induced protein). Na^+ enters the mucosal border of the cell, diffusing down an electrochemical gradient, and is extruded at the serosal border by a pump of which Na^+–K^+ activated ATPase is a major component. AIP might exert its effect on Na^+ transport through one or a combination of the following mechanisms: (1) altered permeability at the apical cell membrane (permease theory); (2) the oxidative pathway leading to the generation of more ATP to drive the pump (energy theory); (3) the Na^+–K^+ ATPase associated with the sodium pump (pump theory).

Pump Theory

Based upon the preferential inhibitory effect of ouabain on aldosterone-stimulated sodium transport by the toad bladder, originally reported by Sharp and Leaf (1966) and later confirmed in our laboratory (Porter, 1970b), Goodman *et al.* (1969) proposed the pump theory for the effect of aldosterone on sodium transport. In essence these authors

(Goodman *et al.*, 1969) concluded that aldosterone either changed the sodium pump, i.e. Na^+–K^+ activated ATPase by activation or synthesis of new ATPase, or altered the local environment of the enzyme thus causing increased net sodium transport. Their conclusions rest heavily on the assumption that ouabain is a specific inhibitor of membrane bound Na^+–K^+ activated ATPase, a supposition which is not universally true according to the findings of Cortas and Walser (1971) alluded to previously. In addition, Hill *et al.* (1973), although unable to exclude a qualitative change in Na^+–K^+ activated ATPase extracted from toad bladder following aldosterone exposure, did exclude a quantitative increase of enzyme activity associated with steroid addition. That the increased sensitivity of aldosterone-stimulated Na^+ transport to mimimal concentrations of ouabain may not be mediated through Na^+–K^+ activated ATPase *per se*, but rather may involve interference with glycolysis, is suggested by work both in our laboratory (1972) and the observations reported by Pelletier (1972). It was demonstrated that the increased ouabain sensitivity of aldosterone-stimulated short-circuit current requires the presence of glucose. Recently, a more rigorous evaluation of the interaction of ouabain and various hormones known to stimulate sodium transport, including aldosterone, has been conducted (Crabbé *et al.*, 1973). From these results we propose an attractive hypothesis for explaining the increased ouabain sensitivity of aldosterone-substrate treated tissue which involves the presence of a new type of ATPase with properties differing from those of the enzyme system operating under non-stimulated conditions. Indirect support for such speculation can be found in the work of Gachelin and Bastide (1968) who reported two membrane-bound Na^+–K^+ activated ATPases in frog urinary bladder. One must be cautious when inferring *in situ* enzyme characteristics from measurements performed on broken cell preparations. If steric orientation of enzyme within the membrane is critical, or if the local environment conditions dictate enzyme activation, then such disparate results as noted above would be expected. Recently, Coplon and Maffly (1972) have re-examined the intracellular action of ouabain inhibition of the short-circuit current in the isolated toad bladder by simultaneously measuring CO_2 production and net Na^+ transport. This study was performed since previous reports were conflicting as to the effect of ouabain on oxygen consumption and pyruvate or acetoacetate utilization by this membrane system (Hallowell *et al.*, 1968; Levy and Richards, 1965; Lipton and Edelman, 1971; Sharp and Leaf, 1965). Based upon the similarity between the effect of ouabain and mucosal sodium removal on both short-circuit current and CO_2 production, the authors concluded that the predominant cause of the reduction of respiration induced by ouabain was due to inhibition of transepithelial sodium transport (Coplon and Maffly, 1972). However, Poole *et al.* (1971), using Ehrlich

ascites tumour cells, have demonstrated a tight linkage between ouabain inhibition of glycolysis and blockade of K^+ transport. Explanations advanced by these authors (Poole et al., 1971) included a secondary accumulation of ATP due to ATPase inhibition, thus interferring
with the glycolytic pathway, or a direct effect of intracellular K^+
depletion which is known to reduce glycolysis. That the action of ouabain in the toad bladder may be mediated through intracellular K^+
depletion seems unlikely since the onset of ouabain inhibition in biological membranes begins within 5 min and is virtually complete by
1 hour, whereas 1 to 1.5 hours of ouabain exposure is needed before a
significant decrease in intracellular K^+ concentration can be measured
in either intact toad bladder (Finn et al., 1966) or isolated epithelial
cells (Handler et al., 1972).

No less confusing is the effect of aldosterone on Na^+–K^+ activated
ATPase, i.e. the pump, in rat kidney preparations. The correlation
between Na^+–K^+ ATPase activity in relation to renal sodium re-
absorption for both adrenalectomized (Landon et al., 1966; Chignell
et al., 1966; Jørgensen, 1969) and non-adrenalectomized rats (Katz
and Epstine, 1967) have been well documented, and this function seems
to reside in the ascending limb and distal tubule of the nephron (Kiil,
1971). Germane to the present subject is the reduced renal Na^+–K^+
ATPase activity associated with adrenalectomy which can be restored
by aldosterone administration (Jørgensen, 1969; 1972). However,
there exists a significant temporal discrepancy between restoration of
enzyme activity, i.e. undetectable prior to 6 hours and maximum after
24 hours (Jørgensen, 1969), and the aldosterone-induced changes in
urinary Na^+K^+ excretion, i.e. evident within 2 hours (Fimognari
et al., 1967a; 1967b). Furthermore, in rats with intact adrenals sodium
depletion does not increase renal Na^+–K^+ ATPase despite increased
aldosterone secretion (DeSanto et al., 1971). Recently, an explanation
for these confusing results has been offered by Hendler et al. (1972).
These workers demonstrated that restoration of Na^+–K^+ ATPase
activity in adrenalectomized rats was dependent on the glucocorticoid
effect of adrenal steroids rather than the mineralocorticoid effect. Thus,
it would appear that although mineralocorticoids may induce qualitative
changes in Na^+–K^+ activated ATPase, quantitative changes are a
manifestation of the glucocorticoid effect.

Permease Theory

That aldosterone acts to increase transepithelial sodium transport
by increasing mucosal entry and thus providing more cation to the
basal–lateral pumps, was first proposed by Crabbé (1963b) and Sharp
and Leaf (1964) and has come to be referred to as the permease theory.
Support for this theory has been marshalled from a variety of experimental procedures. One prominent argument has involved measurement

of the intracellular Na$^+$ pool by radio-sodium and the changes which aldosterone induce in the pool size. As a follow-up of his original report, Crabbé in association with DeWeer (1969), reported extensive experiments in which they demonstrated that aldosterone, vasopressin and cyclic-AMP induced proportional increases in both net sodium transport and transport pool size, whereas insulin preferentially increased net sodium transport relative to sodium pool changes, the latter result being most compatible with a direct effect on the rate of Na$^+$ extrusion. However, results with the former hormones suggested an increased sodium entry. This approach has been modified more recently to involve the direct measurement of intracellular electrolyte content from both intact urinary bladder and scraped epithelial cells. Handler, Preston and Orloff (1972) found that both vasopressin and aldosterone increased cell sodium content and, based upon the three-compartment model of transepithelial sodium movement, i.e. mucosal solution → cell → serosal solution, concluded that hormones stimulated a rate limiting entry step for sodium located at the mucosal border of the cells. Confirmation of the hormonal responsiveness of the isolated cells was assured by demonstrating the characteristic increase in [6-^{14}C] glucose oxidation that occurred following aldosterone addition (Handler et al., 1972). Utilizing a different method for isolating epithelial cells, Lipton and Edelman (1971) were unable to demonstrate any change in the intracellular concentration of Na$^+$, K$^+$ or Cl$^-$ following either vasopressin or aldosterone. Since the isolated cells were shown to react appropriately to hormone addition, as evidenced by increases in oxygen consumption, the latter authors (Lipton and Edelman, 1971) concluded that both vasopressin and aldosterone act simultaneously to facilitate Na$^+$ entry across the apical membrane and active sodium transport across the basal–lateral surface of the epithelial cell. Using a unique approach to the problem by which mucosal sodium entry was eliminated by exposing the mucosal surface to liquid paraffin, Janáček and co-workers (1971) concluded that aldosterone stimulated independently the entry of sodium ions across the apical membrane and the active extrusion across the basal–lateral cell surface. More recently, a combined technique of measuring both chemical and radioisotopic sodium within the scraped toad bladder cells has been employed by Leaf and Macknight (1972) who reported that the intracellular amount of both forms of Na$^+$ were increased as a result of aldosterone addition. A further complication of interpreting the so-called sodium pool results is the statement by Cereijido (1972) that at low external Na$^+$ concentrations, an increase in the "Na$^+$ transporting pool" was not necessary for the effect of vasopressin.

Thus, the accumulated evidence concerning the effect of aldosterone on intracellular electrolyte composition seems to favour an increase in sodium content which would be consistent with increased mucosal

entry. Unfortunately, when interpreted in terms of the kinetics which
relate active sodium transport as a function of external sodium concentration, the data do not represent *prima facie* evidence concerning
the mechanism of action of aldosterone on active sodium transport.
According to data reported by Crabbé (1972), the half maximum velocity of active sodium transport is achieved at an external mucosal concentration of ∼15 mEq/l; whereas, the intracellular Na^+ concentrations which have been measured in the absence of aldosterone stimulation, i.e. 28 mEq/l cell water (Handler *et al.*, 1972) and 65 mEq/kg
cell water (Lipton and Edelman, 1971), approach or exceed the concentration of Na^+ necessary to induce a maximum velocity of sodium
transport. However, as can be appreciated from the results shown
in Fig. 4.1, following aldosterone addition the active sodium transport
rate in isolated toad bladder may more than double. It is quite conceivable, although unproven to date, that specific compartmentalization
of intracellular electrolytes might exist which would allow a concentration-dependent mechanism to operate and still be compatible with
the values reported.

An additional technique which has been used to dissociate permeability properties from effects on the sodium pump involves measurement of electrical resistance. Again, opposite conclusions have been
reached. Cuthbert and Painter (1969), utilizing a combination of ouabain and pyruvate, failed to find any change in the electrical resistance
of aldosterone-treated tissues and concluded that aldosterone acted
on the active process associated with sodium transport. Conversely,
Civan and Hoffman (1971) demonstrated a moderate decline in electrical resistance associated with aldosterone exposure and suggested that
at least part of the steroid's effect was mediated through a change in a
parallel leak channel. An alternative method of dissociating passive
permeability and active extrusion was the voltage clamp experiments of
Fanestil *et al.* (1967). Based upon changes in Na^+ flux ratio, these
authors concluded that aldosterone increased active sodium extrusion
independently of an effect on mucosal sodium entry.

A somewhat different approach to the problem of aldosterone's
action on mucosal permeability was employed by Dalton and Snart
(1967b; 1968). Assuming that transepithelial sodium transport involved a single rate-limiting step and, furthermore, that the effect of
changing environmental temperature would be limited to the permeability step for such transport, Arrhenius plots were derived from toad
bladders with and without exposure to steroid and, based upon the
different slopes derived, they interpreted their results as being most
compatible with a hormonal-induced increase in mucosal sodium
permeability. However, based upon the detailed study between the
interaction of temperature and active sodium transport reported from
my laboratory (Porter, 1970a), such a simplistic interpretation is seen

to be inappropriate. Specifically, we concluded that owing to the complex character of the temperature dependence of the active Na^+ flux, the transport response to increments in temperature does not provide useful information in the study of mineralocorticoid action.

Experiments using various inhibitors to dissociate passive Na^+ entry from active Na^+ extrusion have been performed to define the site of aldosterone action. Most notable has been the use of the acyl-guanidine Amiloride, which interferes with mucosal entry of Na^+ (Bentley, 1968; Crabbé et al., 1971). The observation that aldosterone-treated tissues are less responsive to addition of Amiloride suggested a site of action at the mucosal surface (Crabbé and Ehrlich, 1968; Crabbé, 1972). A more unique observation was reported by Goodman et al. (1971) in which they noted that bladders pretreated with phospholipase A_2 had a significantly shortened latent period following aldosterone exposure. When this observation was combined with data on ^{14}C-fatty acid incorporation, they concluded that a fundamental action of aldosterone involved alteration of the fatty acid metabolism of membrane phospholipids. However, in an earlier study, DeGraeff and associates (1965) were unable to define any relationship between sodium transport and either the amount or metabolic turnover rate of phospholipids, including studies involving membranes treated with aldosterone.

Finally, the polyene antibiotic amphotericin B has been used in an attempt to confirm the permease theory. Rationale for its use was based upon the observation that amphotericin B increases the permeability of artificial lipid membranes (Kinsky et al., 1966) and it was shown by Lichtenstein and Leaf (1965) that an increase in Na^+ transport by the toad bladder could be induced by the antibiotic which simulates the aldosterone response, i.e. that substrate addition to hemibladders pretreated with amphotericin B had an immediate curvilinear increase in the short-circuit current similar to the results in aldosterone-pre-treated tissue. Additional evidence suggesting that amphotericin B and aldosterone shared a common site of action on sodium transport, was afforded by elimination of the amphotericin B effect on short-circuit current in bladders previously exposed to aldosterone (Crabbé, 1967). However, differences between the two compounds have also been observed. Although the glucose dependence of aldosterone action in toad bladder is well established, amphotericin B can be shown to increase the short-circuit current in glucose-depleted tissue (Fanestil et al., 1968). Furthermore, in a detailed study of the interrelationship between oxidative metabolism and the action of aldosterone, Fanestil et al. (1968) were able to dissociate the action of aldosterone from that of either vasopressin or amphotericin B, based upon the dependence of the physiological activity of aldosterone on intact NADH-linked electron transport.

Energy Theory

That aldosterone-induced protein might act to increase the supply of energy, i.e. ATP, available to the pump thus causing a stimulation of net sodium transport, was first proposed to explain the effect of aldosterone on transepithelial sodium transport by Edelman and co-workers (1963). This speculation was based on the absolute metabolic substrate dependence of the bioelectric effect of the aldosterone action in isolated toad bladder. From experiments involving Krebs cyclic intermediates, Fimognari et al. (1967a; 1967b) concluded that aldosterone acted by stimulating the tricarboxylic cycle somewhere between condensing enzyme and α-ketoglutarate dehydrogenase. When analysed in the light of mitochondrial electron transport, these observations suggested that NADH could play a critical role in the aldosterone-mediated sodium transport. Evidence in support of such a critical role came from the observation that anaerobic conditions and antimycin A, both inhibitors of electron transport, eliminated aldosterone's effect on the short-circuit current (Fimognari et al., 1967a; 1967b; Sharp et al., 1966a; 1966b). These observations were both confirmed and extended by the detailed evaluation of Fanestil et al. (1968). Direct evidence regarding the role of tricarboxylic acid cycle enzymes in aldosterone-mediated sodium transport came from Kirsten and co-workers (1968) who measured an aldosterone-induced increase in citrate synthetase activity independent of the presence or absence of stimulated sodium transport. However, unresolved is the observation by Handler et al. (1969a) that under anaerobic conditions sufficient to eliminate oxidative metabolism, a persistent incremental difference in short-circuit current could be demonstrated in aldosterone-treated tissue as compared to non-treated controls. Another unresolved metabolic effect of aldosterone on the toad bladder concerns the differential effect on glucose metabolism. As originally reported by Kirchberger et al. (1968), aldosterone reduced glucose oxidation through the hexose monophosphate shunt (HMPS) pathway while increasing glucose utilization via the Embden–Meyerhof (E–M) pathway. Furthermore, the effect on C-1 oxidation was not the result of stimulated sodium transport since it could be reproduced in sodium-free solutions (Kirchberger et al., 1968). Although Handler and associates (1969b) concluded that the stimulatory effect on the Embden–Meyerhof pathway was a mineralocorticoid effect, while the decrease in the activity of the hexose monophosphate shunt was due to a glucocorticoid effect, a more recent report from Kirchberger and associates (1971a) disputes this conclusion, since their data indicated that the stimulation of Na^+ transport and inhibition of the hexose monophosphate shunt are properties common to mineralocorticoids. In a recent extension of these studies the latter authors have reproduced the mineralocorticoid effect on glucose oxidation using c-AMP, the so-called second messenger for vasopressin-mediated Na^+ transport in the toad bladder (Kirchberger et al., 1971b); however, they were unable to demonstrate any change in the tissue concentration of c-AMP following aldosterone exposure. These results are different from those reported by Stoff et al. (1972) who found a greater accumulation of c-AMP in epithelial cells taken from toad bladder tissue incubated overnight in the presence of aldosterone as compared to untreated control cells. Finally, Morrison et al. (1972), have reported both increased glucose release and increased glycogen content of aldosterone-treated but not cortisol-treated hemi-bladders during incubation with lactate. Based upon these observations the authors have raised doubts regarding the interpretation of studies which relate changes in the pattern of glucose metabolism to acute hormonal effects.

Recently, Feldman, Funder and Edelman (1972) have suggested a modification of the energy theory based upon the work of De Weer (1972). This model, as shown in Fig. 4.11, proposes that the action of aldosterone on active sodium transport is mediated by changes in the ATP:ADP ratio. During the basal, unstimulated state the ATP:ADP ratio is low, favouring a Na^+:Na^+ exchange, whereas under aldosterone stimulation ADP is reduced, resulting in a higher ATP:ADP ratio causing a predominate Na^+:K^+ exchange resulting in an increased net transport of sodium. From this scheme (Fig. 4.11) certain predictions can be entertained regarding the validity of the metabolic theory of the action of aldosterone. If the primary effect of the aldosterone-induced-protein is to stimulate the availability of ATP to drive the sodium pump, then one would expect to find an increase in either the [ATP]:[ADP] ratio or in the coupled redox pair, i.e. [creatine phosphate]:[creatine]; whereas, if an independent mechanism were stimulating Na^+ transport then one might expect the reverse to be true of the ratio of high energy intermediates. Although Handler et al. (1969a) did not detect a change in the [ATP]:[ADP] ratio as a result of the action of aldosterone, they did detect a fall in the [creatine phosphate]:[creatine] ratio due to an approximate 10% decline in phosphocreatine and a reciprocal 10% rise in creatine. This evidence, interpreted in light of the above model, would indicate that the primary action of aldosterone was not on the metabolic component of the sodium pump but rather at some other site, with the metabolic events being a secondary response. However, as the authors themselves point out (Handler et al., 1969a), measurement of whole cell concentrations may not reflect events occurring in an isolated, compartmentalized portion of the cell.

At the present there seems no compelling reason for selecting one of the three alternative theories proposed. One cannot help but speculate that the eventual explanation of the intracellular events associated with the action of aldosterone will incorporate parts of each theory as our knowledge concerning transepithelial sodium transport matures.

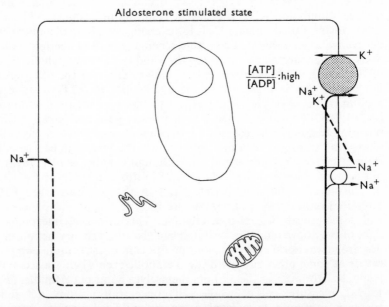

Fig. 4.11. Model proposed by De Weer (1972) and modified by Feldman, Funder and Edelman (1972). The central theme concerns the [ATP]: [ADP] ratio and how its change is coupled to modulating electrolyte transport. In the resting state (upper half) when the [ATP]:[ADP] ratio is low the $Na^+:Na^+$ exchange is favoured, resulting in little net transport. In sharp distinction is the situation in the aldosterone-stimulated state (lower half), where the [ATP]:[ADP] ratio is high which favours the $Na^+:K^+$ exchange pump resulting in an increase in net Na^+ transport.

4.7. UNANSWERED QUESTIONS

Although available evidence supports the concept of the action of aldosterone being mediated via a genetic mechanism in responsive cells, a completely integrated scheme of these subcellular events is not possible at present. It is quite likely that some form of cytoplasmic transfer complex is characteristic of all steroid hormones, although confirmation of this in amphibian tissue is lacking. That such confirmation is desirable is obvious, since many of our current concepts regarding mineralocorticoid action have been derived from this tissue. Extension of the transcriptional phase of aldosterone's action have kept pace with our knowledge regarding this fundamental process and this aspect of the problem appears to be yielding to experimentation. However, definition of the role of translation in the action of aldosterone has been disappointing and has handicapped identification of the mechanism by which decoded genetic information is coupled to or signals the increase of sodium transport. It seems unlikely that any substantial progress in resolving this aspect of the problem will precede the characterization of aldosterone-induced protein.

REFERENCES

Alberti, K. G. M. M. and Sharp, G. W. G. (1969). *Biochim. Biophys. Acta*, **192**, 335
Alberti, K. G. M. M. and Sharp, G. W. G. (1970). *J. Endocrinol.*, **48**, 563
Asano, Y., Tashima, Y., Matsui, H., Nagano, K. and Nakao, M. (1970). *Biochim. Biophys. Acta.*, **219**, 169
Ausiello, D. A. and Sharp, G. W. G. (1968). *Endocrinol.*, **82**, 1163
Baumann, E. J. and Kurland, J. (1927). *J. Biol. Chem.*, **71**, 281
Bentley, P. J. (1968). *J. Physiol. (London)*, **195**, 317
Bishop, W. R., Mumback, M. W. and Scheer, B. T. (1961). *Am. J. Physiol.*, **200**, 451
Bonting, S. L. and Canady, M. R. (1964). *Am. J. Physiol.*, **207**, 1005
Bricker, N. S., Biber, T. and Ussing, H. H. (1963). *J. Clin. Invest.*, **42**, 88
Burden, C. F. (1956). *Biol. Bull.*, **110**, 8
Cartensen, H., Burgers, A. C. J. and Li, C. H. (1961). *Gen. Comp. Endocrinol.*, **1**, 37
Cerijido, M. (1972). Cited in discussion of paper by Leaf, A. and Macknight, A. D. C., *J. Steroid. Biochem.*, **3**, 245
Chignell, C. F., Roddy, P. M. and Titus, E. O. (1966). *J. Biol. Chem.*, **241**, 5083
Choi, J. K. (1963). *J. Cell. Biol.*, **16**, 53
Cirne, B. and Malnic, G. (1972). *Biochim. Biophys. Acta*, **274**, 171
Civan, M. M. and Hoffman, R. E. (1971). *Am. J. Physiol.*, **220**, 324
Cofre, G. and Crabbé, J. (1967). *J. Physiol. (London)*, **188**, 177
Congote, L. F. and Trachewsky, D. (1972). *Biochem. Biophys. Res. Comm.*, **46**, 957
Coplon, N. S. and Maffly, R. H. (1972). *Biochim. Biophys. Acta*, **282**, 250
Cortas, N. and Walser, M. (1971). *Biochim. Biophys. Acta*, **249**, 181
Crabbé, J. (1961a). *Endocrinol.* **69**, 673
Crabbé, J. (1961b). *J. Clin. Invest.*, **40**, 2103
Crabbé, J. (1963a). *The Sodium Retaining Action of Aldosterone*, Presses Acad. Europ. SC, Brussels
Crabbé, J. (1963b). *Nature*, **200**, 787

Crabbé, J. (1964). *Endocrinol*, **75**, 809
Crabbé, J. (1967). *Arch. Int. Physiol Pharmacol*, **75**, 342
Crabbé, J. (1972). *J. Steroid. Biochem.*, **3**, 557
Crabbé, J., Decoene, A. and Ehrlich, E. N. (1971). *Arch. Internat. Physiol. Biochem.*, **79**, 805
Crabbé, J. and De Weer, P. (1964). *Nature*, **202**, 298
Crabbé, J. and De Weer, P. (1969). *Pflügers Arch.*, **313**, 197
Crabbé, J. and Ehrlich, E. N. (1968). *Pflügers Arch.*, **304**, 284
Crabbé, J., Fanestil, D. D., Pelletier, M. and Porter, G. A. (1973). Submitted for publication
Cuthbert, A. W. and Painter, E. (1969). *Nature*, **222**, 280
Dalton, T. and Snart, R. S. (1967a). *Biochim. Biophys. Acta*, **135**, 1062
Dalton, T. and Snart, R. S. (1967b). *Biochim. Biophys. Acta*, **135**, 1059
Dalton, T. and Snart, R. S. (1968). *Comp. Biochem. Physiol.*, **27**, 591
De Graeff, J., Dempsey, E. F., Lameyer, L. D. F. and Leaf, A. (1965). *Biochim. Biophys. Acta*, **106**, 155
De Santo, N. G., Ebel, H. and Hierholzer, K. (1971). *Pflügers Arch.*, **324**, 26
De Weer, P. (1972). Cited in discussion of paper by Kirsten, E. *et al.*, *J. Steroid Biochem.*, **3**, 173
De Weer, P. and Crabbé, J. (1968). *Biochim. Biophys. Acta*, **155**, 280
Di Bona, D. R. (1972). *Nature New Biol.*, **238**, 179
Dicker, S. E. (1970). *Hormones*, **1**, 352
Edelman, I. S. (1968). In *Functions of the Adrenal Cortex*, vol. 1, p. 79 (McKerns, K. W., ed.), Appleton-Century-Crofts; New York
Edelman, I. S. (1669). In *Renal Transport and Diuretics*, p. 139 (Thurau, K. and Jahrnärker, H., eds.), Springer-Verlag; Berlin and New York
Edelman, I. S., Bogoroch, R. and Porter, G. A. (1963). *Proc. Nat. Acad. Sci. (U.S.A.)*, **50**, 1169
Edelman, I. S., Bogoroch, R. and Porter, G. A. (1964). *Trans. Assn. Amer. Phys.*, **77**, 307
Edelman, I. S. and Fanestil, D. D. (1970). In *Biochemical Action of Hormones*, vol. 1, p. 321 (Litwack, G., ed.), Academic Press; New York
Edelman, I. S., Young, H. L. and Harris, J. B. (1960). *Amer. J. Physiol.*, **199**, 666
Edmonds, C. J., (1972). *J. Steroid. Biochem.*, **3**, 143
Erhlich, E. N. and Crabbé, J. (1968). *Pflügers Arch.*, **302**, 79
Fanestil, D. D. (1969). *Ann. Rev. Med.*, **20**, 223
Fanestil, D. D. (1972). Presented at IV Internat. Cong. Endocrinol., Washington, D.C.
Fanestil, D. D. and Edelman, I. S. (1966a). *Fed. Proc.*, **25**, 912
Fanestil, D. D. and Edelman, I. S. (1966b). *Proc. Nat. Acad. Sci. (U.S.A.)*, **56**, 872
Fanestil, D. D., Herman, T. S., Fimognari, G. M. and Edelman, I. S. (1968). In *Regulatory Functions of Biological Membranes*, p. 193 (Jarnefelt, J., ed.), Elsevier; Amsterdam
Fanestil, D. D. and Ludens, J. H. (1973). Personal communication
Fanestil, D. D., Porter, G. A. and Edelman, I. S. (1967). *Biochim. Biophys. Acta*, **135**, 74
Feldman, D., Funder, J. W. and Edelman, I. S. (1972). *Amer. J. Med.*, **53**, 545
Ferguson, D. R., Handler, J. S. and Orloff, J. (1968). *Biochim. Biophys. Acta*, **163**, 150
Ferguson, D. R. and Smith, M. W. (1972). *J. Endocrinol.*, **55**, 195
Fimognari, G. M., Fanestil, D. D. and Edelman, I. S. (1967a). *Amer. J. Physiol.*, **213**, 954
Fimognari, G. M., Porter, G. A. and Edelman, I. S. (1967b). *Biochim. Biophys. Acta*, **135**, 89
Finn, A. L., Handler, J. S. and Orloff, J. (1966). *Amer. J. Physiol.*, **210**, 1279

Forte, L. and Landon, E. J. (1968). *Biochim. Biophys. Acta.*, **157**, 303
Frazier, H. S., Dempsey, E. F. and Leaf, A. (1962). *J. Gen. Physiol.*, **45**, 529
Frazier, H. S. and Hammer, E. I. (1963). *Amer. J. Physiol.*, **205**, 718
Frazier, H. S. and Leaf, A. (1963). *J. Gen. Physiol.*, **46**, 491
Gachelin, G. and Bastide, F. (1968). *Compt. Rend.*, **267**, 906
Gatzy, J. T. and Clarkson, T. W. (1965). *J. Gen. Physiol.*, **48**, 647
Gaunt, R. and Chart, J. J. (1962). *Handbk Exp. Pharmacol.*, **14**, 514
Goodman, D. B. P., Allen, J. E. and Rasmussen, H. (1969). *Proc. Nat. Acad. Sci.* (*U.S.A.*), **64**, 330
Goodman, D. B. P., Allen, J. E. and Rasmussen, H. (1971). *Biochem.*, **10**, 3825
Grantham, J., Cuppage, F. E. and Fanestil, D. D. (1971). *J. Cell. Biol.*, **48**, 695
Hallowell, J. G., Frazer, J. W. and Gardner, L. T. (1968). *J. Clin. Endocrinol.*, **28**, 492
Handler, J. S., Preston, A. S. and Orloff, J. (1969a). *J. Biol. Chem.*, **244**, 3194
Handler, J. S., Preston, A. S. and Orloff, J. (1969b). *J. Clin. Invest.*, **48**, 823
Handler, J. S., Preston, A. S. and Orloff, J. (1972). *Amer. J. Physiol.*, **222**, 1071
Harrop, G. A., Nicholsen, W. M. and Strauss, M. (1936). *J. Exp. Med.*, **64**, 233
Hendler, E. F., Torretti, J., Kupoe, L. and Epstein, F. H. (1972). *Amer. J. Physiol.*, **222**, 754
Herman, T. S., Fimognari, G. M. and Edelman, I. S. (1968). *J. Biol. Chem.*, **243**, 3849
Herman, T. S. and Stiles, J. W. (1972). *Abst. V Internat. Cong. Nephrol.*, p. 79
Herrera, F. C. (1966). *Amer. J. Physiol.*, **210**, 980
Hersey, S. J. (1969). *Biochim. Biophys. Acta*, **183**, 155
Hill, J. H., Cortas, N. and Walser, M. (1973). *J. Clin. Invest.*, **52**, 185
Holmes, W. N. (1959). *Acta Endocrinol.*, **31**, 587
Huf, E. G. and Wills, J. (1953). *J. Gen. Physiol.*, **36**, 473
Huiid-Larsen, E. (1972). *J. Steroid. Biochem.*, **3**, 111
Hutchinson, J. H. and Porter, G. A. (1970a). *Biochem. Med.* **3**, 498
Hutchinson, J. H. and Porter, G. A. (1970b). *Res. Comm. Chem. Path. Pharmacol.*, **1**, 363
Hutchinson, J. H. and Porter, G. A. (1972a). *Res. Comm. Chem. Path. Pharmacol.*, **3**, 585
Hutchinson, J. H. and Porter, G. A. (1972b). *Biochim. Biophys. Acta*, **281**, 55
Hutchinson, J. H. and Porter, G. A. (1973a). Unpublished observations
Hutchinson, J. J. and Porter, G. A. (1973b). *Clin. Res.*, **21**, 228
Hutchinson, J. H. and Porter, G. A. (1973c). *Fed. Proc.*, **32**, 218 abs
Imamura, A. and Sasaki, N. (1962). *Seita. No. Kagaku.*, **13**, 73
Janáček, K., Rybouá, R. and Slaviková, (1971). *Pflügers Arch.*, **326**, 316
Jørgensen, C. B. (1947), *Nature* (*London*), **160**, 872
Jørgensen, P. L. (1969). *Biochim. Biophys. Acta*, **192**, 326
Jørgensen, P. L. (1972). *J. Steroid. Biochem.*, **3**, 181
Kallus, F. T., Vanatta, J. C., Burke, W. H. and Hetherington, E. A. (1971). *Proc. Soc. Exp. Med. Biol.*, **136**, 1245
Kalra, J. and Wheldrake, J. F. (1972). *Febs. Lett.*, **25**, 298
Katz, A. L. and Epstein, F. H. (1967). *J. Clin. Invest.*, **46**, 1999
Keller, A. R. (1963). *Anat. Record*, **147**, 367
Kiil, F. (1971). *Scand. J. Clin. Lab. Invest.*, **28**, 375
Kinsky, S. C., Luse, S. A. and van Deenen, L. L. M. (1966). *Fed. Proc.*, **25**, 1503
Kirchbergern, M. A., Martin, D. G., Leaf, A. and Sharp, G. W. G. (1968). *Biochim. Biophys. Acta*, **165**, 22
Kirchberger, M. A., Chen, L. C. and Sharp, G. W. G. (1971a). *Biochim. Biophys. Acta*, **241**, 861
Kirchberger, M. A., Witkum, P. and Sharp, G. W. G. (1971b). *Biochim. Biophys. Acta*, **241**, 876

Kirsten, E., Kirsten, R., Leaf, A. and Sharp, G. W. G. (1968). *Pflügers Arch.*, **300**, 213
Kirsten, E., Kirsten, R. and Sharp, G. W. G. (1970a). *Pflügers Arch.*, **316**, 26
Kirsten, R., Brinkoff, B. and Kirsten, E. (1970b). *Pflügers Arch.*, **314**, 231
Koefoed-Johnson, V. and Ussing, H. H. (1949). *Acta Physiol. Scand.*, **17**, 38
Koefoed-Johnson, V. and Ussing, H. H. (1958). *Acta Physiol. Scand.*, **42**, 298
Landon, E. F., Jazeb, N. and Forte, L. (1966). *Amer. J. Physiol.*, **211**, 1050
Leaf, A. (1965). *Rev. Physiol.*, **56**, 216
Leaf, A., Anderson, J. and Page, L. B. (1958). *J. Gen. Physiol.*, **41**, 657
Leaf, A. and Dempsey, E. F. (1960). *J. Biol. Chem.*, **235**, 2160
Leaf, A. and Macknight, A. D. C. (1972). *J. Steroid. Biochem.*, **3**, 237
Leaf, A., Page, L. B. and Anderson, J. (1959). *J. Biol. Chem.*, **234**, 1625
Levy, J. V. and Richards, V. (1965). *Proc. Soc. Exp. Biol. Med.*, **118**, 501
Lichtenstein, N. S. and Leaf, A. (1965). *J. Clin. Invest.*, **44**, 1328
Liew, C. C., Liu, O. K. and Gornall, A. G. (1972). *Endocrinol.*, **90**, 488
Lipton, P. and Edelman, I. S. (1971). *Amer. J. Physiol.*, **221**, 733
Loeb, R. F., Atchley, D. W., Benedict, E. M. and Leland, L. (1933). *J. Exp. Med.*, **57**, 775
Lucas, G. H. W. (1926). *Amer. J. Physiol.*, **77**, 114
Ludens, J. H. and Fanestil, D. D. (1971). *Biochim. Biophys. Acta*, **244**, 360
Ludens, J. H. and Fanestil, D. D. (1972). *Amer. J. Physiol.*, **223**, 1338
McAfee, R. D. and Locke, W. (1961). *Amer. J. Physiol.*, **200**, 797
McClane, T. K. (1965). *J. Pharmacol. Exp. Therap.*, **148**, 106
McDougal, B. and Sullivan, L. P. (1971). *Proc. Soc. Exp. Biol. Med.*, **136**, 871
Maetz, J., Jard, S. and Morel, F. (1958). *Compt. Rend.*, **247**, 516
Maffly, R. H. and Coggins, C. H. (1965). *Nature*, **206**, 197
Maffly, R. H. and Edelman, I. S. (1963). *J. Gen. Physiol.*, **46**, 733
Mamelak, M. and Maffly, R. H. (1969). *Abst. Amer. Soc. Nephrol.*, p. 44
Marenzi, A. D. and Fustinoni, O. (1938). *Rev. Soc. Argent. Biol.*, **14**, 118
Marine, D. and Baumann, E. J. (1927). *Amer. J. Physiol.*, **81**, 86
Marver, D., Goodman, D. and Edelman, I. S. (1972). *Kidney Internat.*, **1**, 210
Mishra, R. K., Wheldrake, J. F. and Feltham, L. A. W. (1971). *FEBS Lett.*, **23**, 176
Mishra, R. K., Wheldrake, J. F. and Feltham, L. A. W. (1972). *FEBS Lett.*, **24**, 106
Morrison, A. D., Goodman, D. B. P., Rasmussen, H. and Winegrad, A. I. (1972). *Biochim. Biophys. Acta*, **273**, 122
Nagel, W. and Dörge, A. (1971). *Pflügers Arch.*, **324**, 267
Nandi-Majumdar, A. P. and Trachewsky, D. (1971). *Canad. J. Biochem.*, **49**, 501
Nielsen, R. (1969). *Acta Physiol. Scand.*, **77**, 85
O'Malley, B. W. (1971). *Metabol.*, **20**, 981
Pak Poy, R. F. K. and Bentley, P. J. (1960). *Exp. Cell. Res.*, **20**, 235
Park, Y. (1973). Personal communication
Peachey, L. D. and Rasmussen, H. (1961). *J. Biophys. Biochem. Cytol.*, **10**, 529
Pelletier, M., Ludens, J. H. and Fanestil, D. D. (1972). *Arch. Intern. Med.*, **129**, 248
Phillips, J. G., Holmes, W. N. and Bondy, P. K. (1959). *Endocrinol*, **65**, 811
Phillips, J. G. and Mulrow, P. J. (1959). *Proc. Soc. Exp. Biol. Med.*, **101**, 262
Poole, D. T., Butler, T. C. and Williams, M. E. (1971). *J. Membrane Biol.*, **5**, 261
Porter, G. A. (1968). *Mol. Pharmacol.*, **4**, 224
Porter, G. A. (1970a). *Biochim. Biophys. Acta*, **211**, 487
Porter, G. A. (1970b). Cited by Edelman, I. S. and Fanestil, D. D. in *Biochemical Actions of Hormones*, vol. 1, p. 338 (Letwack, G., ed.), Academic Press; New York
Porter, G. A. (1971a). *Gen. Comp. Endocrinol.*, **16**, 443
Porter, G. A. (1971b). In *Proceedings of the III International Congress on Hormonal Steroids*, p. 1016 (James, V. H. T. and Martini, L., ed.), Excerpta Medica; Amsterdam

Porter, G. A. (1972). *Abst. V Internat. Cong. Nephrol.*, p. 143
Porter, G. A. and Edelman, I. S. (1964). *J. Clin. Invest.*, **43**, 611
Porter, G. A., Bogoroch, R. and Edelman, I. S. (1964). *Proc. Nat. Acad. Sci. (U.S.A.)*, **52**, 1326
Roeder, R. G. and Rutter, W. J. (1970). *Proc. Nat. Acad. Sci. (U.S.A.)*, **65**, 675
Ross, E. J. (1959). In *Aldosterone in Clinical and Experimental Medicine*, Blackwell Scientific Publ.; Oxford
Rousseau, G. and Crabbé, J. (1968). *Biochim. Biophys. Acta*, **157**, 25
Rousseau, G. and Crabbé, J. (1972). *European J. Biochem.*, **25**, 550
Samuels, H. H. and Tomkins, G. M. (1970). *J. Molec. Biol.*, **52**, 57
Scheer, B. T., Mumback, M. W. and Cox, B.L. (1961). *Fed. Proc.*, **20**, 177
Sharp, G. W. G., Coggins, C. H., Lichenstein, N. S. and Leaf, A. (1966a). *J. Clin. Invest.*, **45**, 1640
Sharp, G. W. G., Komack, C. L. and Leaf, A. (1966b). *J. Clin. Invest.*, **45**, 450
Sharp, G. W. G. and Leaf, A. (1964). *Nature*, **202**, 1185
Sharp, G. W. G. and Leaf, A. (1965). *J. Biol. Chem.*, **240**, 4816
Sharp, G. W. G. and Leaf, A. (1966). *Physiol. Rev.*, **46**, 593
Skou, J. C. (1957). *Biochim. Biophys. Acta*, **23**, 394
Smith, D. C. W. (1956). *Mem. Soc. Endocrinol.*, **5**, 83
Snart, R. C. (1967). *Biochim. Biophys. Acta*, **135**, 1056
Stoff, J. S., Handler, J. S. and Orloff, J. (1972). *Proc. Nat. Acad. Sci. (U.S.A.)*, **69**, 805
Swaneck, G. E., Highland, E. and Edelman, I. S. (1969). *Nephrol.*, **6**, 297
Swaneck, G. E., Chu, L. L. H. and Edelman, I. S. (1970). *J. Biol. Chem.*, **245**, 5382
Taubenhaus, M., Fritz, I. B. and Marten, J. V. (1956). *Endocrinol.*, **59**, 458
Thorn, G. W., Garbutt, H. R., Hitchcock, F. A. and Hartman, F. A. (1936). *Proc. Soc. Exp. Biol. Med.*, **35**, 247
Trachewsky, D. and Cheah, A. M. (1971). *Canad. J. Biochem.*, **49**, 496
Trachewsky, D. and Lawrence, S. (1972). *Proc. Soc. Exp. Biol. Med.*, **141**, 14
Trachewsky, D., Nandi-Majumdar, A. P. and Congote, L. F. (1972). *Eur. J. Biochem.*, **26**, 543
Urakabe, S., Handler, J. S. and Orloff, J. (1970). *Amer. J. Physiol.*, **218**, 1179
Ussing, H. H. (1966). *Ann. New York Acad. Sci.*, **137**, 543
Ussing, H. H. and Zerahn, K. (1951). *Acta Physiol. Scand.*, **23**, 110
Vančura, P., Sharp, G. W. G and Malt, R. A. (1971). *J. Clin. Invest.*, **50**, 543
Voûte, C. L. (ed.) (1972). *Symposium on Aldosterone and Active Sodium Transport Across Epithelia, J. Steroid Biochem.*, **3**, 105
Voûte, C. L., Dirix, R., Nielsen, R. and Ussing, H. H. (1969). *Exp. Cell. Res.*, **57**, 448
Voûte, C. L., Hänni, S. and Ammann, E. (1972). *J. Steroid Biochem.*, **3**, 161
Walser, M. (1971). *Biochim. Biophys. Acta*, **225**, 64
Williams, M. W. and Angerer, C. A. (1959). *Proc. Soc. Exp. Biol. Med.*, **102**, 112
Williamson, H. E. (1963). *Biochem. Pharmacol.*, **12**, 1449

Chapter 5

Enzyme Induction by Steroid Hormones with Reference to Cancer

D. C. Williams and M. Smethurst
The Marie Curie Memorial Foundation,
The Chart, Oxted, Surrey

5.1. INTRODUCTION

Mechanism of Action of Steroid Hormones

The gross functional and morphological changes brought about by steroid hormones in their respective target tissues are well documented. Within hours of administration of a steroid hormone characteristic changes in the chemical composition and metabolism of the target cells occur, eventually leading to hypertrophic and hyperplastic growth. Investigation of the biochemical basis for these changes has established that many of the major responses to androgens and oestrogens are associated with the activation of the genetic apparatus of the cell, leading to the synthesis of nucleic acids and the subsequent formation of new protein (Hamilton, 1964; Williams-Ashman *et al.*, 1964). Considerable attention has been given as to how these steroid molecules achieve their specific responses and exhibit such striking organ selectivity. The presence of receptor proteins which specifically bind oestradiol in oestrogen-dependent target tissues was indicated in experiments using radioactive oestradiol. It was demonstrated that particular organs of the immature rat have a striking affinity for this hormone resulting in the uterus, vagina and anterior pituitary taking up and retaining oestradiol to reach concentrations considerably higher than that in the blood (Jensen and Jacobson, 1960). Subsequently, a two-stage system has been demonstrated in which the steroid intially binds to a cytoplasmic receptor protein which is later detected in the nucleus. This transfer to the nucleus is accompanied by a change of the receptor complex from a 4S to a 5S form. Oestrogen-receptor interaction has been extensively reviewed elsewhere and will not be discussed further (Jensen, *et al.*, 1971; Jensen and DeSombre, 1972). The existence of specific protein-receptor systems for androgen and progestogen target tissues has also been demonstrated (Fang and Liao, 1971; Aakvaag *et al.*, 1972; O'Malley, *et al.*, 1970; Terenius, 1972) and it would appear likely that these systems might represent a general pattern for steroid hormone action in mammalian cells.

Activation of the genetic system in steroid hormone target tissues appears to involve the interaction of the nuclear steroid-receptor complex with chromatin (Shyamala and Gorski, 1969; King and Gordon, 1972). The culmination of this interaction is a rapid stimulation in the synthesis of RNA, which the use of inhibitors of RNA synthesis has shown to be important in the hormonal regulation of protein synthesis and growth. Obviously, the induction of RNA synthesis and the transfer of RNA and ribonucleoprotein particles to the cytoplasm constitutes a major basis for any gross change in the pattern of protein synthesis and enzyme induction that may occur with steroid hormones. Hence, the events leading to the activation of the genetic system represent an important aspect in the understanding of steroid hormone action. However, it is also of considerable biochemical and physiological importance to consider the extent and nature of the individual enzyme systems which are induced as a result of the initial activation of the nucleus.

Steroid Hormones in Relation to Cancer

It has been recognized for many years that, at least in their initial stages, some human tumours which arise from hormone-sensitive tissues retain some of the original characteristics of the cells from which they were derived. In certain cases, tumours arising from organs which are normally dependent upon hormones for growth and function may themselves be influenced by these substances until this facility is eventually lost by the normal de-differentiation process which appears to be common to all malignant growth. However, in many cases there is a considerable period of hormone sensitivity as the tumour develops and this effect can often be exploited as the basis for therapy. Ultimately, this advantage is usually lost and the tumour becomes autonomous. In general, the better differentiated the tumour and the more hormone sensitive the original parent tissue, the more likely is a favourable response to hormone therapy.

Endocrine treatment appears most effective in those patients having slow growing tumours. Tumours which have well-differentiated structures, which is sometimes a sign of retained hormone sensitivity, offer a better prognosis than do the rapidly metastasizing anaplastic tumours. A long period between original diagnosis and recurrence of symptoms is also generally a good sign. Beyond these broad generalizations very little information can be gained, either from histological or clinical examination, as to the likely response of a particular tumour to therapy. It would seem desirable that some form of biochemical test might be evolved for this purpose, but, although much work has been done in this direction, there is no reliable test at the present time (Williams, 1973). The best methods for patient evaluation at present appear to be based on clinical experience supported by histological data and results from previous therapy. The specific estimation of levels of hormone secretion or of enzymes involved in hormone metabolism have so far been disappointing (Forrest and Roberts, 1973).

The problem of the interplay between hormone-sensitive tumours and the endocrine status of the host is a fascinating one which remains to be resolved. There is, however, a large body of data which relates to experimental tumours in animals and which suggests that tumours having different hormone responses can co-exist in the same species and even in the same animal (e.g. see Gardner, 1964). It has also been shown that mammary tumours, differing in hormonal specificity, can be induced by the same chemical carcinogen (Williams, 1965). It seems likely that, in a proportion of cases, tumours consist of a mixed population of cells some of which are hormone sensitive, and can be controlled by hormone treatment, but that after prolonged treatment the insensitive cells become increasingly dominant so that the tumour eventually becomes autonomous. It follows that the induction of enzyme systems by hormones must play a vital part in the development of this disease. Indeed, this interaction is of importance in a number of quite different ways in the overall understanding and management of hormone sensitive cancer.

(1) The complex and still ill-understood mechanism by which hormones stimulate, amongst other things, nucleic acid and protein synthesis in the developing cell appears to be mediated through a system of enzyme induction and repression. In this context enzyme repression may presumably be regarded as "negative induction", that is to say a balance between the two form a so-called "negative feedback" reaction which controls the proliferation of the cells.

Since one of the main definitive properties of cancer is the lack of control of cell proliferation it seems likely that the control of enzyme induction within the cell could bring about a repression of malignant properties and possibly even a reversion of malignant transformation which can itself be brought about by steroid hormones in sensitive tissues. It is a matter of clinical experience that the growth of hormone-sensitive tumours, for example breast tumours, may be stimulated or suppressed by either androgen or oestrogen depending on the tumour/host relationship in that particular case. Some aspects of this control are discussed later in this chapter.

At a molecular level it appears that cancer is a genetic disfunction and that the deviation of the cancer cell from normal may well be expressed on the nucleic acid template in which case the controlled induction of enzyme systems, in particular those responsible for nucleic acid repair, offers a possible method of therapy.

(2) The stimulatory effect of, in particular, testosterone on the enzyme systems of prostatic and other cancers has been used as a method of introducing cytotoxic radioactive phosphorus (as phosphate) into the tumour. It has long been known that phosphorus is differentially incorporated into fast-growing prostatic tumours, especially in the case of bone metastases. Radioactive phosphorus was used in the treatment of leukaemia and similar conditions by Lawrence and Tobias (1956) and then extended by these workers to the treatment of cancer of the prostate and breast. It was shown by Hertz (1950) that simultaneous treatment with testosterone resulted in a concentration of radioactive phosphorus in neoplastic tissue of 5 to 10 times that occurring in normal tissue. This technique has been applied clinically by several groups of workers all of whom appear to have achieved valuable remissions.

(3) Enzyme stimulation by hormones and other substances offers the possibility of early diagnostic screening tests for cancer. It has been shown that raised enzyme levels are often obtained very early in the development of the tumour, which in theory should provide a clue as to the existence, and possibly to the site of, the early primary tumour. The difficulty with such a method is that, in general, it has not been possible to distinguish between isoenzymes from malignant and normal tissues so that, although local enzyme concentrations may be very much raised in malignant tissue, the amount excreted is still small compared with that from the normal tissues. Recent developments in our own and other laboratories, however, give reason to hope that there may be immunological differences between these enzymes which may be exploited for diagnostic purposes and also possibly for treatment (Field, 1973; Harris, 1973).

(4) There is a growing body of evidence that the induction of enzyme systems may play an important part in the carcinogenic process. It is well known that certain established carcinogens, which may include some steroid hormones, are detoxified and excreted as glucuronides, sulphates or phosphates. The induction, therefore, of local concentrations of the corresponding hydrolytic enzymes would lead to the prolonged retention of the active carcinogen with a consequent potentiation of its activity. It is interesting to note that these enzymes, in particular β-glucuronidase, are often raised in inflamed tissue so that inflammation itself may play a part in the carcinogenic process. It is tempting to speculate that the well-known co-carcinogenic effect of croton oil in mouse skin may be due to the demonstrated increase in β-glucuronidase levels in treated tissues (Allen et al., 1957).

In this paper, enzyme induction by steroid hormones will be discussed in terms of events in both the nucleus and cytoplasm, and where applicable, data will be considered in relation to neoplastic disease in steroid hormone-sensitive tissues.

5.2. RNA SYNTHESIS

Early investigations into the biochemical effects of steroid hormones established that these compounds had a marked stimulatory effect on the synthesis of RNA and protein. Androgens were shown to stimulate directly RNA synthesis in androgen-dependent tissues such as the seminal vesicles (Wicks and Kenney, 1964) and the ventral prostate (Liao et al., 1965). Similarly, RNA synthesis in rat uteri following the administration of oestrogens to ovariectomized animals is also well documented. Further work indicated that the RNA formed was predominantly ribosomal RNA (Liao et al., 1966) and that there was an acceleration in the appearance of new ribosomes in the cytoplasm (Moore and Hamilton, 1964). The use of inhibitors of RNA and protein synthesis also established that many of the early effects of steroid hormones are highly dependent upon RNA and protein synthesis (Ui and Mueller, 1963; Mueller et al., 1961).

An increase in the rate of RNA synthesis may result from a number of factors including (a) a variation in enzyme levels or enzyme efficiency, (b) an alteration in the template capacity of the chromatin, (c) an increase in the synthesis of essential precursors or (d) a combination of these factors. The early reports described above do not indicate whether the increase in RNA is due to direct effects on the enzyme or on the DNA template. However, it is generally implied that changes result from an activation of the polymerase enzyme (Liao et al., 1965; 1966; Gorski, 1964).

A difference in the extent of activation of RNA polymerase when assayed with Mg^{2+} and in low and high ionic strength solutions was also demonstrated (Liao et al., 1965; Gorski, 1964; Breuer and Florini 1966; Widnell and Tata, 1966a). Two separate DNA-dependent RNA polymerase reactions in rat liver nuclei were identified by Widnell and Tata (1966b), one requiring Mg^{2+} (RNA polymerase I) and the other requiring Mn^{2+} and ammonium sulphate (RNA polymerase II). These reactions were subsequently found to have different intra-nuclear locations, with the Mg^{2+}-activated polymerase present in the nucleolus and the Mn^{2+}-activated reaction to be found primarily in the extra-nucleolar chromatin (Maul and Hamilton, 1967). The former reaction synthesized RNA of a ribosomal nature and the latter a DNA-like RNA. The presence of two inducible RNA polymerase systems in steroid hormone target tissues was shown by Hamilton et al. (1968). Oestrogen treatment of ovariectomized rats caused activation of the Mg^{2+}-dependent RNA polymerase reaction in uterine nuclei, which was detectable at about 1 hr after oestrogen treatment and reached a maximum of 125% of the control value at about 2 hr. The assay of RNA polymerase in the presence of Mn^{2+} and ammonium sulphate revealed no detectable change before 12 hr after oestrogen treatment but activity then increased by about 60%.

A major problem in the assessment of the induction of RNA polymerase by steroid hormones is the degree of purity of the enzyme preparation, especially freeing the enzyme from the endogenous DNA template. Care must therefore be taken in any conclusion drawn from the experiments described above as whole nuclei were used in the assays. The increased synthesis of RNA does not therefore necessarily indicate increased levels of the RNA polymerase protein as other factors such as the priming ability of endogenous DNA may have been changed by the steroid hormones. In fact Mangan et al. (1968) demonstrated an increased template capacity in prostate nuclei from testosterone-treated castrated rats when RNA synthesis was assayed using an Escherichia coli RNA polymerase preparation. Similarly, an increase in the template capacity of chromatin isolated from oestradiol stimulated rat uteri was demonstrated (Barker and Warren, 1966), also using an RNA polymerase preparation from E. coli. This was confirmed by Teng and Hamilton (1968) who found a stimulation of RNA formation of about 25% in rat uteri 30 min after oestradiol treatment in vivo and after 8 hr the template activity was increased by about 100% of the control level. Church and McCarthy (1970) used a partially purified RNA polymerase system prepared from normal endometrial nuclei of rabbit to assay the template activity of chromatin from the endometrium of oestradiol-treated, ovariectomized rabbits. An even greater and more rapid stimulation than that reported above for rat uteri was observed. Thus, there is good evidence that increased RNA synthesis following steroid hormone administration can be partly attributed to an increased template capacity of the chromatin.

The complete and reproducible solubilization of RNA polymerase from nuclei of rat liver cells and sea urchin embryos was achieved by Roeder and Rutter (1969; 1970a; 1970b) and the methods used were applied to the analysis of RNA polymerase enzymes from rat prostate gland by Mainwaring et al. (1971). RNA polymerase was entirely dependent upon added DNA for full activity, and examination of the effects of ammonium sulphate, and the optima for DNA template activity and pH indicated that specific Mn^{2+} and Mg^{2+}-activated enzymes were associated with both the nucleolar and extra-nucleolar regions of prostatic nuclei. Castration caused a pronounced loss of Mg^{2+}-dependent RNA polymerase activity in both the nucleolus and extra-nucleolar regions, but this could be totally restored by administration of testosterone phenyl propionate in vivo. This increase was clearly apparent by 2 hr after androgen treatment, at which time there was little change in the template activity of the chromatin, hence confirming the activation of the Mg^{2+}-enzyme as distinct from the RNA polymerase system as a whole. The Mn^{2+}-dependent RNA polymerase activity was not significantly altered by castration or androgen administration. Dihydrotestosterone and other androgens had no stimulatory effects on either polymerase system when added directly to the assay. Thus the direct involvement of a steroid hormone molecule in the induction of RNA polymerase activity could not be demonstrated and other associated factors, such as specific androgen receptors, were suggested as playing a part in the activation process.

Testosterone metabolism and subsequent formation and retention of dihydrotestosterone in androgen target tissues appears to be a necessary process in the mediation of androgenic effects (Fig. 5.1). It is therefore of interest that Alfheim and Unhjem (1971) found that dihydrotestosterone added *in vitro* had no effect on the RNA polymerase activity of rat prostate tissue whether assayed in homogenates, tissue slices, or nuclei preparations. However, Davies *et al.* (1972) observed stimulation of the RNA polymerase system when androgens were added to the polymerase assay using rat and dog prostatic nuclei. Again this assay only measured the rate of incorporation of nucleotides into whole nuclei and there was no attempt to assay the RNA polymerase enzyme free from its endogenous template. Preparations from both rat and dog were stimulated by dihydrostestosterone (Table 5.1) but testosterone only stimulated both systems to a small extent, perhaps because of conversion of testosterone to dihydrotestosterone. The rat RNA polymerase system was also activated by both 5α-androstane-$3\alpha,17\beta$-diol and 5α-androstane-$3\beta,17\beta$-diol. However, 5α-androstane-3α, 17β-diol had no effect on the RNA polymerase system from dog, whereas 5α-androstane-$3\alpha,17\alpha$-diol had a slight stimulatory effect on the dog prostate preparation but not on that from the rat.

Unlike the androgens, the stimulation of oestrogen-dependent tissues with oestradiol does not require the formation of steroid metabolites. The action of oestradiol on the uterus has therefore been widely used for analysis of the early synthetic events of RNA and protein synthesis which occur in oestradiol-stimulated tissue. The initial enhancement of RNA synthesis occurs within minutes of the administration of oestradiol, followed by the increased synthesis of ribosomal RNA detectable at 30 min, transfer RNA at 60 min and DNA-like RNA after 120 min (Hamilton *et al.*, 1965; 1968; Trachewsky and Segal, 1967; Billing *et al.*, 1968; 1969). Contrary to this, the synthesis of non-ribosomal high molecular weight RNA within 30 min of oestradiol administration has been reported (Knowler and Smellie, 1971) and, more recently, the activation by oestradiol of the RNA polymerase enzymes in the uteri of ovariectomized rats has been critically re-examined by Glasser *et al.* (1972). These workers analysed RNA polymerase activities

6

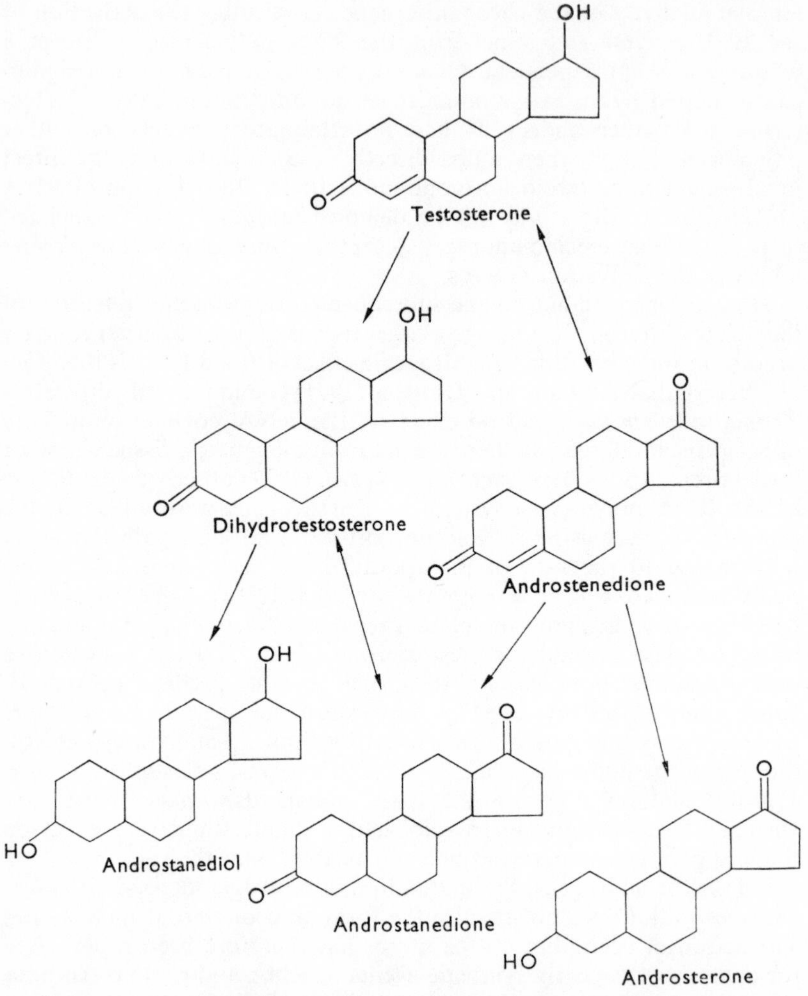

Fig. 5.1. A simplified scheme for the metabolism of testosterone.

in uterine nuclei using various ionic strengths and α-amanitin which selectively inhibits the majority of the RNA polymerase II activity. Within 15 min of the injection of 17β-oestradiol an increase in RNA polymerase II activity occurred (Fig. 5.2), resulting in the synthesis of DNA-like RNA. This stimulation was completely inhibited by α-amanitin. RNA polymerase II activity reached a peak at 30 min and then decreased up to 2 hr with a further increase between 2–12 hr. The RNA polymerase I activity, assayed in low ionic strength conditions,

TABLE 5.1

Biological Activity of Testosterone and Some of Its Metabolites on Rat and Dog Prostatic Tissue

Metabolite	Biological Activity	
	Rat Prostate	Dog Prostate
Testosterone	Slight stimulation of RNA polymerase (a)	Slight stimulation of RNA polymerase (a)
	Maintains secretion (c)	Stimulation of DNA polymerase (b)
Dihydrostestosterone	Stimulation of RNA polymerase (a)	Stimulation of RNA polymerase (a)
	Hyperplasia (c)	
	Maintains secretion (c)	
5α-Androstane-3α,17β-diol	Stimulation of RNA polymerase (a)	No stimulation of RNA polymerase (a)
	Slight hyperplasia (c)	Stimulation of DNA polymerase (b)
5α-Androstane-3β,17β-diol	Stimulation of RNA polymerase (a)	Slight stimulation of RNA polymerase (a)
	Stimulates secretion (c)	No stimulation of DNA polymerase (b)
5α-Androstane-3α,17α-diol	No stimulation of RNA polymerase (a)	Slight stimulation of RNA polymerase (a)
		Stimulation of DNA polymerase (b)

(a) Davies, *et al.*, 1972.
(b) Harper, *et al.*, 1970b.
(c) Robel, *et al.*, 1971.

was stimulated 30–60 min after oestradiol injection and was accompanied by an increase in the template capacity of the chromatin. The authors speculate that the initial synthesis of DNA-like RNA is a prerequisite for the second increase in RNA polymerase II activity and also the initial rise in RNA polymerase I activity located within the nucleolus. It is also suggested that the early synthesis of RNA after 15 min might be a necessary process for the stimulation of protein synthesis detectable after 30–45 min in the uteri of immature or ovariectomized rats (Notides and Gorski, 1966; De Angelo and Gorski, 1966; Mayol and Thayer, 1970).

None of the experiments described above gives any indication as to the manner by which steroid hormones induce RNA polymerase activity. However, there is evidence indicating the involvement of specific steroid hormone receptors in this stimulation. The direct treatment of the uterine nuclei with oestradiol does not result in a

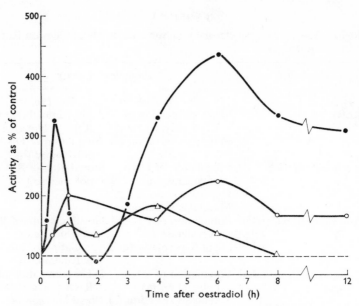

Fig. 5.2. Effects of 17β-oestradiol on uterine RNA polymerase activity and chromatin template capacity (after Glasser *et al.*, 1972).

 Endogenous RNA polymerase activity in uterine nuclei was assayed in high salt conditions (●——●), or low salt conditions +α-amanitin (○——○); template chromatin (△——△).

stimulation of RNA synthesis, but exposure of nuclei to the oestradiol-receptor complex prepared from uterine cytosol does cause an activation of the RNA polymerase I (Mohla *et al.* 1972). Furthermore, this stimulation will only occur after the cytoplasmic receptor has undergone conversion from the 4S form to the 5S nuclear form, this transformation being dependent upon association with oestradiol. However, DeSombre *et al.* (1972) have shown that precipitation of the cytoplasmic oestrogen receptor protein from calf uterus with ammonium sulphate at 2°C will cause its conversion to the 5S nuclear form in the absence of oestradiol. These workers have also demonstrated that this preparation will stimulate RNA synthesis whether complexed with oestradiol or not. Thus it would appear that the function of the oestradiol is to facilitate the conversion of the cytoplasmic receptor to an active form in the nucleus and it is this protein that stimulates RNA polymerase activity.

 In conclusion, current evidence indicates that the initial action of steroid hormones could well involve the induction of RNA synthesis such that small amounts of specific RNA species are formed which regulate the synthesis of a small amount of protein. Conceivably, the latter might include the synthesis of important synthesizing enzymes such as RNA polymerase; however, it is unclear at what stage the synthesis of new RNA polymerase molecules occurs as distinct from an activation of the existing enzyme. It is only after these initial synthetic events that a general increase in the levels of ribosomes and total proteins occurs.

5.3. DNA SYNTHESIS

The growth of steroid hormone target tissues which results when a steroid hormone is administered to endocrine-ablated animals involves DNA synthesis and cell proliferation. The initial response of the tissue is one of hypertrophy with a large increase in RNA and protein synthesis, the synthesis of DNA and cell proliferation occurring at a later time. For example, DNA replication in the uterus of oestrogen-treated, ovariectomized rats does not occur until at least 36 hr after subcutaneous injection of the hormone, long after the RNA and protein contents have increased. Similarly, the ability of the rat prostate to incorporate deoxyribonucleotides into DNA following the administration of testosterone to castrated animals shows little alteration until 24–48 hr after treatment and the rate of incorporation does not reach a maximum until 3–5 days (Kosto et al., 1967). This testosterone-induced increase in DNA synthesis as measured by thymidine incorporation in vivo is paralleled by a large increase in the soluble DNA polymerase activity, assayed using calf thymus DNA as primer (Coffey et al., 1968). The peaks for the rate of DNA synthesis and polymerase activity both occurred 3–6 days following the commencement of daily androgen treatment. An increased DNA polymerase activity almost immediately before the hormone-induced rise in DNA synthesis was also shown for mouse mammary gland following the daily injection of oestradiol and progesterone (Banerjee et al., 1971).

These experiments all measure the in vitro incorporation of radioactive deoxyribonucleotides into DNA using calf thymus DNA as a primer and a soluble tissue extract as the polymerase preparation. However, there is no evidence to suggest that the enzyme system assayed in vitro represents the DNA polymerase involved in the premitotic replication of DNA. Also, no indication can be given as to whether the changes in DNA polymerase activity following steroid hormone admininstration in vivo represent alterations in the rates of synthesis or degradation of the polymerase proteins or a change in the catalytic activity of the pre-existing enzyme. However, the possible dependence of DNA polymerase activity on protein synthesis has been indicated by Chung and Coffey (1971) using testosterone-stimulated rat prostate tissue; the maximum rate of nuclear protein synthesis was shown to precede the peak of DNA synthesis by 24–48 hr and inhibitors of protein synthesis markedly reduced the rate of DNA formation.

The action of testosterone metabolites on various forms of prostatic tissue has been examined by Harper *et al.* (1970a; 1970b) who observed a stimulation of the DNA polymerase from human hyperplastic tissue by both testosterone and dihydrotestosterone *in vitro*, but these androgens did not stimulate all preparations from hyperplastic tissue and there was no effect on the human neoplastic tissue studied. Examination of the normal canine prostate *in vitro* revealed that only testosterone, 5α-androstane-3α,17β-diol and 5α-androstane-3β,17α-diol, out of 13 androgen metabolites studied, stimulated DNA polymerase activity (Table 5.1), and surprisingly dihydrotestosterone had no effect. This is in marked contrast to RNA polymerase which is induced by dihydrotestosterone (Davies *et al.*, 1972). It is of interest that the same workers have found a selective accumulation of 5α-androstane-3α,17α-diol in the nuclei of the canine prostate following studies on testosterone metabolism in this tissue (Harper *et al.*, 1971). This might indicate an important stimulatory role for this metabolite in the dog prostate.

Experiments using testosterone metabolites in prostate organ culture have shown that individual metabolites vary as to their effects on the morphology and secretory activity of the tissue (Robel *et al.*, 1971; Gittinger and Lasnitski, 1972), and that the final expression of androgenic stimulation is dependent upon the balance of metabolites present (Table 5.1). As the induction of both RNA and DNA synthesis in the prostate varies with different metabolites it is conceivable that changes in the ability of the tissue to metabolize testosterone may cause an alteration in its growth pattern by influencing the RNA polymerase and DNA polymerase systems. Changes of this nature might well be the prerequisite for the induction of benign hyperplasia and adenocarcinoma in the human prostate—in fact, high levels of dihydrotestosterone in benign hyperplastic prostatic tissue have been reported (Siiteri and Wilson, 1970). It is of considerable interest that many compounds used in the treatment of prostatic disease cause alterations in the uptake and metabolism of testosterone by the prostate, and other compounds inhibit the induction of the nucleic acid polymerase enzymes. Mangan *et al.* (1967) demonstrated that diethylstilboestrol inhibited the induction of RNA polymerase activity in isolated rat prostate nuclei. This was confirmed in rat and dog prostatic preparations by Davies *et al.* (1972) who also demonstrated inhibition of RNA polymerase activity by 17β-oestradiol, hexoestrol, and three stilboestrol derivatives. The same stilboestrol derivatives also inhibited the DNA polymerase system from human hyperplastic and neoplastic prostatic tissue (Harper *et al.*, 1971), and also from canine prostatic tissue (Robel *et al.*, 1971).

5.4. DRUG METABOLIZING ENZYMES

The hepatic microsomal enzymes which metabolize steroids are themselves induceable by the steroid hormones. The metabolism of steroids in the liver is also influenced when the enzymes of the microsomal mixed function oxidase system are induced by various drugs. Similarities between the two systems have led to the suggestion that both drugs and steroids are substrates for the same enzymes (Kuntzman et al., 1964; 1966). It has also been proposed that the steroid hormones are the normal physiological substrates for these enzymes (Conney, 1967; Tephly and Mannering, 1968). However, measurement of hydroxylase activities in the liver of developing rats indicates that the typical sex difference that develops for drug metabolizing ability is evident two weeks earlier than the appearance of a sex difference in the 16α-hydroxylation of dehydroepiandrosterone (Heinrichs et al., 1966). Whether the enzymes are the same or not, it has been established that in certain species there is a sex difference in the activity of the hepatic microsomal enzymes and evidence indicates that the male and female sex hormones are important in their induction. It follows that age, castration and hormone administration are all significant factors and this has been well demonstrated with studies on a number of enzyme systems, mostly from the rat. These include the steroid hydroxylases (Jacobson and Kuntzman, 1969; Schriefers et al., 1972; Ghraf et al., 1972; Inano et al., 1973), the steroid Δ^4-5α-hydrogenase (Schriefers et al., 1972; Rubin and Strecker, 1961; Schriefers, 1969), and also the steroid 20-keto-reductase (Ghraf et al., 1972; Schriefers, 1969).

Although caution must be used when extrapolating data from one species to another, the fact remains that the induction of hepatic microsomal enzymes by steroid hormones and by drugs is important to the human both pharmacologically and physiologically. Enzyme induction of this nature will lead to an accelerated metabolism of both drugs and steroids and therefore alter the intensity and duration of action of these compounds. This has been shown by a number of workers, including Jacobson and Kuntzman (1969) who demonstrated an increased rate of formation of hydroxylated testosterone derivatives following testosterone propionate administration to castrated male rats. This was a selective effect, as the induction of the 16α-steroid hydroxylase was greater than that of the 6β- and 7α-steroid hydroxylases. Reactions of this nature which may involve a variety of steroid hormone substrates can therefore influence the levels of circulating steroid hormones, and in turn may well affect the function of the secondary sexual tissues and the steroid-producing glands. Thus, in terms of the regulation of hormonal levels, the liver may be considered as the antagonist to the endocrine glands.

The administration of steroid hormones to neonatal rats also influences the activities of the hepatic microsomal steroid-metabolizing enzymes in the adult. Ghraf et al. (1972) demonstrated that treatment of the neonatal male rat with oestradiol benzoate caused changes in the steroid hydroxylase activities such that the adult rat had levels typical of those in the female. In similar work (Schriefers *et al.*, 1972) it was shown that although testosterone propionate and oestradiol benzoate treatment of neonatal female rats did not alter the activities of the steroid Δ^4-5α-hydrogenase, steroid hydroxylases, hydroxysteroid-dehydrogenases and 20-keto-steroid-reductase, oestradiol benzoate administered to male neonatal rats caused a persistent feminization of enzyme levels.

The activities of these hepatic enzymes are therefore not only dependent upon existing steroid hormone levels but also on the hormonal status in the early stages of life. If this is applicable to the human then, the prenatal and neonatal administration of large amounts of steroid hormones and certain drugs may influence enzyme levels and the final hormonal status when maturity is reached. The importance of the prenatal environment is indicated by the association of maternal stilboestrol therapy with the appearance of adenocarcinoma of the vagina in young women (Herbst *et al.*, 1971).

5.5. ENZYMES OF CARBOHYDRATE METABOLISM

Since Warbug reported that cancer tissues show a faster rate of aerobic glycolysis than normal tissues, the enzymes of carbohydrate metabolism have been the source of much study in relation to cancer. Important rate-limiting enzymes include glucose-6-phosphate dehydrogenase, hexokinase, phosphofructokinase and pyruvate kinase (Fig. 5.3) and all show induction by steroid hormones.

Glucose-6-phosphate dehydrogenase is a particularly important enzyme in terms of NADPH production and the formation of ribose. Induction of this enzyme occurs in the seminal vesicles of castrated animals following testosterone administration (Singhal and Ling, 1969) and this is prevented by inhibitors of RNA and protein synthesis. However, testosterone had little effect on glucose-6-phosphate dehydrogenase induction in the prostate (Santti and Villee, 1971). Oestradiol promotes induction of this enzyme in the uteri of ovariectomized rats and induction is also effected by NADP administration directly to the uterus (Barker, 1967; Moulton and Barker, 1971). Both forms of induction are inhibited by cycloheximide but only the former is inhibited by actinomycin D.

Typical inductions are also shown for hexokinase in the uteri of ovariectomized rats following oestradiol administration (Valadares *et al.*, 1968) and in the male secondary sexual tissues following testosterone administration (Singhal and Ling, 1969; Singhal *et al.*, 1968). Again both inductions are prevented by inhibitors of RNA and protein synthesis. However, Santti and Villee (1971) could find no induction after testosterone treatment of rats castrated 12–15 days previously, when calculated on the basis of activity per mg protein, and they suggest that the increases shown by Singhal's group represent the synthesis of

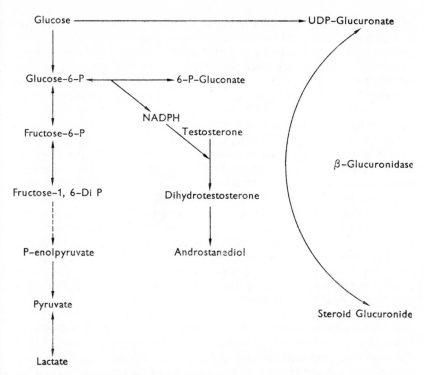

Fig. 5.3. Connection between carbohydrate and steroid metabolism (modified from Schriefers, 1969).

all cell constituents. However, a marked induction of hexokinase on the basis of activity per mg protein was shown following testosterone treatment of rats castrated 48 hr previously (Santti and Villee, 1971).

Phosphofructokinase is also a key enzyme of glycolysis and this can be induced in the prostate and seminal vesicles of castrated rats by testosterone treatment (Singhal and Valadares, 1968), and in the uteri of ovariectomized rats by oestradiol (Singhal et al., 1967). The activity of pyruvate kinase shows similar induction by testosterone and oestradiol in their respective target tissues (Santti and Villee, 1971; Singhal and Valadares, 1970). The induction of both enzymes is again prevented by inhibitors of RNA and protein synthesis.

The conversion of lactate to pyruvate, catalysed by lactate dehydrogenase, is an important process during anaerobic metabolism. Evidence relating to the induction of this enzyme by steroid hormones is sketchy but increased activity in the uterus, prostate and seminal vesicles of immature rats following treatment with the appropriate sex hormone was shown by Goodfriend and Kaplan (1964). Lactate dehydrogenase has been widely studied in relation to cancer especially in respect of the pattern of isoenzymes, and it is of interest that for many of the enzymes described above, induction is frequently accompanied by a change in the isoenzyme pattern. The isoenzymes of lactate dehydrogenase are designated I to V with types II, III and IV representing hybrids of the two pure isoenzymes I and V. The type I enzyme is predominant in tissues where glycolysis is primarily aerobic and type V where glycolysis is relatively anaerobic. In benign prostatic tissue the type I is more active than the type V enzyme, whereas in malignant tissue the reverse is apparent (Oliver et al., 1970). Inhibition of lactate dehydrogenase activity occurs in tissue extracts from both benign and malignant prostates following incubation with diethylstilboestrol; greater inhibition occurs in the malignant tissue with the type V enzyme showing the greatest sensitivity (Belitsky et al., 1970). This illustrates the antagonistic effects of an oestrogen on an androgen-controlled tissue and such effects are apparent in a number of cases of enzyme induction. The use of diethylstilboestrol in vivo also causes a reduction in lactate dehydrogenase activity in malignant prostatic tissues (Miller et al., 1972) but in this case activities in both the malignant and treated-malignant cases were below the activity in normal prostate.

Analysis of the hexokinase isoenzymes by starch gel electrophoresis revealed that the type II isoenzyme was predominant in the prostate and the type I enzyme predominant in the seminal vesicles. Castration and testosterone treatment indicated that the type II isoenzyme, but not the type I, was induceable by androgens (Santti and Villee, 1971). However, since both organs are target tissues of testosterone, the ratio of the isoenzymes may not be an important factor in the regulation of hexokinase activity by testosterone.

The effects of oestradiol on the isoenzyme pattern of glucose-6-phosphate dehydrogenase were examined by Hilf et al. (1972). Separation of the enzymes by polyacrylamide gel electrophoresis indicated that induction in ovariectomized rat uteri was largely due to an increase in the type I isoenzyme, with only a slight increase in type II and a decrease in type III. The use of an anti-oestrogen (U 11, 100A) in these experiments prevented the oestrogen-induced changes in the type I enzyme. The induction of hexokinase in the prostate and seminal vesicles by testosterone can also be prevented, and simultaneous administration of oestradiol stopped the induction of this enzyme but oestradiol alone had no effect. This antagonistic action between the sex hormones is especially important in relation to the role of these enzymes in malignant tissues. Miller et al. (1972) demonstrated that in carcinoma of the prostate there were elevated levels of glucose-6-phosphate dehydrogenase, but only small increases were observed in cases of benign hypertrophy. However, a reduction in enzyme activity to near normal values in both malignant and benign prostates was achieved by diethyl-stilboestrol therapy, indicating an inhibitive effect of an oestrogen in an androgen-dependent tissue.

The metabolism of testosterone to dihydrotestosterone by 5α-reduction is a necessary process in the mediation of many androgenic effects. This conversion requires NADPH, and since glucose-6-phosphate dehydrogenase is involved in the production of this coenzyme, a decrease in the activity of this enzyme might cause a significant reduction in androgenic activity by reducing dihydrotestosterone formation (Fig. 5.3). It is thus of interest whether glucose-6-phosphate is induced by testosterone or whether this is mediated by the metabolite, dihydrotestosterone.

5.6. LYSOSOMAL ENZYMES

One of the earliest papers associating lysosomal enzymes and steroid hormones was published by Kerr et al. (1949) who used β-glucuronidase as an index of growth in the uterus and other organs of the mouse. The general conclusions drawn were that: (a) in the uterus, changes in glucuronidase activity reflect changes in growth, and the action of oestrone on the enzyme is antagonised by testosterone; (b) oestrone produces marked increases in glucuronidase activity and cell division in the liver of ovariectomized mice. This action, which is also seen in normal and castrated males, but not in normal females, is antagonized by testosterone.

The classic work of Fishman and his colleagues in the 1950s (Fishman, 1951; Fishman et al., 1950; 1953; 1955; 1958; Kasdon et al., 1951a; 1951b; Sie and Fishman, 1953; Riotton and Fishman, 1953) showed that certain hydrolytic enzymes, in particular renal β-glucuronidase and hence urinary β-glucuronidase, could be markedly increased in certain animal species by androgens (Sie and Fishman, 1953); a marked sex difference was also apparent in these animals (Fig. 5.4). Elevation of the enzyme activity of several hundred-fold was obtained and was maintained for a prolonged period (Fishman et al., 1955; Fishman and Lipkind, 1958). This response was specific for androgens and could be decreased by oestrogen treatment and/or castration of male animals. Lesser effects produced by corticosteroids were considered to be secondary to the metabolism of these compounds.

The connection between cell proliferation and tissue β-glucuronidase levels was observed by several groups of workers and from there it was a comparatively short step to demonstrate the enhancement of β-glucuronidase activity in malignant tissues from both human and animal sources. The marked sex difference in the incidence of human bladder cancer led Boyland and his colleagues to investigate the excretion of hydrolytic enzymes in this disease (Boyland and Williams,

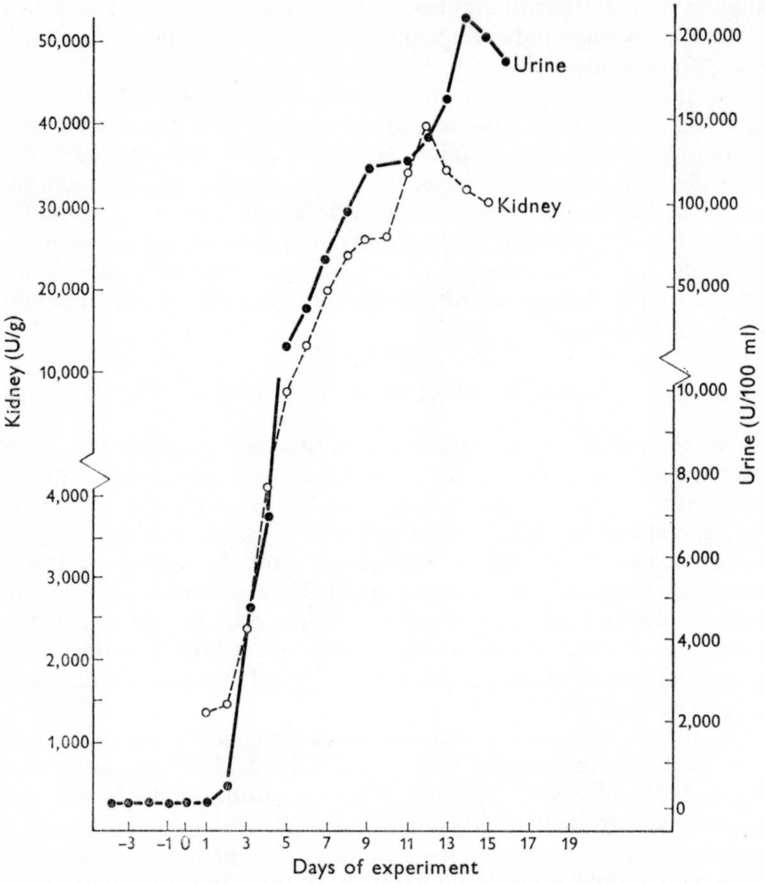

Fig. 5.4. Kidney and urinary β-glucuronidase activities in DBA female mice injected daily for 16 days with 2.5 mg testosterone (after Riotton and Fishman, 1953).

1956; 1957; Boyland *et al.*, 1955a; 1955b; 1956; 1957a; 1957b), a study which, in turn, led to investigation of other forms of human cancer (Boyland *et al.*, 1956) and also to the relationship between the levels of these enzymes and the metabolism of tryptophan in such patients (Boyland and Williams, 1956; 1957). The modification of the induced enzyme patterns with enzyme inhibitors was also examined clinically (Boyland *et al.*, 1957a).

The significance of serum β-glucuronidase activity in patients with liver disease has been studied by Pineda *et al.* (1959) who reflected that the precise function of β-glucuronidase in living cells had not been clearly defined but could possibly be both hydrolytic and conjugative. It has been observed that glucuronidase participates in the synthesis of steroid glucuronides and therefore may play a significant role in steroid metabolism. Further evidence for this theory had been obtained from observations that serum glucuronidase is elevated in late pregnancy, in the immediate post partum period, in postmenopausal women receiving oestrogens, and in patients with cancer of the breast, cervix, and uterus following oestrogen or androgen therapy. In the early stages of liver disease resulting from metastases, biliary obstruction or hepatitis, serum β-glucuronidase levels increase, whereas in advanced disease the normal to low levels of serum glucuronidase observed probably mirrors a decrease in the number of hepatic cells capable of synthesizing the enzyme. However, the observations of liver function were not correlated with changes in the hormone balance of the patients.

Further evidence for a role for the coupling of steroid hormone/ hydrolytic enzyme activities (though not induction by steroid hormones), has been provided by Fishman et al., 1968. They reported that an alkaline phosphatase in serum and tumour cells of a patient with bronchogenic carcinoma had the characteristics of placental alkaline phosphatase rather than those of alkaline phosphatase from the tissue of origin (lung) of the tumour. It is significant that both adrenal glands were completely replaced by tumour but the authors did not attempt to correlate hormone imbalance with the presence of a foetal-type iso-enzyme.

The actual mechanism by which hormones induce increased activity of the lysosomal enzymes is still not completely clear although it seems likely that it is a complex one. Initially it was tacitly assumed by many groups that the increased enzyme activity produced by the action of hormones was a simple de novo increase in protein synthesis. However, the extensive work of Szego and her co-workers, and others (Hechter, 1955; Roberts and Szego, 1953; Szego and Davis, 1967) has shown that, a large part of the effect, particularly in the very early stages, is due to the release of the bound enzyme from cell membranes in the target cell. Thus "within some 15 seconds after the administration of oestrogens to ovariectomized rats secondary reflexions of the interaction of the hormone with receptor sites in the uterus were already evident in release of sequestered amines and in the stimulation of membrane-bound adenylate cyclase" (Szego and Davis, 1969). These effects could be modified by β-blocking agents and by glucocorticosteroids. These observations and others led to the elegant experiments of Szego et al. (1971) on the effects of both oestrogens and androgens on a number of membrane-bound acid hydrolases in rats which seem worthy of closer examination. 17β-Oestradiol, testosterone, or diethylstilboestrol were injected into rats which had previously been ovariectomized and, after short intervals of time, purified lysosomal fractions were obtained from preputial gland tissues. These specimens were then treated under conditions, such as detergent-treatment, autolysis, and mechanical stress, which are known to release all remaining hydrolytic enzyme from lysosomal membranes. Diminished "structural latency" was observed on examining the specimens for several acid hydrolases including acid phosphatases, β-glucuronidase and acid ribonuclease. Similar results were obtained from uterine fractions under similar conditions (Szego, 1972). Pre-treatment with stilboestrol caused membrane destabilizing effects in preputial gland samples from orchidecto-mized rats, but the membrane stability of specimens from uterus was unaffected by pretreatment with testosterone. The authors conclude that these experiments provide direct evidence that modifications in lysosomal function are brought about by selective protein interaction with specific gonadal hormones at specific target sites in steroid sensitive tissues. They also suggest that these are physiologically significant general mechanisms for mediation of secondary biochemical transformations in the cell. This work has shed considerable light on the vexed question of the mechanism of steroid-mediated enzyme induction and interaction in hormone sensitive cells which may now be seen as a two (or more) stage process involving both *de novo* synthesis and membrane modifications. Much work remains before the interplay of these mechanisms is fully appreciated.

Scherstén *et al* (1969) have investigated the lysosomal enzyme activities in liver tissue from patients with renal carcinoma with a view to shedding some light on the question of whether malignancy in man induces increased activity of hepatic lysosomal enzymes and if some property of the primary tumour could be correlated with hepatic hydrolytic enzyme activity. They found that the liver tissue of patients with malignant renal tumour had significantly higher protein contents and also significantly higher free and total activities of aryl sulphatase, cathepsin and β-glucuronidase. Their results did not permit any conclusion about the mechanism of the increase of lysosomal enzyme activities.

In a further paper Scherstén and Lundholm (1972) undertook studies to determine lysosomal enzyme activity in muscle tissue from patients with malignant tumour and whether this activity changes with further propagation of the tumour. Observations in some patients suggested a connection between the course of the tumour disease and changes of the hydrolytic enzyme activities in muscle tissue, implying that the increased activities of these hydrolytic enzymes and their association with anabolic steroid levels may be of importance for the development of cachexia in patients with malignant tumour.

The influence of age and sex on plasma acid hydrolases of apparently healthy humans has been reported by Erikson *et al.* (1972). The activities in plasma of seven acid hydrolases were determined and three different patterns of variation with age were noted. High values in children were found for acid phosphatase and β-galactosidase. Levels increasing with age were noted for N-acetyl-β-glucosaminidase, β-glucuronidase and α-fucosidase, whereas β-glucosidase and α-mannosidase showed similar values in children and adults of all ages. No significant differences were found between the values in the two sexes.

An interesting aspect of the induction and/or synthesis of acid hydrolases is raised by the suggestion that the induction of these enzymes may be a function of a particular cell type within a gland and that cells of variable response may co-exist. Such a system is postulated by Ballantyne and Guillon (1971). These workers point out that the work of Nicol and others (Nicol et al., 1964; 1965; 1967) has led to the suggestion that 17β-oestradiol in particular may be regarded, in some ways, as a natural stimulant of the body's defences. Since oestrogens are in general transported in a soluble form, often as a conjugate with glucuronic acid, and since this is, in turn controlled by β-glucuronidase, which is situated within the target cell (although probably mostly in an unactive form on the lysosomal membranes), it is of interest to consider the relationship between oestradiol and β-glucuronidase in lymphoid tissues. These authors, working with the rabbit, have shown that female rabbits have considerably higher β-glucuronidase levels than males and that treatment with oestradiol stimulates the enzyme in the nodes of male animals only. This leads them to postulate two different types of cell containing β-glucuronidase in the nodes. One, in the medulla, is sensitive to oestradiol and the other, in the cortex, is much less sensitive. It is further suggested that these regional differences may result in locally enhanced reticuloendothelial activity which may play a part in the mechanism of phagocytosis. Such a system, if proved, must surely have an important role in the little-understood mechanism by which hormone-sensitive cancer may sometimes be controlled for prolonged periods with steroid hormones and also in the vexed question of the advisability of the surgical removal of the lymph nodes in such patients.

Two recent papers concern themselves with the role of lysosomes in tumour regression in hormone-dependent tumours of the rat. In the first of these papers, Nicholson et al. (1973) investigated the activities of five acid hydrolases together with protein, DNA and RNA, in growing tumours and in tumours from intact animals and from animals five days after ovariectomy. The general trend was an increase in the specific activity of the acid hydrolases, particularly an increase in activity of acid proteinase. An explanation for the increase in specific activities is the synthesis of lysosomal enzymes by macrophages after oestrogen withdrawal. It was thought that, since acid hydrolases are specific marker enzymes for lysosomes, the investigations would yield information on the role of lysosomes in the regression of hormone-dependent tumours. If this situation is operative in the mammary carcinoma it could account for the known increase in lysosomal enzymes under such conditions. Their general conclusion was that the pattern observed for the regressing tumour is similar to that observed for normal hormone-dependent tissue undergoing comparable physiological changes.

Cutts (1973) has investigated the activities of three lysosomal enzymes in regressing oestrone-induced mammary tumours of the rat. Chemical and histochemical studies of acid phosphatase, β-glucuronidase and cathepsin showed that these enzymes increase during regression of oestrone-induced mammary tumours. One week after removing the oestrone pellet, the tumour had regressed by 5% but acid phosphatase had increased by 67% cathepsin by 111% and β-glucuronidase by 138%. At week 2 all three enzymes had decreased but were still elevated above normal. From these observations Cutts inferred that lysosomal enzymes play a role in tumour regression. Apart from the association with tumour regression the changes in enzyme activity could not be related to either the age of the animal at the time of tumour appearance, the rapidity of tumour growth, the total number of tumours in the animal, or the location of the tumour in a particular mammary gland. Further experimentation of this type would appear to offer hope of an improvement in our understanding of the mechanism by which tumours regress, sometimes apparently spontaneously.

5.7. CONCLUSION

The control of enzyme induction is important in the design and development of carcinostatic and antineoplastic drugs, and in particular to those drugs aimed at the control of cancer arising from hormone-sensitive tissues. The normal cell has a complex system of enzyme regulation which leads to the efficient detoxication of the vast majority of chemicals, both natural and synthetic, with which it may be challenged. These systems often play an even more striking role in the faster proliferating and frequently drug-adapted malignant tissues. It is probable that many of the already effective anti-cancer drugs can be improved still further by combining treatment with an appropriate enzyme inhibitor or, alternatively, by treating with different drugs in sequence so as to minimize the development of adaptive detoxication enzymes.

A further approach to this problem is the use of synthetic rather than natural hormone preparations for use against hormone-sensitive cancer, because these synthetic substances are often retained in the tissues for long periods. For this reason, as mentioned previously (Herbst et al., 1971), synthetic hormones are themselves not without carcinogenic risk and must of course be used with circumspection. The basic problem in cancer chemotherapy is that, so far, no chemical or enzyme system (either normal or induced) has been found which is unique to the malignant cell. Should such a difference be discovered it must be of the very greatest importance in the chemical control or cancer.

It seems likely that, even with the present cytostatic drugs, considerable advances in cancer therapy can be expected by an improved understanding of the interaction between the drug and the enzyme systems which may be induced within the malignant cell.

REFERENCES

Aakvaag, A., Tveter, K. J., Unhjem, O. and Attramadal, A. (1972). *J. Steroia Biochem.*, **3**, 375
Alfheim, I. and Unhjem, O. (1971). *Acta Endoc.*, **68**, 567
Allen, M. J., Boyland, E. and Williams, D. C. (1957). *Repts. Inst. Cancer Res.*, **25**
Ballantyne, B. and Guillon, P. J. (1971). *Microbios.*, **3**, 23
Banerjee, D. N., Banerjee, M. R. and Wagner, J. (1971). *J. Endocr.*, **51**, 259
Barker, K. L. (1967). *Endocr.*, **81**, 791
Barker, K. L. and Warren, J. C. (1966). *Proc. Nat. Acad. Sci. U.S.*, **56**, 1298
Belitsky, P., El Hilali, M. M. and Oliver, J. A. (1970). *J. Urol.*, **104**, 453
Billing, R. J., Barbiroli, B. and Smellie, R. M. S. (1968). *Biochem. J.*, **109**, 705
Billing, R. J., Barbiroli, B. and Smellie, R. M. S. (1969). *Biochem. J.*, **112**, 563
Boyland, E. Gasson, J. E. and Williams, D. C. (1956). *Lancet*, **2**, 975
Boyland, E., Wallace, D. M. and Williams, D. C. (1955a). *Brit. J. Cancer*, **9**, 62
Boyland, E., Wallace, D. M. and Williams, D. C. (1955b). *Brit. J. Urol.*, **27**, 11
Boyland, E., Wallace, D. M. and Williams, D. C. (1957a). *Brit. J. Cancer*, **12**, 578
Boyland, E., Gasson, J. E. and Williams, D. C. (1957b). *Brit. J. Cancer*, **11**, 120
Boyland, E. and Williams, D. C. (1956). *Biochem. J.*, **64**, 578
Boyland, E. and Williams, D. C. (1957). *Proc. Roy. Soc. Med.*, **50**, 451
Breuer, C. B. and Florini, J. R. (1966). *Biochemistry*, **5**, 3857
Chung, L. W. K. and Coffey, D. S. (1971). *Biochim. Biophys. Acta*, **247**, 584
Church, R. H. and McCarthy, B. J. (1970). *Biochim. Biophys. Acta*, **199**, 103
Coffey, D. S., Shimazaki, J. and Williams-Ashman, H. G. (1968). *Arch. Biochem. Biophys.*, **124**, 184
Conney, A. H. (1967). *Pharmacol. Rev.*, **19**, 317
Cutts, J. H. (1973). *Cancer Res.*, **33**, 1235
Davies, P., Fahmy, A. R., Pierrepoint, C. G. and Griffiths, K. (1972). *Biochem. J.*, **129**, 1167
De Angelo, A. B. and Gorski, J. (1970). *Proc. Nat. Acad. Sci. U.S.*, **66**, 693
DeSombre, E. R., Mohla, S. and Jensen, E. V. (1972). *Biochem. Biophys. Res. Commun.*, **48**, 1601
Erikson, O., Ginzburg, B. E., Hultber, B. and Ocherman, P. A. (1972). *Clin. Chim. Acta*, **40**, 181
Fang, S. and Liao, S. (1971). *J. Biol. Chem.*, **246**, 16
Field, E. J. (1973). In *Modern Trends in Oncology*, vol. I (1), p. 183 (ed. Raven, R. W.), Butterworths; London
Fishman, W. H. (1951). *Ann. N.Y. Acad. Sci.*, **54**, 548
Fishman, W. H., Artenstein, M. and Green, S. (1955). *Endocrinol.*, **57**, 646
Fishman, W. H. and Farmelant, M. H. (1953). *Endocrinol.*, **52**, 536

Fishman, W. H., Inglis, N. I., Stolbach, L. L. and Krant, M. J. (1968). *Cancer Res.*, **28,** 150

Fishman, W. H. and Lipkind, J. B. (1958). *J. Biol. Chem.*, **232,** 729

Fishman, W. H., Odell, L. D., Gill, J. E. and Christensen, R. A. (1950). *Amer. J. Obst. & Gynecol.*, **59,** 414

Forrest, A. P. M. and Roberts, M. M. (1973). In *Modern Trends in Oncology*, I (2), p. 109 (Ed. Raven, R. W.), Butterworths; London

Gardner, W. U. (1964). In *Biological Interactions in Normal and Neoplastic Growth*, p. 391 (Eds. Brennan, M. J. and Simpson, W. L.), Boston, Little Brown & Co.; New York

Ghraf, R., Lax, E. R. and Schriefers, H. (1972). *Acta Endocr.*, **71,** 781

Gittinger, J. W. and Lasnitzki, I. (1972). *J. Endocr.*, **52,** 459

Glasser, S. R., Chytil, F. and Spelsberg, T. C. (1972). *Biochem. J.*, **130,** 947

Goodfriend, T. L. and Kaplan, N. O. (1964). *J. Biol. Chem.*, **239,** 130

Gorski, J. (1964). *J. Biol. Chem.*, **239,** 889

Hamilton, T. H. (1964). *Proc. Nat. Acad. Sci. U.S.*, **51,** 83

Hamilton, T. H., Widnell, C. C. and Tata, J. R. (1965). *Biochim. Biophys. Acta*, **108,** 168

Hamilton, T. H., Widnell, C. C. and Tata, J. R. (1968). *J. Biol. Chem.*, **243,** 408

Harper, M. E., Fahmy, A. R., Pierrepoint, C. G. & Griffiths, K. (1970a). *Steroids*, **15,** 89

Harper, M. E., Pierrepoint, C. G., Fahmy, A. R. and Griffiths, K. (1970b). *Biochem. J.*, **119,** 785

Harper, M. E., Pierrepoint, C. G., Fahmy, A. R. and Griffiths, K. (1971). *J. Endocr.*, **49,** 213

Harris, R. G. (1973). Personal communication

Hechter, O. (1955). *Vitamins & Hormones*, **13,** 293

Heinrichs, W. L., Feder, H. H. and Colas, A. (1966). *Steroids*, **7,** 91

Herbst, A. L., Ulfelder, H. and Poskanger, D. C. (1971). *New Eng. J. Med.*, **284,** 878

Hertz, S. (1950). *J. Clin. Invest.*, **29,** 821

Hilf, R., McDonald, E., Sartini, J., Rector, W. D. and Richards, A. H. (1972) *Endocr.*, **91,** 280

Inano, H., Suzuki, K., Wakabayashi, K. and Tamaoki, B. (1973). *Endocr.*, **92,** 22

Jacobson, M. and Kuntzman, R. (1969). *Steroids*, **13,** 327

Jensen, E. V. and DeSombre, E. R. (1972). *Ann. Rev. Biochem.*, **41,** 203

Jensen, E. V. and Jacobson, H. I. (1960). In *Biological Activities in Relation to Cancer.* p. 161 (Ed. Pincus, G. and Vollmer, E. P.) Academic Press; London and New York

Jensen, E. V., Numata, M., Brecher, P. I. and DeSombre, E. R. (1971). In *The Biochemistry of Steroid Hormone Action*, p. 133 (Ed. Smellie, R. M. S.) Academic Press; London and New York

Kasdon, S. C., McGowan, J., Fishman, W. H. and Homburger, F. (1951a). *Amer. J. Obst. & Gynecol.*, **61,** 647

Kasdon, S. C., Romsey, A. B., Homburger, F. and Fishman, W. H. (1951b). *Amer. J. Obst. & Gynecol.*, **61,** 1142

Kerr, L. M. H., Campbell, J. G. and Levvy, G. A. (1949). *Biochem. J.*, **44,** 487

King, R. J. B. and Gordon, J. (1972). *Nature New Biol.*, **240,** 185

Knowler, J. T. and Smellie, R. M. S. (1971). *Biochem. J.*, **125,** 605

Kosto, B., Calvin, H. I. and Williams-Ashman, H. G. (1967). In *Advances in Enzyme Regulation*, vol. 5, p. 25 (Ed. Weber, G.), Pergamon Press; London

Kuntzman, R., Jacobson, M., Schneidmen, K. and Conney, A. H. (1964). *J. Pharmacol. Exp. Therap.*, **146,** 280

Kuntzman, R., Welch, R. and Conney, A. H. (1966). In *Advances in Enzyme Regulation*, **4,** 149 (Ed. Weber, G.), Pergamon Press, London

Lawrence, J. H. and Tobias, C. A. (1956). *Cancer Res.*, **16,** 185

Liao, S., Barton, R. W. and Lin, A. H. (1966). *Proc. Nat. Acad. Sci. U.S.*, **55,** 1593

Liao, S., Leininger, K. R., Sagher, D. and Barton, R. W. (1965). *Endocrinology,* **77,** 763

Mainwaring, W. I. P., Mangan, F. R. and Peterken, B. M. (1971). *Biochem. J.*, **123,** 619

Mangan, F. R., Neal, G. E. and Williams, D. C. (1967). *Biochem. J.*, **104,** 1075

Mangan, F. R., Neal, G. E. and Williams, D. C. (1968). *Arch. Biochem. Biophys.*, **124,** 27

Maul, G. G. and Hamilton, T. H. (1967). *Proc. Nat. Acad. Sci. U.S.*, **57,** 1371

Mayol, R. F. and Thayer, S. A. (1970). *Biochemistry*, **9,** 2484

Miller, H. C., Rector, W. and Hilf, R. (1972). *Invest. Urol.*, **10,** 1

Mohla, S., DeSombre, E. R. and Jensen, E. V. (1972). *Biochem. Biophys. Res. Commun.*, **46,** 661

Moore, R. J. and Hamilton, T. H. (1964). *Proc. Nat. Acad. Sci. U.S.*, **52,** 439

Moulton, B. C. and Barker, K. L. (1971). *Endocr.*, **89,** 1131

Mueller, G. C., Gorski, J. and Aizawa, Y. (1961). *Proc. Nat. Acad. Sci. U.S.*, **47,** 164

Nicholson, R. I., Bagnall, I. and Davies, M. (1973). *Europ. J. Cancer*, **9,** 313

Nicol, T. and Vernon-Roberts, B. (1965). *J. Reticuloendothelial Soc.*, **2,** 15

Nicol, T., Bilbey, D. L. J., Charles, L. M., Cordingley, J. L. and Vernon-Roberts, B. (1964). *J. Endocrinol.*, **30,** 277

Nicol, T., Quantoch, D. C. and Vernon-Roberts, B. (1967). *Adv. Exp. Med. Biol.*, **I,** 221

Notides, A. and Gorski, J. (1966). *Proc. Nat. Acad. Sci. U.S.*, **56,** 230

Oliver, J. A., El Hilali, M. M., Belitsky, P. and MacKinnon, K. J. (1970). *Cancer*, **25,** 863

O'Malley, B. W., Sherman, M. R. and Toft, D. O. (1970). *Proc. Nat. Acad. Sci. U.S.*, **67,** 501

Pineda, E. P., Goldberge, J. A., Banks, B. M. and Rutenburg, A. M. (1959). *Gastroenterology*, **36,** 222

Riotton, G. and Fishman, W. H. (1953). *Endocrinol.*, **52,** 692

Robel, P., Lasnitzki, I. and Baulieu, E. E. (1971). *Biochimie*, **53,** 81

Roberts, S. and Szego, C. M. (1953). *Physiol Rev.*, **33,** 593

Roeder, R. G. and Rutter, W. J. (1969). *Nature, Lond.*, **224,** 234

Roeder, R. G. and Rutter, W. J. (1970a). *Proc. Nat. Acad. Sci. U.S.*, **65,** 675

Roeder, R. G. and Rutter, W. J. (1970b). *Biochemistry*, **9,** 2543

Rubin, B. L. and Strecker, H. J. (1961). *Endocr.*, **69,** 257

Santti, R. S. and Villee, C. A. (1971). *Endocr.*, **89,** 1162

Scherstén, T. and Lundholm, K. (1972). *Cancer*, **30,** 1246

Scherstén, T., Wahlqvist, L. and Johansson, L. G. (1969). *Cancer*, **23,** 608

Schriefers, H. (1969). In *Advances in the Biosciences 2*, p. 69 (Ed. Raspé, G.) (Pergamon Press; London

Schriefers, H., Hoff, H., Graf, R. and Ockenfels, H. (1972). *Acta Endocr.*, **69,** 789

Shyamala, B. and Gorski, J. (1969). *J. Biol. Chem.*, **244,** 1097

Sie, H. and Fishman, W. H. (1953). *Cancer Res.*, **13,** 590

Siiteri, P. K. and Wilson, J. D. (1970). *J. Clin. Invest.*, **49,** 1737

Singhal, R. L. and Ling, G. M. (1969). *Canad. J. Physiol. Pharmacol.*, **47,** 233

Singhal, R. L. and Valadares, J. R. E. (1968). *Biochem. J.*, **110,** 703

Singhal, R. L. and Valadares, J. R. E. (1970). *Am. J. Physiol.*, **218,** 321

Singhal, R. L., Valadares, J. R. E. and Ling, G. M. (1967). *J. Biol. Chem.*, **242,** 2593

Singhal, R. L., Wang, D. and Ling, G. M. (1968). *Proc. Canad. Fedn. Biol. Soc.*, **II,** 146

Szego, C. M. (1972). In *The Sex Steroid; Molecular Mechanisms* (Ed. McKerns, K. W.), Appleton-Century-Crofts; New York
Szego, C. M. and Davis, J. S. (1967). *Proc. Nat. Acad. Sci.*, **58**, 1711
Szego, C. M. and Davis, J. S. (1969). *Life Sci.*, **8**, 1109
Szego, C. M., Seeler, B. J., Steadman, R. A., Hill, D. F., Kimura, A. K. and Roberts, J. A. (1971). *Biochem. J.*, **123**, 523
Teng, C. S. and Hamilton, T. H. (1968). *Proc. Nat. Acad. Sci. U.S.*, **60**, 1410
Tephly, T. R. and Mannering, G. J. (1968). *Molec. Pharmac.*, **4**, 10
Terenius, L. (1972). *Steroids*, **19**, 787
Trachewsky, D. and Segal, S. J. (1967). *Biochem. Biophys. Res. Commun.*, **27**, 588
Ui, H. and Mueller, G. C. (1963). *Proc. Nat. Acad. Sci. U.S.*, **50**, 256
Valadares, J. R. E., Singhal, R. L. and Parulekar, M. R. (1968). *Science, N.Y.*, **159**, 990
Wicks, W. D. and Kenney, F. T. (1964). *Science, N.Y.*, **144**, 1346
Widnell, C. C. and Tata, J. R. (1966a). *Biochem. J.*, **98**, 621
Widnell, C. C. and Tata, J. R. (1966b) *Biochim. Biophys. Acta*, **132**, 478
Williams, D. C. (1965). *Europ. J. Cancer*, **I**, 115
Williams, D. C. (1973). In *Modern Trends in Oncology*, Vol. 1, p. 209 (Ed. Raven, R. W.), Butterworths; London
Williams-Ashman, H. G., Liao, S., Hancock, R. L., Jurkowitz, L. and Silverman, D. A. (1964). *Recent Prog. Horm. Res.*, **20**, 247

Chapter 6

The Control of Tryptophan Metabolism

G. Curzon
Department of Neurochemistry, Institute of Neurology,
Queen Square, London, W.C.1., England

6.1. INTRODUCTION

This volume as a whole describes enzyme induction mechanisms and the influences of enzyme induction upon metabolism and its control. The present chapter deals with various ways in which the metabolism of a single amino acid tryptophan is controlled. While processes influencing this include induction, the *de novo* synthesis of enzymes which metabolize tryptophan is only one of a variety of ways in which its metabolism is apportioned between various pathways. For example, altered activity of existing enzyme molecules and altered availability of plasma tryptophan for metabolism are also important. The roles of these factors will therefore be discussed as well as that of induction. Mechanisms of a less specific nature such as altered availability of tissue amino acid pools are described in detail elsewhere (Munro, 1970) and will not be dealt with here.

Figure 6.1 shows the principal pathways of tryptophan metabolism in animals. A quantitatively important route is initiated by the action of liver tryptophan pyrrolase (or oxygenase), an enzyme that is rapidly induced and whose activity is readily demonstrated. These properties have been used in many classical enzyme induction studies. Tryptophan pyrrolase splits the pyrrole ring of tryptophan to formylkynurenine and this, through the action of formylase, is converted to kynurenine (Mehler and Knox, 1950). As formylase is present in the liver in large excess, pyrrolase is the rate limiting enzyme for kynurenine formation.

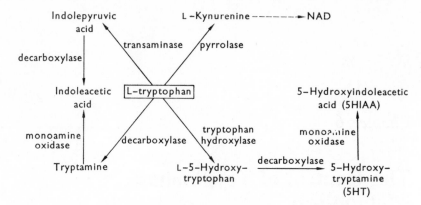

Fig. 6.1. Main pathways of tryptophan metabolism.

The pathway is involved in nicotinamide synthesis as kynurenine, through a complex sequence of reactions, gives rise to NAD (Nishizuka and Hayaishi, 1963).

The tryptophan-5-hydroxylase pathway is quantitatively minor but physiologically important because it represents the rate limiting step (Moir and Eccleston, 1968) in synthesis of the neurotransmitter 5-hydroxytryptamine (5HT). Liver, mast cells, intestinal mucosa and brain all contain tryptophan hydroxylase activity and can thus convert tryptophan to 5-hydroxytryptophan which is then decarboxylated to 5HT by excess aromatic amino acid decarboxylase. It is noteworthy that 5HT does not penetrate effectively to the brain which must therefore synthesize its own requirements.

Relatively little attention has been paid to the two alternative pathways which lead to indole-acetic acid as their terminal metabolite. The first proceeds via decarboxylation to tryptamine which is converted to indole-acetic acid by monoamine oxidase (Weissbach et al., 1959). An alternative route to indole-acetic acid can proceed via transamination to indole-pyruvic acid (Semba and Civen, 1970) and subsequent oxidative decarboxylation. Interest in tryptamine has recently been stimulated by its detection in brain (Saavedra and Axelrod, 1972).

A number of important questions arise concerning the routes of metabolism shown in Fig. 6.1.

1. What part do they play in the control of tryptophan levels?
2. What are the relative roles of enzyme and substrate concentrations in determining formation of the products NAD and 5HT?
3. Can altered activity on one route change tryptophan levels so that metabolism on another route is influenced?

6.2. CONTROL OF PLASMA TRYPTOPHAN CONCENTRATION

That plasma tryptophan levels are under specific control is suggested by the different pattern of diurnal variation of rat plasma tryptophan from that of most other amino acids (Fernstrom *et al.*, 1971). Most amino acids are in phase with maximum and minimum concentrations at about 6 a.m. and 6 p.m. respectively. Trytophan is exceptional with corresponding concentrations at 2 a.m. and 10 a.m. Dietary intake of tryptophan can hardly be invoked to explain the specific nature of the diurnal tryptophan pattern. Three other possibilities may be considered as specific controls or influences upon the diurnal pattern and absolute concentrations of plasma tryptophan. These are: (*a*) destruction of tryptophan by pyrrolase, (*b*) uptake by tissues, and (*c*) protection from (*a*) and (*b*) by the binding of tryptophan to plasma albumin.

Much attention has been paid to the possibility that pyrrolase activity may be a dominating influence on plasma tryptophan levels. Pyrrolase activity alters readily in response to changes of milieu. For example, Fig. 6.2 shows the increase of rat liver pyrrolase activity during immobilization stress. These changes appear to be mediated by adrenocortical secretions in so far as they can be produced by cortisol injection and are prevented by adrenalectomy. The idea that they have a physiological function related to tryptophan metabolism, or at least a physiological effect on the latter, is attractive. As well as adrenal corticoids, exogenous tryptophan itself increases pyrrolase activity (Knox and Auerbach, 1955) which suggests that in some circumstances pyrrolase might be involved in homeostasis of plasma tryptophan level. It is of interest that the relative inductive potencies of corticoids and tryptophan are grossly different in different species. Thus the

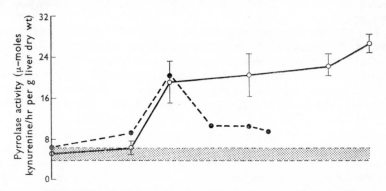

Fig. 6.2. Effects of immobilization and of cortisol on rat liver pyrrolase. ○, immobilization; ●, after 5 mg/kg hydrocortisone i.p. Points represent means ± one S.D. on at least 4 rats (Curzon and Green, 1969).

gerbil (Baughman and Franz, 1971) and guinea-pig (Hvitfelt and Santii, 1972) enzymes respond to substrate but are resistant to corticoids, while the rat and mouse enzymes respond to both stimuli.

Plasma corticoids and liver pyrrolase also both show well-defined diurnal rhythms, that of pyrrolase activity in the mouse roughly following that of plasma corticosterone but about six hours later. However, the rhythm of tryptophan concentration in whole blood shows no clear temporal relationship to that of pyrrolase activity (Hardeland and Rensing, 1968; Rapoport et al., 1966). Furthermore, adrenalectomy results in the pyrrolase rhythm being almost completely flattened out but a tryptophan rhythm still remains strikingly apparent though absolute levels change. Another negative suggestion is provided by the finding that the tryptophan rhythm but not the pyrrolase rhythm is reversed by changing the period of food intake (Ross et al., 1972). Moreover, even when pyrrolase activity is increased considerably by cortisol injection or by stress, plasma tryptophan may show little change. Thus there was only a small and transient fall of plasma tryptophan after injecting rats intraperitoneally with 5 mg/kg cortisol, even though pyrrolase activity increased three-fold (Green and Curzon, 1970). Similarly, after three hours' immobilization (Knott and Curzon, 1972) plasma tryptophan fell by an insignificant extent (15%), though under these conditions pyrrolase rose four-fold (Curzon and Green, 1969).

Still greater pyrrolase increases occur on injecting α-methyltryptophan. This is not a substrate and massively elevates pyrrolase activity not by inducing new enzyme but by an action on the enzyme already formed (Civen and Knox, 1960). Its administration markedly decreases blood, brain and liver tryptophan (Sourkes et al., 1970). Results in general, therefore, suggest that only in extreme circumstances is increased pyrrolase activity associated with a well-defined fall of plasma tryptophan.

The interaction between pyrrolase and tryptophan is complex (Feigelson and Maeno, 1967; Greengard et al., 1966a; Knox, 1966; Schimke, et al., 1964) and while the commonly used methods for determination of pyrrolase activity, which involve the in vitro action on tryptophan, may give good estimates of the total number of pyrrolase molecules, these may not necessarily be proportional to the activity of the enzyme in vivo.

Pyrrolase influences tryptophan concentrations more strikingly following tryptophan administration. Thus the rate of disappearance of injected tryptophan from rat plasma is proportional to liver pyrrolase activity (Knox, 1966), and conversely, when pyrrolase is decreased by adrenalectomy then plasma tryptophan concentration remains high after tryptophan injection much longer than in intact rats (Powanda and Wannemacher, 1971). The protective action of pyrrolase against the toxic effects of administered tryptophan is consistent with this finding. Thus, the toxicity of tryptophan towards adrenalectomized rats is much greater than towards intact animals or adrenalectomized rats previously given cortisol (Knox, 1966).

The dependence of the fate of exogenous tryptophan upon pyrrolase activity could be of therapeutic relevance. This amino acid has been used in the treatment of depressive illness (Coppen et al., 1967) and many depressives appear to have high pyrrolase activity, as after a standard oral dose of tryptophan they excrete large amounts of its metabolites formed subsequent to pyrrolase action (Curzon and Bridges, 1970). These findings suggest that in some patients high pyrolase activity might oppose the therapeutic efficiency of administered tryptophan.

The above provides evidence that pyrrolase can influence tryptophan levels in extreme and non-physiological situations; for example, when pyrrolase is massively increased by α-methyltryptophan or when tryptophan is administered. Evidence that it can alter physiological tryptophan levels is as yet less convincing. However, many past findings on the relationship between pyrrolase and tryptophan levels now require reconsideration because the binding of tryptophan by plasma albumin was not taken into account. It is important to realize that most of the tryptophan in plasma is bound to albumin (McMenamy and Oncley, 1958) and is therefore not immediately available for transport to tissues and metabolism. Only the so-called free tryptophan with which the albumin-bound tryptophan is in equilibrium can be metabolized. The free tryptophan can be estimated using dialysis or more conveniently by centrifugal ultrafiltration. Unfortunately, though it is more than a decade since plasma tryptophan was found to be largely albumin-bound, almost all reported determinations until recently were of total plasma tryptophan only.

The equilibrium between bound and free tryptophan is very different under different conditions. Thus free tryptophan concentrations are approximately doubled by food deprivation for 24 hours and are significantly increased by 3 hours' immobilization (Table 6.1). Under these circumstances total plasma tryptophan only changes slightly if at all. It is of interest that in the first experiment shown in Table 6.1, in which the control animals were relatively undisturbed, they had particularly low plasma-free tryptophan concentrations, but in the other experiments in which controls had been either re-caged in isolation, or injected with saline and therefore exposed to some stress, their plasma-free tryptophan concentrations were higher. In general, these results suggest that the amount of plasma tryptophan available for metabolism is far more sensitive to stress, nutrition and plasma composition than is the total plasma tryptophan.

An important factor influencing free tryptophan concentration is

TABLE 6.1

Effects of Various Procedures on the Disposition of Rat Plasma Tryptophan

		Plasma tryptophan (μg/ml)		
		Total	Free	
Expt. 1.	Controls (20)	21.2 ± 2.8	1.2 ± 0.35	} $P < 0.001$
	Starved (20)	20.6 ± 5.0	3.1 ± 1.2	
Expt. 2.	Controls (8)	23.3 ± 2.9	2.4 ± 0.85	} $P < 0.05$
	Immobilized (8)	19.8 ± 3.3	3.6 ± 1.1	
Expt. 3.	Controls (5)	18.6 ± 2.5	2.2 ± 0.3	} $P < 0.001$
	Heparin* (5)	15.2 ± 3.8	4.1 ±0.9	

* Killed 15 min after 500 I.U./kg heparin i.v from Knott and Curzon (1972)
Results ±1 S.D.

plasma unesterified fatty acid concentration. The latter are almost completely bound to albumin and addition *in vitro* of physiologically occurring fatty acids to rat and human plasma within the range of physiological concentration displaces bound tryptophan and thus elevates the free fraction (Curzon *et al.*, 1973a). Plasma fatty acids and free tryptophan both increase upon deprivation, immobilization or heparin injection (Knott and Curzon, 1972). These findings suggested a mechanism by which the many hormonal agents which alter lipolysis and fatty acid levels through influencing fat-cell cyclic AMP (Robison *et al.*, 1971) could also alter plasma-free tryptophan concentration.

Thus isoprenaline and aminophylline which increase fat-cell cyclic AMP, by activating adenyl cyclase and inhibiting cyclic AMP destruction, respectively, both increase plasma unesterified fatty acid and free tryptophan, while insulin and nicotinic acid, which decrease fat-cell cyclic AMP, cause changes in the opposite direction (Curzon and Knott, 1973). The decrease of both plasma unesterified fatty acid and free tryptophan concentrations after an oral glucose load (Lipsett *et al.*, 1973) is also consistent with these findings. Again the increase of brain tryptophan after large doses of L-dopa (Weiss *et al.*, 1971) is associated with increased plasma fatty acid and free tryptophan (Curzon and Knott, 1973). A pathological situation in which elevated fatty acid and free tryptophan concentrations occur is acute hepatic failure (Curzon *et al.*, 1973).

Increased availability of tryptophan for metabolism following the increase of free tryptophan upon food deprivation could be a physiologically important control mechanism opposing deficiencies which would otherwise occur more rapidly, e.g. NAD deficiency. Also, as tryptophan availability may have a limiting role in protein synthesis, especially in young animals (Wunner *et al.*, 1966; Aoki and Siegel, 1970) the increase of free tryptophan would tend to oppose development of defective protein synthesis.

Determination of free tryptophan changes may well lead to reappraisal of many earlier physiological, pharmacological and pathological findings on tryptophan metabolism. For example, the diurnal and other relationships between plasma tryptophan and pyrrolase must be reconsidered. Also as aspirin (McArthur *et al.*, 1971) can affect tryptophan metabolism by directly displacing it from plasma albumin it seems likely that many other drugs will be found to have a similar effect.

6.3. METABOLISM ON THE PYRROLASE PATHWAY

The relative importance of pyrrolase induction by adrenocorticoids, and of availability of tryptophan in the control of the amount of flux of tryptophan metabolism on the pyrrolase pathway, must be considered. In the previous section, evidence of the effect of the pyrrolase route on the disposal of a tryptophan load was described. The positive correlation between pyrrolase activity of human liver biopsy material and urinary excretion of kynurenine after an oral tryptophan load (Altman and Greengard, 1966) is consistent with these findings. However, results under loading conditions are not necessarily capable of extrapolation to physiological conditions.

This problem has been studied by Powanda and Wannemacher (1971) who determined NAD as an index of tryptophan flux through the pathway subsequent to pyrrolase action. They found a significant positive correlation between the dose of tryptophan injected and the eventual amount of NAD present in mouse liver. Thus NAD synthesis may be increased by increasing available substrate for pyrrolase. Therefore, it may well be increased either when tryptophan is given therapeutically or when the plasma tryptophan equilibrium is shifted in the direction of free tryptophan. However, when pyrrolase activity was increased five-fold by cortisol, no increase was found in the NAD determined one hour after the induced pyrrolase maximum. This suggests that flux of tryptophan towards NAD is not readily altered by increasing pyrrolase activity. Earlier work showed that dietary NAD deficiency disease was prevented by administration of corticoids but that increased pyrrolase was not required as the effect was obtained even using corticoids which did not induce pyrrolase (Greengard *et al.*, 1966b) and was associated with increased plasma tryptophan (Greengard *et al.*, 1968). A somewhat ambiguous finding is the slight increase of NAD when pyrrolase activity was increased three- to four-fold by α-methyltryptophan (Powanda and Wannemacher, 1971), as

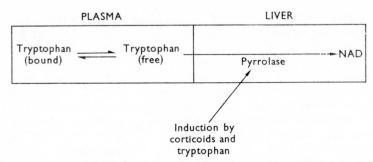

Fig. 6.3. Tryptophan disposition and metabolism on the pyrrolase pathway.

available tryptophan may well have fallen markedly under these conditions (Oravec and Sourkes, 1970), in which circumstance the results would indicate a relative increase of metabolism towards NAD associated with increased pyrrolase.

Previously, Kim and Miller (1969) had used a different approach, injecting rats with small amounts of ^{14}C-labelled tryptophan and determining the radioactivity of the exhaled $^{14}CO_2$ as a measure of flux specifically on the pyrrolase step. Although cortisol caused large increases of liver pyrrolase as measured *in vitro*, the output of $^{14}CO_2$ only increased appreciably if unlabelled carrier tryptophan was also given. This suggested that changes induced by cortisol which were demonstrable *in vitro* might have little relevance to physiological conditions. However, others have found an appreciable increase in $^{14}CO_2$ output from a tracer dose of labelled tryptophan after cortisol treatment (Chytil, 1971; Joseph, 1972).

A general scheme consistent with most of the studies discussed above is shown in Fig. 6.3. It is suggested that cortisol (under some circumstances at least) increases the flux of tryptophan through the pyrrolase pathway by inducing new enzyme. Increased tryptophan availability, resulting either from tryptophan administration or from decreased binding to plasma albumin, may also increase net flux and this might be further enhanced by an effect of tryptophan upon pyrrolase activity. As the plasma-free tryptophan pool available for metabolism is normally in equilibrium with a much larger albumin-bound pool, pyrrolase activity changes could have relatively small effects on total plasma tryptophan, except in extreme circumstances, e.g. when α-methyltryptophan or a large tryptophan load is given. Conversely, decreased pyrrolase activity due to adrenalectomy sharply increases the plasma half-life of a tryptophan load.

Competition Between Pathways

In what circumstances can tryptophan be diverted from one of its pathways by increasing its metabolism on another? One example is carcinoid disease, in which there is a malignancy of the cells in the intestine which produce 5HT so that a large fraction of the whole body tryptophan metabolism is diverted to 5HT synthesis (Sjoerdsma et al., 1956). It is therefore not too surprising that pellagra symptoms, apparently due to resultant nicotinamide deficiency, have been described in some carcinoid patients (Thorson et al., 1954).

A further possibility which has been the subject of some speculation (Curzon, 1969; Lapin and Oxenkrug, 1969) is the opposite diversion of tryptophan away from brain 5HT synthesis because of increased pyrrolase activity. This is of some pathological interest as evidence suggests both impaired brain 5HT synthesis and high cortisol levels in depressive illness (see Curzon, 1969 and Coppen, 1972a for reviews). A clear example of diversion of tryptophan away from 5HT synthesis is provided by the action of α-methyltryptophan. Sourkes et al. (1970) showed that this substance caused prolonged and extremely high pyrrolase activity, a large fall of tryptophan levels and also significant decreases of brain 5HT and of its metabolite 5-hydroxyindole-acetic acid (5HIAA).

Other procedures which increase pyrrolase activity may also lead to decreased brain 5HT, e.g. injection of adrenal corticoids and related substances (Curzon and Green, 1968; De Schaepdryver et al., 1969). Time relationships between the pyrrolase and 5HT changes were consistent with a causal role of pyrrolase. Unlike the situation in the case of α-methyltryptophan a comparable diversion of tryptophan was not demonstrable, so that the mechanism by which brain 5HT falls after cortisol injection remains unclear. As injection of the tryptophan metabolites, kynurenine, hydroxykynurenine or hydroxyanthranilic acid also resulted in decreased brain 5HT (Green and Curzon, 1970), it was suggested that increased synthesis of these substances is responsible, although subsequent work failed to show that plasma kynurenine rose after cortisol injection (Joseph 1972).

6.4. THE CONTROL OF TRYPTOPHAN METABOLISM ON THE 5HT PATHWAY IN THE BRAIN

As tryptophan hydroxylase, the rate limiting enzyme for brain 5HT synthesis, is normally unsaturated with its substrate tryptophan, brain 5HT synthesis can alter as a result of altered availability of tryptophan to the brain (Eccleston et al., 1965). Thus, Tagliamonte and co-workers (1971) showed that when rats are given various drugs such as amphetamine, lithium salts, dibutyryl cyclic AMP, bulbocapnine, probenecid or γ-aminobutyric acid (GABA), or when they are exposed to a 40° environment, then brain tryptophan and 5HT turnover both increase. Food deprivation or immobilization have similar effects (Curzon *et al.*, 1972) which are not merely a reflection of non-specific brain amino acid changes (Knott *et al.*, 1973). Although the increased brain tryptophan is presumably derived from plasma, the brain and plasma concentrations of this amino acid are not significantly correlated, only the plasma-free tryptophan (see page 173) being available to the brain (Curzon *et al.*, 1972; Tagliamonte *et al.*, 1973).

It seems likely that the effects of many drugs on brain 5HT level or turnover are due to plasma-free tryptophan changes, produced either by the drug directly displacing tryptophan from plasma albumin (e.g. salicylates and other antirheumatic drugs—McArthur *et al.*, 1971) or by it changing plasma fatty acid levels so that albumin binding is altered. The latter mechanism appears to be involved in the increases after aminophylline or amphetamine injection and in hepatic failure (Curzon *et al.*, 1973b) and the decrease after nicotinic acid injection (Curzon and Knott, 1973).

While availability of tryptophan to the brain is a function of the plasma-free tryptophan the latter must compete for transport to brain with other amino acids. Thus the ratio of plasma tryptophan to a group of competing plasma neutral amino acids (tyrosine, phenylalanine, leucine, isoleucine and valine) is also a determinant of brain tryptophan (Fernstrom and Wurtman, 1972). This was used to explain why, although a specific increase of plasma tryptophan within the physiological range results in increases of brain tryptophan and of 5HT and its turnover (Fernstrom and Wurtman, 1971), the increases upon ingestion of protein-containing diets have little effect. High concentrations of members of the group of competing amino acids may well have pathological consequences, the resultant decrease of brain 5HT or protein synthesis especially at a vulnerable stage of development possibly leading to permanent brain damage, e.g. in phenylketonuria, tyrosinaemia, leucinosis and leucine pellagra. It is relevant that inhibition of rat brain 5HT synthesis from the 7th to the 15th day leads to subsequent impairment of brain growth but similar treatment of older rats is without permanent effect (Hole, 1972). Also hyperphenylalaninaemia in young rats decreases brain protein synthesis and this is prevented by tryptophan administration (Aoki and Siegel, 1970).

The qualitative importance of brain tryptophan concentration as a determinant of 5HT synthesis is clear but its quantitative significance is less well defined as the latter is synthesized specifically within 5HT neurones and there is as yet no way in which their tryptophan content can be determined. It is known that an active transport mechanism exists by which tryptophan is transported into neuronal preparations (Grahame-Smith and Parfitt, 1970). These synaptosomal preparations are, however, derived not specifically from 5HT neurones but from neurones of various types.

Turnover on the brain 5HT pathway may be influenced not only by tryptophan availability but also by altered synthesis, or activity, of tryptophan hydroxylase. The activity of this enzyme appears to be under the influence of the adrenal cortex as rat mid-brain tryptophan hydroxylase activity was reported to be decreased by adrenalectomy and partly restored by corticosterone (Azmitia and McEwen, 1969). However, the rate of transport of newly synthesized tryptophan hydroxylase to nerve endings appears too slow for it to be able to deal with acute demands for increased 5HT synthesis (Meek and Neff, 1972). These demands are thus presumably met by mechanisms such as allosteric changes of the tryptophan hydroxylase molecule or increased availability of its pteridine cofactor, or of tryptophan.

Homeostasis of 5HT following increased breakdown related to neuronal firing probably occurs via allosteric change of tryptophan hydroxylase. Thus electrical stimulation of the cell bodies of brain 5HT neurones *in vivo* led to increased 5HT breakdown and increased synthesis of 5HT from tracer amounts of labelled tryptophan (Shields and Eccleston, 1972). As brain tryptophan did not increase and as 5HT synthesis increased very rapidly an allosteric mechanism is suggested. The newly synthesized 5HT was probably within a distinct small rapidly turning over pool as the total rate of accumulation was unaltered.

Various and conflicting studies have been made on feedback control of 5HT synthesis in which net brain 5HT was increased by monoamine oxidase inhibition (e.g. Macon *et al.*, 1971; Millard *et al.*, 1972). However, as after inhibition the 5HT concentration is still able to rise far above physiological levels such a mechanism must be of limited efficiency, at least in the presence of the inhibitory drug. Indeed evidence indicates that normal control of 5HT synthesis is lost after monoamine oxidase inhibition (Meek and Fuxe, 1971).

Other studies suggest that 5HT concentration specifically at the synapse may influence 5HT synthesis. Thus, chlorimipramine which increases synaptic 5HT by preventing its neuronal re-uptake from the synaptic cleft also decreases its turnover (Meek and Werdinius, 1970). A related finding is the decreased 5HT synthesis observed (Lin *et al.*, 1969; Carlsson and Lindqvist, 1972) following administration of lysergic acid diethylamide (LSD) which may activate 5HT receptors (Anden *et al.*, 1968). Pre-synaptic activity of 5HT neurones is not obligatory for 5HT synthesis, as hydroxylation of tryptophan in the brain is only moderately affected by axonal lesions of 5HT neurones although synthesis in spinal neurones is rapidly decreased by axotomy (Carlsson *et al.*, 1972).

Another factor to be considered is the evidence of an effect of 5HT concentration on the rate of firing of 5HT neurones. Thus monoamine oxidase inhibitors rapidly depress firing (Aghajanian, 1972). Tryptophan also rapidly blocks firing, although 5HT synthesis may not be involved here as the blockade was not prevented by treatment with *p*-chloro-phenylalanine (an inhibitor of 5HT synthesis from tryptophan). An important question is whether depression of firing can occur when brain tryptophan increases following either administration of tryptophan in the treatment of depression or increased availability of plasma tryptophan due to physiological changes.

In studying the relationships between 5HT synthesis and brain activity the ideal experimental design is elusive. The physiological relevance of experiments with drugs can always be questioned and though direct neuronal stimulation is superficially ideal the manipulations required may well result in even control animals being in an abnormal state which could have effects obscuring physiological mechanisms operating within narrow limits. An alternative approach is to compare brain 5HT metabolism of control animals disturbed as little as possible and of animals for which environmental or nutritional parameters are altered. The advantage of such experiments is their high physiological relevance. A disadvantage is the complexity of the changes which may influence 5HT metabolism and their debatable relationship to brain 5HT function, i.e. to the activity of 5HT neurones.

For example, the imposition of immobilization leads to two opposing influences upon the availability of tryptophan to the brain. Thus, pyrrolase induced through increased adrenocortical activity might tend to divert tryptophan from 5HT synthesis (Curzon and Green, 1969), while this will be opposed by increased availability of plasma trypto-phan to the brain mediated via sympathetic activity, cyclic AMP and increased plasma fatty acids (Knott and Curzon, 1972; Curzon *et al.*, 1973a).

The above opposing effects may well summate to give qualitative differences in different stresss situations or in animals of different strains or different previous life experiences. This may explain the widely different effects of stresses upon brain 5HT metabolism which have been reported. They may also provide a framework within which to consider the various indications of defective brain 5HT synthesis in endogenous depression (Coppen, 1972a). Thus, low availability of plasma tryptophan (Coppen, 1972b), high pyrrolase activity (Curzon and Bridges, 1970) and diminished response of the sympathetic system to stress (Perez-Reyes, 1969) would all tend to decrease the availability of tryptophan for brain 5HT synthesis.

REFERENCES

Aghajanian, G. K. (1972). Chemical feedback regulation of serotonin containing neurones in brain. *Ann. N.Y. Acad. Sci.*, **193**, 86–94

Altman, K. and Greengard, O. (1966). Correlation of kynurenine excretion with liver tryptophan pyrrolase levels in disease and after hydrocortisone induction. *J. Clin. Invest.*, **45**, 1527–34

Anden, N. E., Corrodi, H., Fuxe, K. and Hokfelt, T. (1968). Evidence for a central 5-hydroxytryptamine receptor stimulation by lysergic acid diethylamide. *Brit. J. Pharmac.*, **34**, 1–7

Aoki, K. and Siegel, F. L. (1970). Hyperphenylalaninemia: disaggregation of brain polyribosomes in young rats. *Science*, **168**, 129–30

Azmitia, E. C. and McEwen, B. S. (1969). Corticosterone regulation of tryptophan hydroxylase in midbrain of the rat. *Science*, **166**, 1274–76

Baughman, K. L. and Franz, J. M. (1971). Control of tryptophan oxygenase and formamidase activity in the gerbil. *Int. J. Biochem.*, **2**, 201–11

Carlsson, A. and Lindqvist, M. (1972). The effect of tryptophan and some psychotropic drugs on the formation of 5-hydroxytryptophan in the mouse brain *in vivo*. *J. Neural Transmission*, **33**, 23–43

Carlsson, A., Bedard, P., Lindqvist, M. and Magnusson, T. (1972). The influence of nerve-impulse flow on the synthesis and metabolism of 5-hydroxytryptamine. *Biochem. J.*, **128**, 70–71P

Chytil, F. (1971). Contribution to discussion. *Amer. J. Clin. Nutr.*, **24**, 709–10

Civen, N. and Knox, W. E. (1960). The specificity of tryptophan analogues as inducers, substrates, inhibitors and stabilizers of liver tryptophan pyrrolase. *J. Biol. Chem.*, **235**, 1716–18

Coppen, A. (1972a). Biogenic amines and affective disorders. *J. psychiat. Res.*, **9**, 163–75

Coppen, A., Eccleston, E. G. and Peet, M. (1972b). Total and free tryptophan concentration in the plasma of depressive patients. *Lancet*, **ii**, 1415–16

Coppen, A., Shaw, D. M., Herzberg, B. and Maggs, R. (1967). Tryptophan in the treatment of depression. *Lancet*, **ii**, 1178–80

Curzon, G. (1969). Tryptophan pyrrolase—a biochemical factor in depressive illness? *Brit. J. Psychiat.*, **115**, 1367–74

Curzon, G. and Bridges, P. (1970). Tryptophan metabolism in depression. *J. Neurol. Neurosurg. Psychiat*, **33**, 698–704

Curzon, G. and Green, A. R. (1968). Effect of hydrocortisone on rat brain 5-hydroxytryptamine. *Life Sci.*, **7**, 657–63

Curzon, G. and Green, A. R. (1969). Effects of immobilization on rat liver tryptophan pyrrolase and brain 5-hydroxytryptamine metabolism. *Brit. J. Pharmac.*, **37**, 689–97

Curzon, G. and Knott, P. K. (1973). Drugs influencing plasma and brain tryptophan. *Brit. J. Pharmac.*, **48**, 253–353P

Curzon, G., Joseph, M. H. and Knott, P. J. (1972). Effects of immobilization and food deprivation on rat brain tryptophan metabolism. *J. Neurochem.*, **19**, 1967–74

Curzon, G., Friedel, J. and Knott, P. J. (1973a). The effect of fatty acids on the binding of tryptophan to plasma protein. *Nature*, **242**, 198–200

Curzon, G., Kantamaneni, B. D., Winch, J., Rojas-Bueno, A., Murray-Lyon, I. M. and Williams, R. (1973b). Plasma and brain tryptophan changes in experimental acute hepatic failure. *J. Neurochem.*, **21**, 137–145

De Schaepdryver, A. F., Preziosi, P. and Scapagnini, U. (1969). Brain monoamines and stimulation or inhibition of ACTH release. *Arch. Int. Pharmacodyn.*, **180**, 11–18

Eccleston, D., Ashcroft, G. W. and Crawford, T. B. B. (1965). 5-Hydroxyindole metabolism in rat. A study in intermediate metabolism using the technique of tryptophan loading II. *J. Neurochem.*, **12**, 493–503

Feigelson, P. and Maeno, H. (1967). Studies on enzyme-substrate interactions in the regulation of tryptophan oxygenase activity. *Biochem. Biophys. Res. Commun.*, **28**, 289–293

7

Fernstrom, J. D. and Wurtman, R. J. (1971). Brain serotonin content: physio-logical dependence on plasma tryptophan levels. *Science*, **173**, 149–52

Fernstrom, J. D. and Wortman, R. J. (1972). Brain serotonin content: physio-logical regulation by plasma neutral amino acids. *Science*, **178**, 414–16

Fernstrom, J. D., Larin, F. and Wurtman, R. J. (1971). Daily variations in the concentrations of individual amino acids in rat plasma. *Life Sci.*, **10**, Pt 1, 813–19

Grahame-Smith, D. G. and Parfitt, A. G. (1970). Tryptophan transport across the synaptosomal membrane. *J. Neurochem.*, **17**, 1339–53

Green, A. R. and Curzon, G. (1970). The effect of tryptophan metabolites on brain 5-hydroxytryptamine metabolism. *Biochem. Pharmacol.*, **19**, 2061–18

Greengard, O., Mendelsohn, N. and Acs, G. (1966a). Effect of cytoplasmic par-ticles on tryptophan pyrrolase activity of rat liver. *J. Biol. Chem.*, **241**, 304–8

Greengard, P., Sigg, E. B., Fratta, I. and Zak, S. B. (1966b). Prevention and re-mission by adrenocortical steroids of nicotinamide deficiency disorder. *J. Pharmac.*, **154**, 624–31

Greengard, P., Kalinsky, H., Manning, T. J. and Zak, S. B. (1968). Prevention and remission by adrenocortical steroids of nicotinamide deficiency disease. II. A study of the mechanism. *J. Biol. Chem.*, **243**, 4216–21

Hardeland, R. and Rensing, L. (1968). Circadian oscillation in rat liver tryptophan pyrrolase and its analysis by substrate and hormone induction. *Nature*, **219**, 619–21

Hole, K. (1972). Reduced 5-hydroxyindole synthesis reduces postnatal brain growth in rats. *European J. Pharmac.*, **18**, 361–66

Hvitfelt, J. and Santii, R. S. (1972). Tryptophan pyrrolase in the liver of guinea pig: the absence of hydrocortisone induction. *Biochim. Biophys. Acta*, **258**, 358–65

Joseph, M. H. (1972). *Ph.D. Thesis*. University of London

Kim, J. H. and Miller, L. L. (1969). The functional significance of changes in activity of the enzymes, tryptophan pyrrolase and tyrosine transaminase, after induction in intact rats and in the isolated, perfused rat liver. *J. Biol. Chem.*, **244**, 1410–16

Knott, P. J. and Curzon, G. (1972). Free tryptophan in plasma and brain tryptophan metabolism. *Nature*, **239**, 452

Knott, P. J., Joseph, M. H. and Curzon, G. (1973). Effects of food deprivation and immobilization on tryptophan and other amino acids in rat brain. *J. Neurochem.*, **20**, 249–51

Knox, W. E. (1966). The regulation of tryptophan pyrrolase activity by tryptophan. *Adv. Enz. Reg.*, **4**, 287–97

Knox, W. E. and Auerbach, V. H. (1955). The hormonal control of tryptophan peroxidase in the rat. *J. Biol. Chem.*, **214**, 307–13

Lapin, I. P. and Oxenkrug, G. F. (1969). Intensification of the central seroton-inergic processes as a possible determinant of the thymoleptic effect. *Lancet.* **i**, 132–36

Lin, R. C., Ngai, S. H. and Costa, E. (1961). Lysergic acid diethylamide; role in conversion of plasma tryptophan to brain serotonin (5-hydroxytryptamine). *Science*, **166**, 237–39

Lipsett, D., Madras, B. K., Wurtman, R. J. and Munro, H. N. (1973). Serum tryptophan level after carbohydrate ingestion: selective decline in non-albumin-bound tryptophan coincident with reduction in serum free fatty acids. *Life Sciences*, **12**, Pt 2, 57–64

McArthur, J. N., Dawkins, P. D. and Smith, M. J. H. (1971). The displacement of L-tryptophan and dipeptides from bovine albumin in vitro and from human plasma in vivo by antirheumatic drugs. *J. Pharm. Pharmac.*, **23**, 393–98

Macon, J. B., Sokoloff, L. and Glowinski, J. (1971). Feedback control of brain 5-hydroxytryptamine synthesis. *J. Neurochem.* **18**, 323–31

McMenamy, R. H. and Oncley, J. L. (1958). The specific binding of L-tryptophan to serum albumin. *J. Biol. Chem.*, **233**, 1436–47

Meek, J. L. and Fuxe, K. (1971). Serotonin accumulation after monoamine oxidase inhibition. Effects of decreased impulse flow and of some antidepressants and hallucinogens. *Biochem. Pharmac.*, **20**, 693–706

Meek, J. L. and Neff, N. H. (1972). Tryptophan 5-hydroxylase: approximation of half-life and rate of axonal transport. *J. Neurochem.*, **19**, 1519–25

Meek, J. and Werdinius, B. (1970). Hydroxytryptamine turnover decreased by the antidepressant drug chlorimipramine. *J. Pharm. Pharmac.*, **22**, 141–3

Mehler, A. H. and Knox, W. E. (1950). The conversion of tryptophan to kynurenine in liver. I. The coupled tryptophan peroxidase-oxidase system forming formyl-kynurenine. *J. Biol. Chem.*, **187**, 419–30

Millard, S. A., Costa, E. and Gal, E. M. (1972). On the control of brain serotonin turnover rate by end product inhibition. *Brain Research*, **40**, 545–51

Moir, A. T. B. and Eccleston, D. J. (1968). The effect of precursor loading in the cerebral metabolism of 5-hydroxyindoles. *J. Neurochem.*, **15**, 1093–108

Munro, H. N. (1970). Free amino acid pools and their role. in *Regulation in Mammalian Protein Metabolism* (ed. Munro, H. N.), **4**, 299–387

Nishizuka, Y. and Hayaishi, O. (1963). Studies on the biosynthesis of nicotinamide adenine dinucleotide. *J. Biol. Chem.*, **238**, 3369–77

Oravec, M. and Sourkes, T. L. (1970). Inhibition of hepatic protein synthesis by α-methyl-DL-tryptophan *in vivo*. Further studies on the glyconeogenic action of α-methyltryptophan. *Biochemistry*, **9**, 4458–64

Perez-Reyes, M. (1969). Differences in the capacity of the sympathetic and endocrine systems of depressed patients to react to a physiological stress. *Pharmakopsychiatric-Neuropsychopharmakologie*, **2**, 245–51

Powanda, M. C. and Wannemacher, R. W. (1971). Tryptophan availability as a control of hepatic pyridine nucleotide concentration in mice. *Biochim. Biophys. Acta*, **252**, 239–45

Rapoport, M. I., Feigin, R. D., Bruton, J. and Beisel, W. R. (1966). Circadian rhythm for tryptophan pyrrolase activity and its circulating substrate. *Science*, **153**, 1642–44

Robison, G. A., Butcher, R. W. and Sutherland, E. W. (1971). Cyclic AMP, Academic Press; New York and London

Ross, D. S., Fernstrom, J. D. and Wurtman, R. J. (1972). Difference in the roles of dietary protein in generating the daily rhythms in hepatic tryptophan pyrrolase and tyrosine transaminase. *Fed. Proc.*, **31**, 900

Saavedra, J. M. and Axelrod, J. (1972). A specific and sensitive enzymatic assay for tryptamine in tissues. *J. Pharmac*, **182**, 363–69

Schimke, R. T., Sweeney, E. W. and Berlin, C. M. (1964). An analysis of the kinetics of rat liver tryptophan pyrrolase induction: the significance of both enzyme synthesis and degradation. *Biochem. Biophys. Res. Commun.*, **15**, 214–19

Semba, T. and Civen, M. (1970). Subcellular distribution of aromatic amino acid transaminases in rat brain. *J. Neurochem.*, **17**, 795–800

Shields, P. J. and Eccleston, D. (1972). Effects of electrical stimulation of rat mid-brain on 5-hydroxytryptamine synthesis as determined by a sensitive radio-isotope method. *J. Neurochem.*, **19**, 265–72

Sjoerdsma, A., Weissbach, H. and Udenfriend, S. (1956). A clinical, physiological and biochemical study of patients with malignant carcinoid (agentaffinoma). *Amer. J. Med.*, **20**, 520–32

Sourkes, T. L., Missala, K. and Oravec, M. (1970). Decrease of cerebral serotonin and 5-hydroxyindolylacetic acid caused by (–)-α-methyltryptophan. *J. Neurochem.*, **17**, 111–15

Tagliamonte, A., Tagliamonte, P., Perez-Cruet, J., Stern, S. and Gessa, G. L. (1971). Effect of psychotropic drugs on tryptophan concentration in the rat brain. *J. Pharmac.*, **177**, 475–80

Tagliamonte, A., Biggio, G., Vargiu, L. and Gessa, G. L. (1973). Free tryptophan in serum controls brain tryptophan level and serotonin synthesis. *Life Sciences*, **12**, Pt 2, 277–87

Thorson, A., Biorck, G., Bjorckman, G. and Waldenstrom, J. (1954). Malignant carcinoid of the small intestine. *Amer. Heart J.*, **47**, 795–817

Weiss, B. F., Munro, H. N. and Wurtman, R. J. (1971). L-Dopa: disaggregation of brain polysomes and elevation of brain tryptophan. *Science*, **173**, 833–35

Weissbach, H., King, W., Sjoerdsma, A. and Udenfriend, S. (1959). Formation of indole-3-acetic acid and tryptamine in animals. *J. Biol. Chem.*, **234**, 81–86

Wunner, W. H., Bell, J. and Munro, H. N. (1966). The effect of feeding with a tryptophan-free amino acid mixture on rat-liver polysomes and ribosomal ribonucleic acid. *Biochem. J.*, **101**, 417–28

Chapter 7

The Effect of Drugs on 5-Aminolaevulinate Synthetase and Other Enzymes in the Pathway of Liver Haem Biosynthesis

Francesco De Matteis
Biochemical Mechanisms Section, MRC Toxicology Unit,
Woodmansterne Road, Carshalton, Surrey

7.1. INTRODUCTION

Haem, the iron complex of protoporphyrin, is the essential component of the many cytochromes and haem-containing enzymes involved in the important functions of electron transport, oxidation and hydroperoxidation. Its biosynthetic pathway has been almost completely elucidated and some insight has also been obtained into the control mechanisms which regulate its biosynthesis. Both these aspects have been reviewed extensively (Lascelles, 1964; Marks, 1969; Tait, 1968; Granick and Sassa, 1971; Marver and Schmid, 1972); only those points which will facilitate the description and understanding of the effects of drugs will be summarized here.

The first intermediate in the pathway is an aminoketone, 5-aminolaevulinate (5-ALA) which results from the condensation of glycine with succinyl-CoA carried out by the enzyme 5-ALA synthetase. Two molecules of 5-ALA are then condensed with each other by the enzyme 5-ALA dehydratase to yield the monopyrrolic precursor porphobilinogen (PBG). Four molecules of PBG are then joined together by the uroporphyrinogen synthetase to produce a cyclic tetrapyrrole, uroporphyrinogen (or hexahydro-uroporphyrin), which is then decarboxylated up to the stage of coproporphyrinogen. This undergoes oxidative decarboxylation to protoporphyrin, the substrate for the last enzyme in the biosynthetic sequence, the chelatase, which inserts Fe^{2+} into protoporphyrin to yield haem. From knowledge of the distribution of the enzymes in the various fractions of the liver homogenate, one can postulate intracellular compartmentation of the various synthetic steps (Sano and Granick, 1961). The first reaction leading to the synthesis of 5-ALA and the last two steps leading to protoporphyrin and then to haem take place inside the mitochondrion, while the intermediary steps are apparently carried out in the soluble part of the cytoplasm.

There are two main reasons why the first enzyme of the pathway—the 5-ALA synthetase—is thought to play a key role in the regulation of haem biosynthesis. The first is that its activity appears to be rate-limiting in the overall pathway. This is best illustrated by the observations that an increased supply of 5-ALA will increase both *in vitro* and *in vivo* the liver synthesis of porphyrins and haem (Scott, 1955; Granick, 1966; Doss, 1969; De Matteis and Gibbs, 1972; Druyan and Kelly, 1972), while an increased supply of the precursors of 5-ALA will not. Thus, at least under normal conditions, the rate of haem synthesis appears to be determined mainly by the supply of 5-ALA. A comparison of the activities of the various enzymes of the pathway (as measured with liver preparations *in vitro*) is also compatible with the 5-ALA synthetase being the rate limiting step (Table 7.1). The activity of the uroporphyrinogen synthetase is also fairly low, however, and this may explain why under conditions where the supply of 5-ALA is excessive the uroporphyrinogen synthetase may become limiting and lead to accumulation of PBG (see page 200).

The second reason why 5-ALA synthetase is considered important in the regulation of the pathway is that this enzyme is the site where haem, the end product, exercises a negative feedback control on its own synthesis. It is not yet clear whether the feedback control exercised by haem on liver 5-ALA synthetase involves changes in the amount of enzyme formed by the protein synthesizing apparatus (end-product repression), changes in the activity of the enzyme (end-product inhibition) or both. Another possibility which has been suggested more recently (Kurashima *et al.*, 1970) is that haem may control the migration from the cytosol and assembly in the mitochondrion of a soluble precursor form of the 5-ALA synthetase. Experiments involving the use in chicken embryo liver cell cultures of differential inhibitors of protein synthesis (Sassa and Granick, 1970; Tyrrell and Marks, 1972; Strand *et al.*, 1972) suggest that the most important mechanism is repression of 5-ALA synthetase by haem at the translational stage of protein synthesis. However, a direct inhibition of the 5-ALA synthetase by haem has also been documented with partially purified preparations of the liver enzyme (Scholnick *et al.*, 1969; Whiting and Elliott, 1972). Since the haem-forming enzyme (the chelatase) is in close

TABLE 7.1

Comparison of the *In Vitro* Activities of the Various Enzymes of the Haem
Biosynthetic Pathway in the Liver of Normal Rodents
All enzymic activities are expressed as nmol of 5-ALA equivalents produced
or utilized per min by 1 g wet liver.

Enzyme	Rat	Mouse	References
5-ALA synthetase	0.70*	0.6–1.4†	Strand *et al.*, 1970 Gross and Hutton, 1971
5-ALA dehydratase	60.0*	50–160†	De Matteis and Gibbs, 1972 Doyle and Schimke, 1969
Uroporphyrinogen synthetase	0.7*	<2.0 3.0	Strand *et al.*, 1970 Hutton and Gross, 1970 Miyagi *et al.*, 1971
Uroporphyrinogen decarboxylase	0.8	3.0–7.0	Taljaard *et al.*, 1971 Romeo and Levin, 1971
Coproporphyrinogen oxidase	11.5		Shanley *et al.*, 1970
Ferrochelatase	190.0 67*		Jones and Jones, 1969 De Matteis *et al.*, 1973

* Values obtained with liver preparations from fasted animals.
† Range of values represent variations observed among different strains.

association with the 5-ALA synthetase at or near the mitochondrial
inner membrane (McKay *et al.*, 1969; Zuyderhoudt *et al.*, 1969;
Jones and Jones, 1969), the possibility should be considered that end-
product inhibition may also play a part in the regulation of 5-ALA
synthetase.

From the turnover rates of the haemoprotein haems and from their
respective concentrations, the amount of haem which needs to be synthesized to keep a steady state of haemoproteins can be calculated
(Marver and Schmid, 1972). From these calculations it appears that
the amount of haem required is of the same order as that expected
from the activity of 5-ALA synthetase and also that cytochrome P-450
accounts for more than half the total haem requirement of the liver
cell. This latter point is of particular significance when considering the
effect of drugs on haem metabolism in the liver as many drugs can cause
large and relatively rapid changes in the concentration of cytochrome
P-450 (see page 198).

The utilization of haem for the synthesis of the various haemoproteins must be coordinated in some way with the formation of their
specific protein moieties, but very little is known of the mechanism by

which these two processes are integrated. There is evidence from *in vitro* experiments that, at least in the case of haemoglobin, haem may stimulate the synthesis of the polypeptide chains of globin at the ribosomal site and also assist their assembly (Zucker and Schulman, 1968; Maxwell and Rabinovitz, 1969; Tavill *et al.*, 1972). It does not appear likely, however, that an increased supply of haem is the main controlling factor in stimulating the rate of synthesis of the liver haemoproteins since this would not provide a satisfactory explanation for the selectivity of the response (for example, after administering a drug) which can concern only one or few liver haemoproteins. It is more likely that the main controlling factor is the synthesis of the specific apoproteins and that this in turn will increase haem utilization and therefore haem synthesis. This view appears to be consistent with the finding of a free apoprotein pool which has been reported for cytochrome c (Kadenback, 1970), cytochrome b_5 (Bock and Siekevitz, 1970; Hara and Minakami, 1970; Negishi and Omura, 1970) and catalase (Lazarow and De Duve, 1971).

A point which is not yet clear is whether the utilization of haem for the synthesis of haemoproteins is an essentially irreversible process or whether, and to what extent, exchange between free haem (for which some evidence has recently been obtained (Druyan and Kelly, 1972)) and haemoprotein haem may occur *in vivo*. In animals treated with cycloheximide the labelling from radioactive 5-ALA of total liver haem (Levitt *et al.*, 1968; Garner and McLean, 1969a) and of the haems of pre-existing haemoproteins (Druyan *et al.*, 1973) still occurs when liver protein synthesis is almost completely inhibited. This finding may indicate a rapid exchange between newly synthesized free haem and the haem of pre-existing haemoproteins; although it can also be interpreted to indicate progressive labelling by radioactive haem of the pre-existing apoprotein pools. More direct evidence for *in vivo* haem exchange has been obtained by Bock and Siekevitz (1970) for the microsomal cytochrome b_5.

The ultimate fate of haem is its degradation to bile pigments and other degradation products which are then eliminated from the liver. This organ is the source of a substantial portion of the "early labelled bilirubin" produced from haem by the haem oxygenase system of the microsomes (Tenhunen *et al.*, 1969). It is not yet known to what extent this hepatic "early labelled bilirubin" is produced from turnover of the haemoprotein haem (Schmid *et al.*, 1966) or from the degradation of free haem. There is some evidence that under conditions where supply of liver haem may exceed the requirements for the formation of the haemoproteins (for example, after marked inhibition of liver protein synthesis by cycloheximide or after administering large amounts of exogenous 5-ALA), the excess haem is degraded to bile pigments (Levitt *et al.*, 1968; Song *et al.*, 1971). Therefore, degradation of haem can also be visualized as an "overflow" pathway which ensures that the intrahepatic concentration of free haem will not become too large.

The main regulatory aspects of haem synthesis, utilization and degradation, which have been discussed above can be brought together in a model of the regulation of the pathway (Fig. 7.1) centred on the feedback control exercised by haem at the level of 5-ALA synthetase. On account of the uncertainties which still exist on the detailed mechanism of this feedback control, it is not yet possible to define the identity and intracellular localization of the pool of haem which is responsible

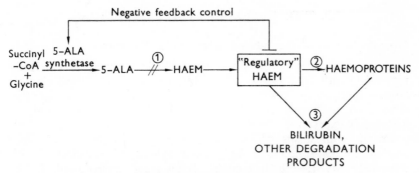

Fig. 7.1. Postulated regulatory mechanism.
(from De Matteis, 1973a).

for the regulation of 5-ALA synthetase. On purely theoretical grounds, this "regulatory" haem can be visualized as a haem pool of relatively small size and rapid rate of turnover into which newly synthesized haem is fed and out of which haem is drawn for either the synthesis of haemoproteins or for degradation. It could be free haem or a pool in rapid equilibrium with free haem. Accordingly, a decrease in the concentration of the "regulatory" haem and a consequent stimulation of 5-ALA synthetase can be brought about by one or more of the following mechanisms: 1. inhibition of the synthesis of haem; 2. increase in the rate of utilization of haem; or 3. increase in the rate of haem degradation. Possible examples of each of these mechanisms will be provided below, in the discussion of the effect of drugs on liver 5-ALA synthetase.

7.2. DRUGS AND HAEM BIOSYNTHESIS
IN THE LIVER OF RODENTS

Many drugs stimulate the activity of the haem biosynthetic pathway in the liver of rodents. A clear distinction can be made between drugs which increase the concentration of haem in the liver (usually the haem of the microsomal cytochromes P-450 and b_5) without causing accumulation of the intermediates of the pathway, and drugs which induce the experimental porphyrias where these intermediates accumulate in excess. These two responses probably reflect different mechanisms of action of the drugs and will be considered separately. The main purpose of this paper is to discuss in the light of recent findings the possible mechanisms involved in these effects of drugs on the liver of the intact animal; the reader is referred to several reviews for a detailed description of the two responses (Conney, 1967; De Matteis, 1967; Goldberg, 1968; Tschudy and Bonkowsky, 1972).

7.3. THE EXPERIMENTAL PORPHYRIAS

Over-production of the Intermediates of the Pathway

Under normal conditions the biosynthetic pathway is efficiently regulated, with little waste of the intermediates. In porphyria this control mechanism breaks down and far more precursors are synthesized than are turned into haem, so that they accumulate and are excreted in excess.

Examples of drugs which induce porphyria in the experimental animal are as follows: 2-allyl-2-isopropylacetamide (AIA), 3,5-di-ethoxycarbonyl-1,4-dihydrocollidine (DDC) and griseofulvin. In the porphyria caused by any of these agents the distribution of the porphyrins and of their precursors within the body, as well as metabolic studies conducted *in vitro* and other considerations (Nakao *et al.*, 1967), indicate that the liver is the main site, if not the only site, of production of these pigments. The effect of feeding griseofulvin to the mouse is shown in Table 7.2. The total amount of intermediates of the pathway which are excreted in 24 hr is some 40-fold in excess of the amount of haem required for the synthesis of the haemoproteins of the liver, indicating clearly that the increased excretion of haem precursors cannot be merely due to a block in their utilization, but must represent a large increase in the rate of their production by the liver.

An essentially similar picture of over-production of pigments in the liver is observed with the other porphyrogenic drugs, with the important difference that the main intermediate which is excreted and accumulates in the liver is protoporphyrin with DDC and griseofulvin, while with AIA, PBG and 5-ALA predominate. The reason for these differences in the biochemical picture will be considered in the discussion of the mechanism of action of DDC and griseofulvin.

In agreement with the concept that in porphyria there is an increased production of intermediates in the liver, Granick and Urata (1963) first showed that the activity of the rate-limiting enzyme of the pathway, 5-ALA synthetase, was markedly increased in the liver of animals treated with DDC; a finding later extended to hepatic porphyria

TABLE 7.2

The Effect of Feeding Griseofulvin (2.5% in the Diet)
on the Porphyrin Metabolism of the Mouse
The amounts of haem precursors excreted on the fifth
day of the experiment are all expressed as haem equivalent in nmol/24 hr per mouse of 25 g body wt.*

	Control	Porphyric
Faeces		
Protoporphyrin	17	2160
Coproporphyrin	8	250
Urine		
Coproporphyrin	0.2	2
Porphobilinogen	1.9	30
5-Aminolaevulinate	12.1	109
Total haem precursors	——	——
excreted in 24 hr	39.2	2551

* Haem required for the synthesis of liver haemoproteins by a normal mouse of 25 g body wt would
be expected to be about 65 nmol/24 hr (Marver and
Schmid, 1972).

caused by 2-allyl-2-isopropylacetamide (Marver *et al.*, 1966a) and
griseofulvin (Nakao *et al.*, 1967). The activity of the two subsequent
enzymes in the pathway, 5-ALA dehydratase and uroporphyrinogen
synthetase, have also been reported to be increased in the liver of
animals rendered porphyric with these drugs (Gibson *et al.*, 1955;
De Matteis and Gibbs, 1972; Miyagi *et al.*, 1971; Strand *et al.*, 1970);
these increases are, however, relatively small and may be secondary to
the stimulation of the 5-ALA synthetase and to consequent accumulation of 5-ALA and PBG in the liver (cf. Onisawa and Labbe, 1962).

Loss of Liver Haem in Porphyria

The mechanism by which porphyrogenic drugs stimulate the activity
of 5-ALA synthetase is not yet known. Granick (1966) suggested that
these drugs increased the activity of 5-ALA synthetase by interfering
in some way with the feedback control of haem. A possible mechanism
by which griseofulvin, AIA and DDC might interfere with the regulation
by haem of 5-ALA synthetase is provided by the recent findings from
several laboratories (Wada *et al.*, 1968; Waterfield *et al.*, 1969) De
Matteis, 1970; Meyer and Marver, 1971; Sweeney *et al.*, 1971; Satyanarayana Rao *et al.*, 1972) that these drugs all cause a transient loss of
microsomal cytochrome P-450 and haem from the liver, coincidental
with the rise in activity of 5-ALA synthetase.

In the case of 2-allyl-2-isopropylacetamide the loss of liver haem is due to increased destruction and conversion into certain unidentified degradation products (De Matteis, 1971a; Meyer and Marver, 1971; Landaw et al., 1970; Levin et al., 1972). The allyl group in the molecule of the drug appears to be important for both destruction of liver haem and stimulation of 5-ALA synthetase activity, probably after conversion, by way of drug metabolism, into a reactive derivative (Abbritti and De Matteis, 1971–72; Levin et al., 1973). Although the microsomal fraction is the main site of haem destruction, a loss of radioactivity is also observed from rapidly labelled haem in the cytosol and in the crude mitochondrial fraction of the liver homogenate of treated animals (De Matteis, 1973a).

In the case of DDC there is also some degree of liver haem destruction (Abbritti and De Matteis, 1973), but inhibition of liver haem synthesis may also be important in causing the loss of liver haem and the stimulation of 5-ALA synthetase activity. A rapid inhibition of the chelatase activity is caused by the drug in liver mitochondria, before any increase in the activity of the synthetase becomes apparent (De Matteis and Gibbs, 1972); mice which are more sensitive than rats to the stimulation of the synthetase by this drug, are also more sensitive to the inhibition of the chelatase (De Matteis et al., 1973). In addition, the oxidative analogue of DDC, 3,5-diethoxycarbonyl-collidine (DC) which had been reported not to cause porphyria in mice (Solomon and Figge, 1960) and in guinea-pigs (Marks et al., 1965), when given to rats does not cause great changes in either 5-ALA synthetase or chelatase activity (Table 7.3).

Inhibition of liver haem synthesis may also be important in the stimulation of 5-ALA synthetase and in the production of porphyria by griseofulvin and related drugs. There exists a good correlation between stimulation of the synthetase and inhibition of the mito-chondrial chelatase after administration of drugs of this group (De Matteis and Gibbs, unpublished). In mice, isogriseofulvin is more active than griseofulvin in inhibiting the chelatase and it also causes a greater stimulation of the synthetase; whereas HET-griseofulvin (the 2'-β-hydroxyethyl thioether analogue of griseofulvin), which is not porphyrogenic, does not inhibit the chelatase. Furthermore, griseo-fulvin inhibits the chelatase much more effectively in the mouse where it causes a marked stimulation of the synthetase, than in the rat where it causes only a slight increase in 5-ALA synthetase activity.

As originally suggested by Onisawa and Labbe (1963) and by Lock-head et al. (1967), a block in haem synthesis at the level of the chelatase provides an explanation for the accumulation of large amounts of protoporphyrin which are seen in DDC and griseofulvin porphyria. This interpretation is strongly supported by the finding in animals treated with either drug of a linear correlation between the ratios of

TABLE 7.3

Effect of Administering a Single Dose of Either 3,5-Diethoxycarbonyl-1,4-Dihydrocollidine (DDC) or of 3,5-Diethoxycarbonyl-Collidine (DC) on the Concentration of Total Porphyrins and on the 5-ALA Synthetase and Chelatase Activities of Rat Liver. Comparison Between Adult and Newborn Rats.

Rats of the Porton strain were injected interperitoneally with either arachis oil or drugs, and killed 4 hr later (exper. a) or 5 hr later (exper. b and c). Adult rats were starved for 24 hr before injections. Newborn rats were less than 24 hr old and the livers from 10 animals were pooled for each observation. Results are expressed as means ±S.E.M. of the number of observations in parenthesis or as the values obtained in individual observations.

Experiment	Age of rat	Treatment and dose (ml/kg or mg/kg)	5-ALA synthetase activity (nmol/min/g liver)	Total porphyrins (nmol/g liver)	Porphyrin–metal chelatase activity (nmol of Co^{2+}-mesoporphyrin formed/min per mg of mitochondrial protein)
(a)	adult (160–180 g)	Oil, 10 ml	1.2, 1.7	0.08, 0.16	0.90, 0.78
		DC, 100 mg	—	<0.06	1.0, 1.0
		DC, 250 mg	2.3, 3.1	0.18, 0.38	0.77, 0.78
		DC, 500 mg†	2.4, 1.5	0.17, 0.09	0.84, 0.84
		DC, 750 mg†	1.9, 2.0	0.18, 0.09	0.88, 0.81
(b)*	adult (160–180 g)	Oil, 10 ml	1.44 ± 0.08 (3)	0.13 ± 0.02 (10)	1.25 ± 0.09 (4)
		DDC, 100 mg	10.31 ± 1.08 (7)	10.68 ± 1.30 (12)	0.26 ± 0.01 (5)
(c)	newborn	Oil, 10 ml	3.51 ± 0.69 (4)	0.18 ± 0.05 (4)	0.94 ± 0.08 (4)
		DDC, 100 mg	2.34 ± 0.32 (4)	0.41 ± 0.06 (4)	0.66 ± 0.02 (4)

* Data from De Matteis et al., 1973. Similar changes in enzyme activities and porphyrin concentration were found 4 hr after DDC.
† These doses caused hypnosis and lowering of body temperature.

the activities of chelatase and 5-ALA synthetase on the one hand and the reciprocal of the liver porphyrin concentration on the other (De Matteis *et al.*, 1973; De Matteis and Gibbs, unpublished).

Possible Role of Drug-metabolism and of Protein Synthesis in Porphyria Caused by Drugs

The stimulation of 5-ALA synthetase caused by DDC can be prevented by prior administration to the rat of either an inhibitor of drug-metabolizing enzymes (SKF 525-A) or of an inhibitor of protein synthesis (cycloheximide). An important difference is that SKF 525-A also prevents the inhibition of the chelatase, whereas cycloheximide does not (De Matteis *et al.*, 1973). Similar results have been obtained with AIA: here again either inhibitor could prevent the stimulation of the 5-ALA synthetase, but only SKF 525-A could also prevent the destruction of liver haem (De Matteis, 1971a; Satyanarayana Rao *et al.*, 1972).

These results suggest that if a decrease in the concentration of liver haem is implicated in the stimulation of 5-ALA synthetase by AIA and DDC, two successive stages can be distinguished in the induction process (Fig. 7.2): a first stage, which is affected by SKF 525-A, leading to

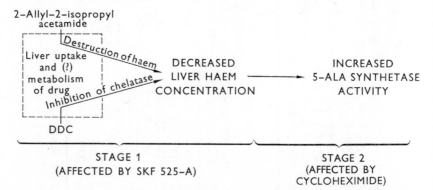

Fig. 7.2. Schematic representation of the induction of 5-ALA synthetase by porphyrogenic drugs. Possible role of drug metabolism and of protein synthesis.

(De Matteis, 1973b).

loss of liver haem through increased destruction and/or inhibition of synthesis; and a second stage, inhibited by cycloheximide, leading to increased activity of 5-ALA synthetase and to porphyria. The effect of SKF 525-A could be explained by an interference with the liver uptake of the porphyria-inducing drugs, but Satyanarayana Rao *et al.* (1972) have shown that, at least in the case of AIA, this does not seem to be the case. Alternatively, the effect of SKF 525-A may indicate that both AIA and DDC need to be converted by way of microsomal drug metabolism to some active derivative in order to cause destruction of liver haem or inhibition of the chelatase, respectively. Compatible with this possibility is some evidence obtained *in vitro* with AIA (De Matteis, 1971a), and in the case of DDC the finding that in newborn rats, which have a low activity of liver drug-metabolizing enzymes (Kato *et al.*, 1964), DDC will not increase the activity of 5-ALA synthetase (see also Woods and Dixon, 1972) and will only cause a slight increase in liver porphyrins and a slight inhibition of the chelatase (Table 7.3). If metabolism of AIA, DDC and griseofulvin to some active derivative was important for the destruction of haem and for the inhibition of the chelatase, then this might provide an explanation for the structural requirements for activity, which have been described in all three classes of compounds (Goldberg and Rimington, 1955; Marks *et al.*, 1965; De Matteis, 1966) and also for the species differences in response to these drugs, which can be very marked (De Matteis *et al.*, 1973). On the basis of results obtained in the chick embryo, Marks and co-workers favour, however, the opposite view that an increased rate of metabolism of either AIA or DDC may result in a decreased (rather than increased) porphyrogenic activity (Racz and Marks, 1972; Marks *et al.*, 1973).

The ability of cycloheximide and of other inhibitors of protein synthesis to prevent the induction of 5-ALA synthetase has been known for some time and the simplest and most likely explanation for this effect (Granick, 1966) is that the changes in 5-ALA synthetase activity reflect changes in the amount of the enzyme protein which is synthesized *de novo* under the influence of the porphyrogenic drugs. It cannot be excluded, however, that protein synthesis may also be related to the activity of the enzyme in some other, less direct way; for example, if 5-ALA synthetase was subjected to direct inhibition by a free haem pool *in vivo*, cycloheximide, by inhibiting the synthesis of the apoprotein of the cytochromes, might conceivably increase the concentration of free haem and lead to an inhibition of the enzyme. Whiting and Elliot (1972) have recently obtained some more direct evidence that the amount of 5-ALA synthetase may be increased in the liver of animals given DDC.

Factors which Influence the Response of 5-ALA Synthetase to Drugs

In the light of the two-stage model of induction of 5-ALA synthetase (illustrated in Fig. 7.2) it is possible to discuss the site of action of certain factors which can influence the degree of induction of the enzyme by drugs. Iron and starvation can potentiate the effect of porphyrogenic drugs (Del Favero *et al.*, 1968; Stein *et al.*, 1970; Rose *et al.*, 1961); glucocorticoid hormones are essential for the induction of 5-ALA synthetase by AIA, both in the whole animal (Marver *et al.*, 1966b) and in the isolated perfused liver (Bock *et al.*, 1971); and glucose in large doses inhibits the induction of the enzyme (Tschudy *et al.*, 1964). Glucocorticoid hormones are known to induce certain liver enzymes and to stimulate the liver formation of both RNA and proteins, whereas glucose (or a high carbohydrate diet) have been reported to inhibit the induction of several liver enzymes. It is therefore likely that both hydrocortisone and glucose act at the second stage of 5-ALA synthetase induction, namely, that requiring synthesis of protein. This is also suggested by the finding (Satyanarayana Rao *et al.*, 1972) that glucose, like cycloheximide, inhibits the induction of 5-ALA synthetase by AIA without preventing the loss of cytochrome P-450. The following lines of evidence suggest, on the other hand, that iron and perhaps also (at least in part) starvation may potentiate the effect of drugs by acting at the first stage of the induction of 5-ALA synthetase, namely, that leading to loss of haem in the liver: (*a*) in rats, the administration of iron dextran is followed 24 hr later by a 35% loss of microsomal haem and by a corresponding three-fold increase in 5-ALA synthetase activity (De Matteis and Sparks, 1973); (*b*) iron can also stimulate *in vitro* the NADPH-dependent loss of microsomal haem, whereas the addition of iron chelators completely prevents the NADPH-dependent loss of haem (Schacter *et al.*, 1972; De Matteis and Sparks, 1973); (*c*) finally, microsomes isolated from starved animals, when incubated in presence of NADPH, degrade both exogenous and endogenous haem at a significantly greater rate than microsomes from fed animals (Bakken *et al.*, 1972; De Matteis and Sparks, 1973).

Conclusions

The results described above are compatible with the hypothesis that a reduction in the concentration of liver haem may, by reducing its feedback control on 5-ALA synthetase, lead to a stimulation of the activity of the enzyme. The main mechanism involved may be inhibition of haem synthesis with DDC and griseofulvin, and increased haem degradation with AIA. As discussed elsewhere (De Matteis, 1973a) the "regulatory" haem need not be cytochrome P-450 haem. Indeed there are cases (for example after simultaneous feeding of DDC and phenylbutazone (De Matteis and Gibbs, 1972)) where the activity of the synthetase is markedly raised but the level of cytochrome P-450 is increased rather than decreased. It appears more likely that some other pool(s) of haem in the liver (perhaps rapidly turning-over free haem) may be concerned with the control of the enzyme and that the early loss of cytochrome P-450 haem described above after administration of AIA and other porphyrogenic drugs may be accompanied by a decrease in the concentration of this "regulatory" pool of haem.

There are two lines of evidence, however, which are difficult to

TABLE 7.4

Effect of a Single Dose of Cobaltous Chloride, Given on Its Own or in Combination with Either Phenylbutazone or with DDC, on the Concentration of Cytochrome P-450 and on the Activity of 5-ALA Synthetase of Rat Liver

Male rats of the Porton strain were fasted for 24 hr, then treated and killed 17 hr later. Cytochrome P-450 was estimated in isolated microsomes and 5-ALA synthetase activity in liver homogenates. Results are expressed as means ±S.E.M. of 3 observations or as the values obtained in individual observations. The dose of cobaltous chloride refers to the hydrated salt $(CoCl_2.6H_2O)$.

Experiment	Administered by S.C. injection (ml/kg or mg/kg)	Administered orally (ml/kg or mg/kg)	Liver wet wt (g/100 g body wt)	5-ALA synthetase (nmol/min) per total liver of 100 body wt	Cytochrome P-450 (nmol) per total liver of 100 body wt
(a)	Saline, 10 ml	Oil, 10 ml	3.4, 3.3	12.5, 10.6	98.0, 97.0
	Saline, 10 ml	phenylbutazone, 150 mg	3.7, 3.7	10.3, 19.0	141.4, 143.8
	$CoCl_2$, 60 mg	Oil, 10 ml	4.3, 3.8	14.8, 8.7	52.7, 55.5
	$CoCl_2$, 60 mg	phenylbutazone, 150 mg	4.1, 4.0	11.0, 12.1	48.0, 33.8
(b)	Saline, 10 ml	Oil, 10 ml	3.25 ± 0.09	10.37 ± 1.35	—
	Saline, 10 ml	DDC, 200 mg	3.51 ± 0.06	38.56 ± 7.6	—
	$CoCl_2$, 60 mg	Oil, 10 ml	4.36 ± 0.14	8.65 ± 1.8	—
	$CoCl_2$, 60 mg	DDC, 200 mg	4.04 ± 0.22	28.16 ± 1.7	—

reconcile with the idea that a decrease in the concentration of liver haem is important for the stimulation of 5-ALA synthetase by porphyrogenic drugs. The first of these is provided by the effect of administering phenobarbitone to rats before AIA. When rats are pretreated with phenobarbitone, AIA causes more degradation of microsomal haem (De Matteis, 1971a) but less stimulation of 5-ALA synthetase (Kaufman *et al.*, 1970). No satisfactory explanation for this discrepancy can as yet be offered.

The other apparent exception is provided by effects on the liver of cobaltous chloride. In agreement with the work of Tephly and Hibbeln (1971), this metal causes a rapid loss of liver cytochrome P-450 and prevents the rise of this haemoprotein in response to an inducing drug, like phenylbutazone (Table 7.4,*a*). Cobalt does not stimulate 5-ALA synthetase activity, however (see also Tephly, 1973), nor does it increase the liver concentration of porphyrins (results not shown). The lack of response of the enzyme cannot be due—at least not entirely—to an aspecific toxic effect on the liver, since cobalt does not prevent the rise in 5-ALA synthetase activity due to DDC (Table 7.4,*b*). If the effect of cobalt on the cytochrome is due to a block in haem synthesis (Tephly *et al.*, 1973) (presumably because of a preferential incorporation of Co^{2+}, rather than Fe^{2+} into protoporphyrin by the chelatase (Labbe and Hubbard, 1961)), then it is difficult to see how a decrease in liver haem concentration large enough to block the synthesis of the major liver haemoprotein does not result in a stimulation of 5-ALA synthetase, unless the cobalt chelate of protoporphyrin is still able to exercise a feedback control on the enzyme even though it cannot be utilized for the formation of the haemoproteins.

For these reasons, although there are several lines of evidence which suggest that loss of liver haem may be implicated in the stimulation of 5-ALA synthetase by AIA and other porphyrogenic drugs, no definite conclusions about this can as yet be drawn.

7.4. THE INDUCTION OF MICROSOMAL CYTOCHROME P-450 BY DRUGS

In clear contrast with the porphyrogenic drugs considered so far, a second group of drugs, including phenobarbitone, phenylbutazone and many other lipid-soluble compounds, increases rather than decreases the concentration of haem in the liver without causing accumulation of porphyrins. The increase in liver haem content is accounted for by an accumulation of haemoproteins, particularly the cytochrome P-450 of the microsomes. Although an increase in the liver content of other haemoproteins has also been described after administration of drugs (a rise in catalase after feeding ethyl-α-p-chlorophenoxyisobutyrate (Reddy *et al.*, 1971) and an increase in the microsomal cytochrome b_5

after administration of several drugs), the accumulation of microsomal cytochrome P-450 is by far the most important response: it is the most rapid and pronounced and, on account of the quantitative importance of this haemoprotein, the most likely to result in an increased demand for haem.

Synthesis of Liver Haem and Control of Cytochrome P-450 Formation

A stimulation of both the activity of 5-ALA synthetase and rate of liver haem synthesis precedes the increase in cytochrome P-450 after phenobarbitone (Marver, 1969; Baron and Tephly, 1969a, 1970). Baron and Tephly (1969b) have also reported that the accumulation of cytochrome P-450 caused by phenobarbitone in rat liver can be prevented by administration of large doses of 3-amino-1,2,4-triazole, an inhibitor of hepatic haem synthesis. These findings have all been interpreted to indicate that an increased supply of haem is the primary event leading to the accumulation of the microsomal haemoprotein (Baron and Tephly, 1970).

As discussed elsewhere (De Matteis and Gibbs, 1972), there are several reasons why this interpretation should be viewed with caution. Firstly, the results obtained by Baron and Tephly (1969b) with 3-amino-1,2,4-triazole may merely indicate that haem, as a constituent of cytochrome P-450, is essential for the formation of the haemoprotein and that, under conditions in which the synthesis of haem is inhibited, haem supply may become rate-limiting. Secondly, after single or repeated doses of drugs a considerable accumulation of cytochrome P-450 can also be observed without great changes in either 5-ALA synthetase activity or the rate of haem synthesis *in vitro* (Bock *et al.*, 1971; De Matteis, 1971b; Song *et al.*, 1971). Finally, it has been found (De Matteis, 1971b; Song *et al.*, 1971; Druyan and Kelly, 1972) that the administration of exogenous 5-ALA to rats does not result in accumulation of liver cytochrome P-450 even though it causes increased formation of haem and bile pigments; this also suggests that an increased supply of 5-ALA and haem is probably not sufficient in itself to stimulate the formation of the cytochrome.

Increased Haem Utilization for the Synthesis of Cytochrome P-450

An alternative explanation is possible for the increased activity of 5-ALA synthetase observed after administration of drugs which stimulate formation of cytochrome P-450. These drugs may act primarily by increasing the amount of the apoprotein of cytochrome P-450 (De Matteis and Gibbs, 1972) and thereby stimulating the rate of haem utilization. In labelling experiments phenobarbitone has been found to increase the radioactivity recovered in the total haem of partially purified preparations of cytochrome P-450 not only from radioactive glycine, but also from radioactive 5-ALA (Song *et al.*, 1971; Levin and Kuntzman, 1969; Garner and McLean, 1969a; 1969b; Greim *et al.*, 1970). Since this latter precursor by-passes the rate-limiting step of the pathway and since the labelling of the haem of cytochrome b_5 was not increased in these experiments (Greim *et al.*, 1970), the greater incorporation of radioactive 5-ALA into cytochrome P-450 haem suggests that phenobarbitone stimulates the utilization of haem for the synthesis of cytochrome P-450.

An increased utilization of haem (mechanism No. 2, in Fig. 7.1) will in turn result in a lowering of the concentration of the "regulatory" pool of haem and therefore in a stimulation of the activity of 5-ALA synthetase. Whether a measurable stimulation of the enzyme will actually occur (after phenobarbitone or a similar drug), and to what extent, will presumably depend on the balance between the rate of haem utilization and the rates of the other two pathways of haem metabolism which control the levels of the "regulatory" haem (Fig. 7.1), namely haem synthesis and degradation. The same stimulus to increased haem utilization by phenobarbitone might be expected to result in a greater depletion of the "regulatory" haem if, at the same time, either the synthesis of haem is impaired or the rate of haem degradation is increased (greater than if both the rate of haem synthesis and rate of haem degradation were normal). There is experimental evidence compatible with the main features of this hypothetical model schematically illustrated in Fig. 7.1.

Under normal conditions, when there is no excessive degradation of haem and the synthesis of haem is unimpaired, a considerable accumulation of cytochrome P-450 is observed after phenylbutazone or phenobarbitone, with either small or no changes in the activity of 5-ALA synthetase. This suggests (De Matteis, 1971b) that the amount of 5-ALA (and haem) which is normally made by the liver is sufficient to meet the increased demand for enhanced P-450 formation (without big changes in the level of "regulatory" haem) or that, if an increased quantity of 5-ALA must be made, it need only be a small amount. When, however, liver haem synthesis is inhibited by DDC, an inducer of cytochrome P-450 (such as phenylbutazone or phenobarbitone) causes a marked stimulation of 5-ALA synthetase, well above that seen when DDC is given on its own (De Matteis and Gibbs, 1972, De Matteis, 1972): the production of 5-ALA is then so large that the two subsequent enzymes, 5-ALA dehydratase and uroporphyrinogen synthetase, become rate-limiting and both 5-ALA and PBG accumulate in excess in the liver (Fig. 7.3). Similarly, phenobarbitone can stimulate markedly the activity of 5-ALA synthetase, when the rate of liver haem destruction is increased by AIA (De Matteis, 1973c). Dose-response experiments suggest that the induction of 5-ALA synthetase by DDC and its potentiation by phenylbutazone may be produced through different

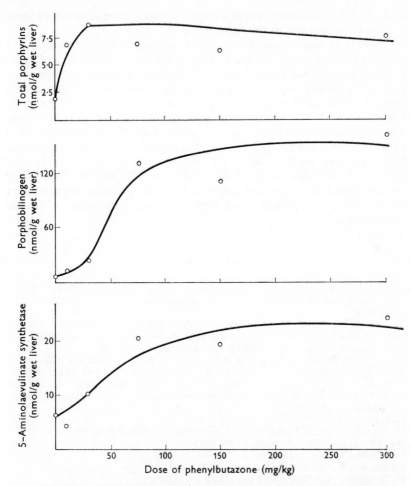

Fig. 7.3. Potentiation by phenylbutazone of porphyria caused by DDC.
Male rats of the Porton strain were fasted for 24 hr, then given DDC
(100 mg/kg) intraperitoneally and at the same time either oil (10 ml/kg)
or increasing doses of phenylbutazone orally. They were killed 15–16 hr
later. Results are the averages of two observations.

mechanisms. In addition, this potentiation effect has recently been
confirmed in the isolated perfused liver (Bock *et al.*, 1973) suggesting
that extra-hepatic factors, like hormones, are not likely to be responsible
for it. As discussed above, the most likely interpretation for these
findings is that drugs can lead to a decrease in the level of "regulatory"
haem and to a stimulation of 5-ALA synthetase by different mechanisms

and that the potentiation effect is obtained when more than one of these mechanisms operates at the same time.

Alternative explanations for this potentiation effect, though less likely, cannot be discounted. For example, the potentiation may be in some way related to the property, which both phenobarbitone and phenylbutazone possess, of increasing the liver size and its protein content (De Matteis and Gibbs, 1972) or to changes caused by these latter drugs in the distribution and metabolism of porphyrogenic compounds. Therefore, even though most of the experimental findings obtained so far are compatible with the model of regulation which has been discussed in this paper, all the evidence in its favour is only indirect and cannot be taken as conclusive. The model should only be considered as a working hypothesis which provides a convenient background for an experimental approach to the many points which are still obscure.

REFERENCES

Abbritti, G. and De Matteis, F. (1971–72). *Chem.-Biol. Interactions*, 4, 281–286
Abbritti, G. and De Matteis, F. (1973). *Hoppe-Seyler's Z. Physiol. Chem.*, 354, 849–850
Bakken, A. F., Thaler, M. M. and Schmid, R. (1972). *J. Clin. Invest.*, 51, 530–536
Baron, J. and Tephly, T. R. (1969a). *Biochem. Biophys. Res. Commun.*, 36, 526–532
Baron, J. and Tephly, T. R. (1969b). *Mol. Pharmacol.*, 5, 10–20
Baron, J. and Tephly, T. R. (1970). *Arch. Biochem. Biophys.*, 139, 410–420
Bock, K. W., Krauss, E. and Fröhling, W. (1971). *Europ. J. Biochem.*, 23, 366–371
Bock, K. W. and Siekevitz, P. (1970). *Biochem. Biophys. Res. Comm.*, 41, 374–380
Bock, K. W., Weiner, R. and Fröhling, W. (1973). *Hoppe-Seyler's Z. Physiol. Chem.*, 354, 857–858
Conney, A. H. (1967). *Pharmacol. Rev.*, 19, 317–366
Del Favero, A., Coli, L. and Abbritti, G. (1968). *Haematologica*, 53, 1109–1114
De Matteis, F. (1966). *Biochem. J.*, 98, 23C–25C
De Matteis, F. (1967). *Pharmacol. Rev.*, 19, 523–557
De Matteis, F. (1970). *FEBS Letters*, 6, 343–345
De Matteis, F. (1971a). *Biochem. J.*, 124, 767–777
De Matteis, F. (1971b). *S. Afr. J. Lab. Clin. Med.*, 17, 126–133
De Matteis, F. (1972). *Biochem. J.*, 127, 21–22P
De Matteis, F. (1973a). *Drug Metab. Dispos.*, 1, 267–272
De Matteis, F. (1973b). *5th Int. Congr. Pharmacology*, San Francisco, 1972, 2, 89–99. S. Karger AG; Basel
De Matteis, F. (1973c). *Enzyme*, 16, 266–275
De Matteis, F., Abbritti, G. and Gibbs, A. H. (1973). *Biochem. J.*, 134, 717–727
De Matteis, F. and Gibbs, A. (1972). *Biochem. J.*, 126, 1149–1160
De Matteis, F. and Sparks, R. G. (1973). *FEBS Letters*, 29, 141–144
Doss, M. (1969). *Z. Klin. Chem. u. Klin. Biochem.*, 2, 133–147
Doyle, D. and Schimke, R. T. (1969). *J. Biol. Chem.*, 244, 5449–5459
Druyan, R., Jakovcic, S. and Rabinowitz, M. (1973). *Biochem. J.*, 134, 377–385
Druyan, R. and Kelly, A. (1972). *Biochem. J.*, 129, 1095–1099
Garner, R. C. and McLean, A. E. M. (1969a). *Biochem. Biophys. Res. Comm.*, 37, 883–887

Garner, R. C. and McLean, A. E. M. (1969b). *Biochem. J.*, **114**, 7P
Gibson, K. D., Neuberger, A. and Scott, J. J. (1955). *Biochem. J.*, **61**, 618–629
Goldberg, A. (1968). *Biochem. Soc. Symp.*, **28**, 35–43
Goldberg, A. and Rimington, C. (1955). *Proc. Roy. Soc. (London), Ser. B*, **143**, 257–280
Granick, S. (1966). *J. Biol. Chem.*, **241**, 1359–1375
Granick, S. and Sassa, S. (1971). *Metab. Pathways*, **5**, 77–141
Granick, S. and Urata, G. (1963). *J. Biol. Chem.*, **238**, 821–827
Greim, H., Schenkman, J. B., Klotzbücher, M. and Remmer, H. (1970). *Biochem. Biophys. Acta*, **201**, 20–25
Gross, S. R. and Hutton, J. J. (1971). *J. Biol. Chem.*, **246**, 606–614
Hara, T. and Minakami, S. (1970). *J. Biochem. (Japan)*, **67**, 741–743
Hutton, J. J. and Gross, S. R. (1970). *Arch. Biochem. Biophys.*, **141**, 284–292
Jones, M. S. and Jones, O. T. G. (1969). *Biochem. J.*, **113**, 507–514
Kadenbach, B. (1970). *Europ. J. Biochem.*, **12**, 392–398
Kato, R., Vassanelli, P., Frontino, G. and Chiesara, E. (1964). *Biochem. Pharmacol.*, **13**, 1037–1051
Kaufman, L., Swanson, A. L. and Marver, H. S. (1970). *Science*, **170**, 320–322
Kurashima, Y., Hayashi, N. and Kikuchi, G. (1970). *J. Biochem. (Japan)*, **67**, 863–865
Labbe, R. F. and Hubbard, N. (1961). *Biochim. Biophys. Acta*, **52**, 130–135
Landaw, S. A., Callahan, E. W., Jun. and Schmid, R. (1970). *J. Clin. Invest.*, **49**, 914–925
Lascelles, J. (1964). *Tetrapyrrole Biosynthesis and Its Regulation*, W. A. Benjamin, Inc.; New York
Lazarow, P. B. and De Duve, C. (1971). *Biochem. Biophys. Res. Comm.*, **45**, 1198–1204
Levin, W., Jacobson, M. and Kuntzman, R. (1972). *Arch. Biochem. Biophys.*, **148**, 262–269
Levin, W., Jacobson, M., Sernatinger, E. and Kuntzman, R. (1973). *Drug Metab. Dispos.*, **1**, 275–284
Levin, W. and Kuntzman, R. (1969). *Mol. Pharmacol.*, **5**, 499–506
Levitt, M., Schacter, B. A., Zipursky, A. and Israels, L. G. (1968). *J. Clin. Invest.*, **47**, 1281–1294
Lockhead, A. C., Dagg, J. H. and Goldberg, A. (1967). *Brit. J. Dermat.*, **79**, 96–102
McKay, R., Druyan, R., Getz, G. S. and Rubinowitz, M. (1969). *Biochem. J.*, **114**, 455–461
Marks, G. S. (1969). *Heme and Chlorophyll*, Van Nostrand; London
Marks, G. S., Hunter, E. G., Terner, U. K. and Schneck, D. (1965). *Biochem. Pharmacol.*, **14**, 1077–1084
Marks, G. S., Krupa, V., Creighton, J. C. and Roomi, M. W. (1973). *Hoppe-Seyler's Z. Physiol. Chem.*, **354**, 856–857
Marver, H. S. (1969). In *Microsomes and Drug Oxidation*, pp. 495–511 (Gillette, J. R., Conney, A. H., Cosmides, G. J., Estabrook, R. W., Fouts, J. R. and Mannering, G. J., eds.), Academic Press; New York
Marver, H. S., Collins, A., Tschudy, D. P. and Rechcigl, M. (1966a). *J. Biol. Chem.*, **241**, 4323–4329
Marver, H. S., Collins, A. and Tschudy, D. P. (1966b). *Biochem. J.*, **99**, 31C–33C
Marver, H. S. and Schmid, R. (1972). *The Porphyrias*. In *The Metabolic Basis of Inherited Disease* 3rd edn., pp. 1087–1140 (Stanbury, J. B., Wyngaarden, J. B. and Fredrickson, D. S., eds.), McGraw-Hill; New York
Maxwell, C. R. and Rabinovitz, M. (1969). *Biochem. Biophys. Res. Comm.*, **35**, 79–85
Meyer, U. A. and Marver, H. S. (1971). *Science*, **171**, 64–66

Miyagi, K., Cardinal, R., Bossenmaier, I. and Watson, C. J. (1971). *J. Lab. Clin. Med.*, **78**, 683–695

Nakao, K., Wada, O., Takaku, F., Sassa, S., Yano, Y. and Urata, G. (1967). *J. Lab. Clin. Med.*, **70**, 923–932

Negishi, M. and Omura, T. (1970). *J. Biochem.* (*Japan*), **67**, 745–747

Onisawa, J. and Labbe, R. F. (1962). *Biochim. Biophys. Acta*, **56**, 618–620

Onisawa, J. and Labbe, R. F. (1963). *J. Biol. Chem.*, **238**, 724–727

Racz, W. J. and Marks, G. S. (1972). *Biochem. Pharmacol.*, **21**, 143–151

Reddy, J., Chiga, M. and Svoboda, D. (1971). *Biochem. Biophys. Res. Comm.*, **43**, 318–324

Romeo, G. and Levin, E. Y. (1971). *Biochim. Biophys. Acta*, **230**, 330–341

Rose, J. A., Hellman, E. S. and Tschudy, D. P. (1961). *Metabolism*, **10**, 514–521

Sano, S. and Granick, S. (1961). *J. Biol. Chem.*, **236**, 1173–1180

Sassa, S. and Granick, S. (1970). *Proc. Nat. Acad. Sci. U.S.A.*, **67**, 517–522

Satyanarayana Rao, M. R., Malathi, K. and Padmanaban, G. (1972). *Biochem. J.*, **127**, 553–559

Schachter, B. A., Marver, H. S. and Meyer, U. A. (1972). *Biochim. Biophys. Acta*, **279**, 221–227

Schmid, R., Marver, H. S. and Hammaker, L. (1966). *Biochem. Biophys. Res. Comm.*, **24**, 319–328

Scholnick, P. L., Hammaker, L. E. and Marver, H. S. (1969). *Proc. Nat. Acad. Sci. U.S.A.*, **63**, 65–70

Scott, J. J. (1955). In *Ciba Foundation Symposium on Porphyrin Biosynthesis and Metabolism*, pp. 43–58 (Wolstenholme, G. E. W. and Millar, E. C. P., eds.) J. & A. Churchill Ltd.; London

Shanley, B. C., Zail, S. S. and Joubert, S. M. (1970). *Brit. J. Haemat.*, **18**, 79–87

Solomon, H. M. and Figge, F. H. J. (1960). *Proc. Soc. Exptl. Biol. Med.*, **105**, 484–485

Song, C. S., Moses, H. L., Rosenthal, A. S., Gelb, N. A. and Kappas, A. (1971). *J. exp. Med.*, **134**, 1349–1371

Stein, J. A., Tschudy, D. P., Corcoran, P. L. and Collins, A. (1970). *J. Biol. Chem.*, **245**, 2213–2218

Strand, L. J., Felsher, B. F., Redeker, A. G. and Marver, H. S. (1970). *Proc. Nat. Acad. Sci. U.S.A.*, **67**, 1315–1320

Strand, L. J., Manning, J. and Marver, H. S. (1972). *J. Biol. Chem.*, **247**, 2820–2827

Sweeney, G. D., Janigan, D., Mayman, D. and Lai, H. (1971). *S. Afr. J. Lab. Clin. Med.*, **17**, 68–72

Tait, G. H. (1968). *Biochem. Soc. Symp.*, **28**, 19–34

Taljaard, J. J. F., Shanley, B. C. and Joubert, S. M. (1971). *Life Sci.*, **10**, 887–893

Tavill, A. S., Vanderhoff, G. A. and London, I. M. (1972). *J. Biol. Chem.*, **247**, 326–333

Tenhunen, R., Marver, H. S. and Schmid, R. (1969). *J. Biol. Chem.*, **244**, 6388–6394

Tephly, T. R. (1973). *Drug Metabol. Dispos.*, **1**, 266

Tephly, T. R. and Hibbeln, P. (1971). *Biochem. Biophys. Res. Comm.*, **42**, 589–595

Tephly, T. R., Webb, C., Trussler, P., Kniffen, F., Hasegawa, E. and Piper, W. (1973). *Drug. Metab. Dispos.*, **1**, 259–265

Tschudy, D. P. and Bonkowsky, H. L. (1972). *Fed. Proc. Fed. Amer. Soc. Exp. Biol.*, **31**, 147–159

Tschudy, D. P., Welland, F. H., Collins, A. and Hunter, G., Jr. (1964). *Metabolism*, **13**, 396–406

Tyrrell, D. L. J. and Marks, G. S. (1972). *Biochem. Pharmacol.*, **21**, 2077–2093

Wada, O., Yano, Y., Urata, G. and Nakao, K. (1968). *Biochem. Pharmacol.*, **17**, 595–603

Waterfield, M. D., Del Favero, A. and Gray, C. H. (1969). *Biochim. Biophys. Acta*, **184**, 470–473

Whiting, M. J. and Elliott, W. H. (1972). *J. Biol. Chem.*, **247**, 6818–6826
Woods, J. S. and Dixon, R. L. (1972). *Biochem. Pharmacol.*, **21**, 1735–1744
Zucker, W. V. and Schulman, H. M. (1968). *Proc. Nat. Acad. Sci. U.S.A.*, **59**, 582–589
Zuyderhoudt, F. M. J., Borst, P. and Huijng, F. (1969). *Biochim. Biophys. Acta.*, **178**, 408

Chapter 8

Induction of the Drug-Metabolizing Enzymes

Dennis V. Parke
Department of Biochemistry, University of Surrey,
Guildford, England

The deactivation of drugs and detoxication of environmental chemicals is brought about by the metabolism of these xenobiotics by enzymes, found primarily in the liver, which decrease the lipid-solubility of these compounds and facilitate their excretion (Parke, 1968). The enzymes of mammalian liver which metabolize drugs and other xenobiotics, such as pesticides, food additives and industrial chemicals, are located predominantly in the endoplasmic reticulum and are known as the "microsomal drug-metabolizing enzymes". These enzymes, which are responsible for oxidation, reduction and conjugation of drugs and other exogenous compounds are integral components of the membranes of the endoplasmic reticulum and although exceptionally difficult to solubilize and purify have been extensively studied using "microsomal suspensions", consisting of the centrifuged 10,000 × g supernatant fraction, or the resuspended 100,000 × g deposit fortified with the necessary co-enzyme requirements. One notable characteristic of these hepatic microsomal enzymes is the facility with which their activities may be enhanced by enzyme induction following pretreatment with a drug or xenobiotic (Conney, 1967; Parke, 1968, 1972). Indeed, the induction of these enzymes has probably been studied more extensively than any other, largely because of the important consequences to therapeutics, oncology and toxicology, and the value of such information in the safety evaluation of drugs and environmental chemicals.

8.1. THE HEPATIC MIXED-FUNCTION OXIDASES AND CYTOCHROME P-450

The most important of the hepatic microsomal drug-metabolizing enzymes are the mixed-function oxidases which oxygenate drugs and are complex, multicomponent systems comprising $NADPH_2$, a phospholipid-protohaeme-sulphide-protein complex known as cytochrome P-450 and a linking electron transport system of cytochrome P-450 reductase, $NADPH_2$-cytochrome c reductase, and possibly cytochrome b_5 (Coon, Strobel and Boyer, 1973; Estabrook et al., 1973). The mechanism of oxygenation (hydroxylation) of drugs and xenobiotics is postulated to be as shown in Fig. 8.1.

It has been suggested that one single enzyme system is responsible for all the various types of hepatic microsomal hydroxylation, which may include the following:

Aliphatic oxygenation $\quad -RCH_3 \xrightarrow{[O]} -RCH_2OH$

Alicyclic oxygenation

Aromatic oxygenation

Epoxidation

O- and N-Dealkylation $\quad \diagup\!\!NCH_3 \xrightarrow{[O]} \diagup\!\!NCH_2OH$

$\longrightarrow \diagup\!\!NH + HCHO$

N-Oxidation $\quad \diagup\!\!NR \longrightarrow \diagup\!\!\overset{O}{\underset{\uparrow}{N}}R$

Sulphoxidation $\quad \diagup\!\!SR \xrightarrow{[O]} \diagup\!\!\overset{O}{\underset{\uparrow}{S}}R$

Desulphuration $\quad \diagup\!\!C{=}S \xrightarrow{[O]} \diagup\!\!C{=}O$

Furthermore, it is considered that this one enzyme system may be responsible for oxygenation of numerous diverse substrates, including drugs, environmental chemicals, steroids, haem and fatty acids.

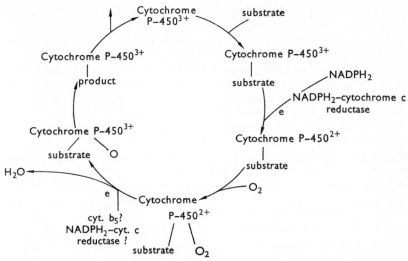

Fig. 8.1. The mechanism of hepatic microsomal hydroxylation by cyto-chrome P-450.

The evidence for the existence of a single terminal oxygenase, namely cytochrome P-450, rests on absorption spectra data (Remmer *et al.*, 1966) and, applied in the strictest sense of one enzyme one protein, is of doubtful validity. Most drugs, including phenobarbital and the mixed function oxygenase inhibitor SKF-525A (β-diethyl-amino-ethyldiphenylpropylacetate hydrochloride) react with hepatic micro-somal preparations of cytochrome P-450 to produce the same characteristic difference spectrum with an absorption minimum at 419–425 nm, and an absorption maximum at 385–390 nm (type I spectrum). Other drugs and chemicals, usually those containing a nitrogen or sulphur atom that could directly ligand with the haem moiety of cytochrome P-450 and which include aniline and the inhibitor DPEA (2,4-dichloro-6-phenylphenoxyethylamine), produce a second type of difference spectrum with an absorption minimum at 390–405 nm, and an absorption maximum at 426–435 nm (type 2 spectrum). This spectral evidence has been interpreted to imply that all the diverse substrates of the hepatic microsomal mixed-function oxidase react with cytochrome P-450 in one of two ways, namely by interaction with the lipoprotein moiety of the cytochrome (type I spectrum) or by forming a ligand complex with the haem moiety (type 2 spectrum). The first mode of reaction, and possibly also the second, could of course involve interactions of the substrates with different areas of the apoenzyme and yet still produce the same spectral shift. This would explain the apparent anomaly of a single enzyme catalysing many differing chemical reactions with a multitude of substrates, yet manifestly preserving a high degree of both substrate and reaction specificity.

Cytochrome P-450 is the most abundant cytochrome present in the liver, and is also found in adrenal mitochondria and in several other tissues which hydroxylate steroids and xenobiotic compounds. In its reduced form cytochrome P-450 forms a ligand complex with carbon monoxide which shows a characteristic absorption maximum at 450 nm, hence its name cytochrome P-450 and its earlier synonym of "carbon monoxide-binding pigment". This cytochrome requires $NADPH_2$ and molecular O_2 for activity, is inactivated by CO, and is induced by pretreatment of animals with phenobarbital and a large number of other drugs and foreign chemicals. Consequently, the accepted criteria for the involvement of cytochrome P-450 in an enzymic reaction are: (a) the 450 nm absorption maximum after reduction and treatment with CO, (b) the requirement of $NADPH_2$ and O_2 for oxygenation, (c) inhibition by CO, and (d) induction by phenobarbital. Induction of the hepatic microsomal enzymes by certain carcinogenic polycyclic hydrocarbons, such as 3-methylcholanthrene, produces a cytochrome with a characteristic CO-difference absorption spectrum at 448 nm, instead of the usual 450 nm, from which it may be inferred that perhaps more than one distinct form of this cytochrome exists (Sladek and Mannering, 1966). In addition to cytochrome P-450 the hepatic endoplasmic reticulum contains another cytochrome, namely cytochrome b_5, which may also be involved in microsomal electron transport (see Fig. 8.1), and this also may be induced by pretreatment with phenobarbital and other chemicals.

Apart from the metabolism of drugs and other xenobiotics, many of which occur in the diet and may be considered to be the natural substrates of the hepatic microsomal enzymes, the natural endogenous steroid hormones are also metabolized by these enzymes and in view of the low values of K_m which characterize the hydroxylation of testosterone, progesterone and oestradiol these steroids may indeed be the most preferred substrates (Kuntzman, Lawrence and Conney, 1965).

In addition to the mixed-function oxygenase activities the hepatic endoplasmic reticulum also contains a number of reductases, certain of which may be mediated through cytochrome P-450, e.g. nitroreductase (Gillette, Kamm and Sasame, 1968), and the UDP-glucuronyl transferases which conjugate drugs and their metabolites with glucuronic acid to form the more polar and water-soluble glucuronides.

8.2. INDUCTION OF THE HEPATIC MICROSOMAL ENZYMES

The increased activities of the microsomal drug-metabolizing enzymes following treatment with a wide variety of drugs, pesticides, food additives, polycyclic hydrocarbons and other xenobiotic compounds, is now well known and extensively documented (Conney, 1967; Conney et al., 1967). Stimulation of these enzymes occurs only when the inducing compounds are administered to the living animal, are perfused through the isolated liver or other organs (Juchau et al., 1965) or are added to mammalian cell cultures in vitro (Nebert and Gelboin, 1968a; 1968b).

The drug-metabolizing enzymes are largely absent or latent in the neonate but develop to normal levels during the first few weeks of life. They may readily be induced in the neonatal animal or in the foetus and indeed may be induced at any age, although the extent of induction falls off with the onset of maturity (Kato and Takanaka, 1968). The postnatal development of the microsomal enzymes might therefore be the result of substrate-mediated induction initiated by the greatly increased level of anutrient compounds ingested in the diet and from the metabolic activity of the newly acquired intestinal microflora.

Classical inducers of the microsomal enzymes, such as phenobarbital, 3-methylcholanthrene or, butylated hydroxytoluene (BHT), administered to rats for several days, increased the activities of the hepatic drug-metabolizing enzymes but generally had no effect on other enzymes of the endoplasmic reticulum, such as the phosphatases, (glucose-6-phosphatase, pyrophosphatase, inosine diphosphatase, guanosine diphosphatase, or uridine diphosphatase), whereas hepatotoxic compounds such as carbon tetrachloride, ethionine, thioacetamide and coumarin, which had little effect on the drug-metabolizing enzymes, produced a marked reduction of the microsomal phosphatases (Feuer and Granda, 1970). Other workers have shown that although the available glucose-6-phosphatase is reduced by phenobarbital pretreatment, the latent enzyme manifested by high pH or deoxycholate treatment of the microsomal membranes is markedly increased, resulting in an overall increase of enzyme activity of about 50%. Phenobarbital treatment does therefore result in induction of this enzyme but this is accompanied by a change in the conformation of the endoplasmic reticulum which masks the enhanced activity (Pandhi and Baum, 1970).

Nature of the Inducing Agents

The drugs and foreign compounds which induce the microsomal enzymes have widely differing pharmacological activities, e.g. phenobarbital (hypnotic), butylated hydroxytoluene (BHT, food antioxidant), DDT (pesticide), 3-methylcholanthrene (carcinogen), and the only features they would all seem to have in common are that (a) they are lipid-soluble and hence become localized in the endoplasmic reticulum of the liver, and (b) they are substrates of, or become bound to, the microsomal drug-metabolizing enzymes. This latter appears to be an important criterion for microsomal enzyme induction, at least at translational level. In a recent study of the inductive effects of a series of barbiturate drugs on rat hepatic microsomal enzymes, an inverse correlation was observed between the rates of metabolism of the barbiturates, as determined by their plasma half-lives, and the extents of induction that they produced (Ioannides, 1973) (see Fig. 8.2).

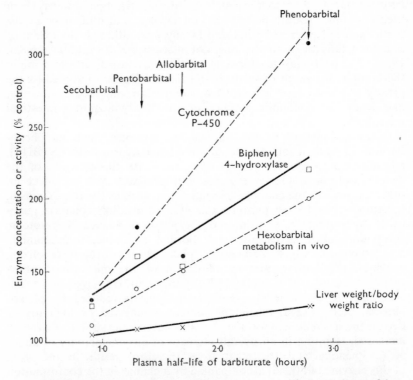

Fig. 8.2. Dependence of enzyme induction on rate of metabolism of barbiturate inducing agents.

The ratio of liver weight/body weight (×——×), rate of hexobarbital metabolism *in vivo* (O – – – O), hepatic microsomal biphenyl 4-hydroxylase activity *in vitro* (□——□), and concentration of hepatic microsomal cytochrome P-450 (● – – – ●) in rats pretreated with secobarbital, pentobarbital, allobarbital or phenobarbital, have been plotted against the plasma half-lives of these barbiturates.

The values for cytochrome P-450 and biphenyl 4-hydroxylase activity after pretreatment with allobarbital are expectedly low because of the destruction of cytochrome P-450 by this barbiturate.

(Ioannides 1973).

It therefore seems reasonable to infer that the longer the substrate remains in the body and hence in contact with the enzyme the greater will be the extent of induction, and indeed this is substantiated by the high level of induction that results with the methylenedioxyaryl compounds, such as safrole and piperonyl butoxide (see page 259) which form stable ligand complexes with cytochrome P-450 (Parke and Rahman, 1970; Philpot and Hodgson, 1971). It is, moreover significant that most compounds studied show initial inhibition of these enzymes which is followed by an increase in enzyme activity some 12–24 hours later.

This phenomenon of induction of the microsomal drug-metabolizing enzymes would thus appear to be an example of substrate-mediated enzyme induction where, because of the unique, multifunctional nature of the enzyme system involved, one substrate is able to increase not only the enzymic activity required for its own metabolism but also that for many others. Many naturally occurring anutrients, e.g. terpenes, coumarins, flavones and caffeine have been shown to stimulate the hepatic drug-metabolizing enzymes thus confirming the fundamental nature of anutrients as natural substrates of these enzymes.

The 3'4'-dichloro analogue of phenobarbital, the classical inducer of the drug-metabolizing enzymes, appears to be an even more potent inducing agent than its parent, possibly because of the substitution of the 4'-position which is the major site of metabolic hydroxylation; hence the molecule reacts with cytochrome P-450 as a potential substrate but is metabolized with difficulty and is consequently an effective inducer at the very low dose level of 2 mg/kg/day (Owen et al., 1971). Recent studies of other potential inducing agents have shown that in rats the tricyclic psychoactive drugs, chlorpromazine and imipramine, were as potent as phenobarbital (Breyer, 1972), but in human patients tricyclic drugs such as amitriptyline, imipramine and chlorimipramine, although enzyme inducers, were less active in this respect than the barbiturates (O'Malley et al., 1973).

The polychlorinated biphenyls, mixtures of various chlorinated biphenyls used as industrial stabilizers and plasticizers, are potent enzyme inducers, more effective than the same dose of phenobarbital and at least as effective as DDT (Litterst and van Loon, 1972). The mixture known as Aroclor 1254, like 3-methylcholanthrene, results in the formation of a haemoprotein with a CO-difference spectrum at 448 nm instead of 450 nm and a substantial increase in the hepatic hydroxylation of benzo[a]pyrene, but unlike methylcholanthrene and like the phenobarbital-type of inducing agent, it also produces a marked increase in ethylmorphine N-demethylase activity. This polychlorinated biphenyl, therefore, appears to possess the enzyme-inducing properties characteristic of both the "drug" and "polycyclic

8

carcinogen" type of inducing agent (Alvares, Bickers and Kappas, 1973a). A new type of polychlorinated biphenyl, Arochlor 1016 (16% chlorine), is a much less potent inducer of hepatic microsomal oxidations than the old product Arochlor 1254 (54% chlorine) (Bickers, et al., 1972), and the extent of induction appears to parallel the chlorine content; the highly polychlorinated triphenyl, Arochlor 5460, administered for 21 days to rats gave rise to marked liver hypertrophy and proliferation of the smooth endoplasmic reticulum in addition to considerable induction of microsomal enzymes such as N-demethylase and nitroreductase (Norback and Allen, 1972). 2,3,7,8-Tetrachlorodibenzo-p-dioxin, a highly toxic contaminant of the herbicide 2,4,5-trichlorophenoxyacetic acid, is also a most potent inducer, producing a marked increase of δ-aminolaevulinic acid synthetase (Poland and Glover, 1973), and an increase of zoxazolamine metabolism (type 2 substrate) but inhibits the metabolism of hexobarbitone (type 1) (Greig, 1972), a similar situation to that observed with the anaesthetic halothane (Brown, 1971).

Effects of Induction on Alternative Pathways of Metabolism

The alternative pathways for metabolism of a drug are not necessarily all equally stimulated by an inducing agent, and the relative extents of induction of different paths may vary with the inducing agent and animal species. Following the pretreatment of rabbits with phenobarbital or DDT, the N-hydroxylation of aniline was markedly stimulated but little or no increase occurred in the para-C-hydroxylation, whereas with N-ethylaniline the p-hydroxylation was markedly stimulated with no increase in N-hydroxylation; moreover, the N-dealkylation of N-ethylaniline was stimulated by phenobarbital but not by DDT (Lange, 1967). Similarly, the prolonged oral dosage of rabbits with 2-acetamidofluorene preferentially induces the N-hydroxylation of this substrate and this selective induction was made use of by Irving (1962) to facilitate the isolation of this metabolite. Pretreatment of rats with either phenobarbital or the steroid-inducing agent, pregnenolone-16α-carbonitrile (PCN), preferentially increases the C-5 hydroxylation of digitoxigenin in vitro (see Table 8.1) (Talcott and Stohs, 1973). Also, after pretreatment of rats with phenobarbital the extent of [3H]-migration in the in vitro 4-hydroxylation of [4-3H]acetanilide (the NIH effect) is less than normal, but after induction with methylcholanthrene or 3,4-benzpyrene [3H]-migration is increased (Daly et al., 1969).

Certain hydroxylations of drugs are known to be stereoselective and enzyme induction is known also to affect this. Treatment of rats with phenobarbital reduces the stereoselectivity of the hydroxylation of ethylbenzene to methylphenylcarbinol (90% D- and 10% L-, is reduced to 77% D-) (McMahon and Sullivan, 1966) and this has been shown

TABLE 8.1

Metabolism of ^{3}H-Digitoxigenin by Liver Microsomes of Rats Pretreated with Phenobarbital or Pregnenolone-16α-carbonitrile (PCN)

Treatment	Polar metabolites (% control)	% Total hydroxylated metabolites			
		1β-hydroxy	5β-hydroxy	12β-hydroxy	16β-hydroxy
None	100	9.6	44	41	6.9
Pheno-barbital	380	6.6	77	7.4	3.9
PCN	600	7.5	82	3.5	4.7

(From Talcott and Stohs, 1973).

to be due to quantitative differences in the pathways of metabolism of ethylbenzene → methylphenylcarbinol → acetophenone → methylphenylcarbinol (Maylin, Cooper and Anders, 1973).

Species and Strain Differences in Hepatic Enzyme Induction

The levels of activity of the various drug-metabolizing enzymes, and the extents to which they may be induced, are both under genetic regulation as may be seen from comparison of different species or of different strains of a given species. Large differences in the basal levels of the hepatic microsomal oxidases and in their induction by phenobarbital have been observed in different strains of rabbits (Cram, Juchau and Fouts, 1965) and rats (see Table 8.2) (Page and Vessell, 1969).

Species differences in induction of the microsomal haem proteins have also been observed (Alvares, Schilling and Levin, 1970). Treatment of mice, guinea-pigs, rabbits and rats with phenobarbital produces a three-fold increase in hepatic benzo[a]pyrene hydroxylase in the first three species but only a 50% increase in rats, nevertheless in all four species the induced CO-binding haemoprotein was cytochrome P-450. In contrast, treatment with 3-methylcholanthrene results in no significant increase of hepatic benzo[a]pyrene hydroxylase in rabbits, a two-fold increase in mice and guinea-pigs, and a five-fold increase in rats, but the induced haemoproteins vary from species to species in their carbon monoxide- and ethyl isocyanide-difference spectra. The induced cytochrome-ethyl isocyanide complex from mice, guinea-pigs and rabbits show increases in both the 430 and 455 nm peaks and differ from that of rat which shows an increase only in the 455 nm peak. Subsequent work has indicated that pretreatment of rabbits and mice with 3-methylcholanthrene results in a change in the ratio of high- and low-spin hepatic CO-binding haemoprotein (types a and b respectively)

TABLE 8.2

The Effect of Phenobarbital Pretreatment on Hepatic Metabolism of Ethylmorphine and Aniline in Various Strains of Male Rats

Strain of rat	Ethylmorphine metabolism (mμ mol HCHO/mg protein/10 min)			Aniline metabolism (mμ mol p-aminophenol/mg protein/15 min)		
	Control	Pretreated	% Induction	Control	Pretreated	% Induction
R. norvegious ACI	15	39	160	9	25	170
R. norvegious Buf.	10	23	120	17	29	70
R. norvegious F344	16	46	180	19	29	50
R. norvegious LEW	13	41	220	10	18	90
Kangaroo rat	0.6	24	4000	4.2	53	1250

(From Page and Vessell, 1969).

but in contrast, rats after pretreatment with methylcholanthrene show no change from the normal ratio of these two forms of the cytochrome (see Table 8.3) (Levin *et al.*, 1973).

Contrary to what happens in the various mammalian systems, pretreatment of house flies with phenobarbital, and naphthalene, induces synthesis of a new species of haemoprotein, namely cytochrome P-448 (Capdevila *et al.*, 1973).

TABLE 8.3

Ratio of High- to Low-spin Hepatic CO-binding Cytochromes after Enzyme Induction

Preparation	Species	Ratio of high spin to low spin (type a/type b)
Partially purified cyt. P-450 (normal)	Rat	0.12
Partially purified cyt. P-450 after phenobarbital	Rat	0.08
Partially purified cyt. P-448 after 3-methylcholanthrene	Rat	0.09
Normal microsomes	Mouse	0.14
"Phenobarbital" microsomes	Mouse	0.17
"3-Methylcholanthrene" microsomes	Mouse	0.37

(From Levin *et al.*, 1973).

8.3. INDUCTION OF UDP-GLUCURONYLTRANSFERASE AND ENZYME LATENCY

This enzyme of the hepatic endoplasmic reticulum which catalyses the conjugation of many xenobiotics and their metabolites with glucuronic acid is induced by many drugs (barbiturates, chlorcyclizine, cinchophen) and by the polycyclic hydrocarbons, 3-methylcholanthrene and 3,4-benzpyrene. However, with phenobarbital and other barbiturate inducers, induction of UDP-glucuronyltransferase (EC 2.4.1.17) is not always directly observed, possibly because these inducing agents strongly increase the latency of this enzyme (Mulder, 1970). Treatment of the hepatic microsomal preparations with the detergents Triton X-100 and digitonin (Wisnes, 1969), or by ultrasonication (Jansen and Henderson, 1972), facilitates the manifestation of this latent enzymic activity, and pretreatment of rats with phenobarbital followed by detergent treatment or ultrasonication of the liver microsomes increases *p*-nitrophenol UDP-glucuronyltransferase activity several-fold (Mulder, 1970). Treatment of liver microsomes from normal

Wistar rats with phospholipase A also reveals this latent enzyme activity and markedly increases o-aminophenyl and p-nitrophenyl UDP-glucuronyl transferases. Phospholipase treatment does not, however, similarly enhance the hepatic UDP-glucuronyl transferase activity in microsomes from the Gunn rat, a strain which normally exhibits very low levels of this enzyme activity (Zakim, Goldenberg and Vessey, 1973). This latency therefore appears to be a feature of the membrane phospholipid and the membrane environment of the enzyme. It may therefore be exacerbated by the relative increase in the lecithin/protein ratio which occurs following phenobarbital treatment (Wisnes, 1972). The degree of latency of UDP-glucuronyltransferase has been shown to vary with age and sex as well as between normal and phenobarbital-treated animals (Wisnes, 1971).

The hepatocarcinogen, diethylnitrosamine, administered to rats *in vivo* or added to rat liver microsomal preparations *in vitro*, also increases the activity of UDP-glucuronyltransferase using o-aminophenol or p-acetamidophenol as substrates. However, this activation of UDP-glucuronyltransferase is distinct from the detergent or phospholipid activation in that (a) it is specific to rat liver and the two substrates mentioned, (b) it is additive to the activation by detergents, and (c) it is especially marked with the Gunn rat (Mowat and Arias, 1970). These observations would suggest that diethylnitrosamine affects the protein rather than the phospholipid moiety of the membrane, and may induce some activating conformational change in the enzyme or in related areas of the endoplasmic reticulum.

8.4. EFFECT OF ENZYME INDUCERS ON HEPATIC HYPERTROPHY

Together with induction of the microsomal enzymes, administration of phenobarbital and other drugs also gives rise to proliferation of the hepatic smooth endoplasmic reticulum and hypertrophy of the liver cells. Even though synthesis of DNA is also activated the increase in liver weight is due mostly to increase in cell size (hypertrophy) rather than to hyperplasia. However, phenobarbital does have a moderate growth-promoting action and in both normal and regenerating liver gives rise to a mild degree of parenchymal cell hyperplasia (Japundzić *et al.*, 1967), accompanied by a marked increase in arginase activity just prior to mitosis (Japundzić *et al.*, 1969). The microsomal enzyme inhibitors, SKF 525-A and CFT 1201 (β-diethylaminoethylphenyldiallyl acetate) inhibit the hyperplastic response of the liver to phenobarbital and butylated hydroxytoluene (BHT), but not the hypertrophic response (Schulte-Hermann *et al.*, 1972). Induction of hepatic microsomal enzymes and liver hypertrophy are seen to be related but independent phenomena; halothane gives rise to hypertrophy without enzyme induction whereas 3-methylcholanthrene and other carginogenic inducers lead to enzyme induction without hypertrophy (Seifert and Vácha, 1970).

The enhancement of drug metabolism by phenobarbital treatment is much more marked when observed *in vivo* than *in vitro*, and results in a 12- to 15-fold decrease in the plasma half-lives of many drugs yet increases the rates of *in vitro* enzyme activity by only two- to four-fold. These differences have been attributed to the marked increase in hepatic blood flow (30–175%) observed in rats after phenobarbital treatment (Ohnhaus *et al.*, 1971) which may be associated with the hypertrophy of the liver. This is supported by the fact that antipyrine also results in an increased hepatic blood flow, whereas benzo[a]pyrene does not.

8.5. MECHANISMS OF INDUCTION OF HEPATIC DRUG-METABOLIZING ENZYMES BY PHENOBARBITAL

The precise mechanism of induction of the microsomal enzymes by drugs and other xenobiotics has not yet been fully elucidated and may involve enhanced synthesis of the enzyme protein, reduced degradation, or both. Conney and Gilman (1963) have shown that the increased enzyme activity was inhibited by puromycin and actinomycin D and therefore probably involved synthesis of new enzyme protein, and this was confirmed by Kato and his co-workers (Kato, Loeb and Gelboin, 1965) who found that phenobarbital increased the incorporation, both *in vitro* and *in vivo*, of labelled amino acids into liver microsomal protein.

The Role of Transcription

As the phenobarbital-induced enzyme activity is inhibited by actinomycin D, it has been suggested that induction may be the result of enhanced DNA–RNA polymerase activity due to genomal derepression (Orrenius, Ericson and Ernster, 1965); both phenobarbital and methylcholanthrene administered to rats were found to stimulate hepatic RNA-polymerase activity (Gelboin, Wortham and Wilson, 1967) and give rise to nuclear chromatin which is a more effective template for RNA synthesis than that isolated from control animals. It has also been shown that phenobarbital administration to rats induces the synthesis of a specific class of acidic hepatic nuclear protein, which may play a role in increased transcription (Ruddon and Rainey, 1970). Other workers have suggested that phenobarbital primarily affects the transcription of ribosomal RNA, whereas methylcholanthrene produces an increase in the transcription of messenger RNA (Wold and Steele, 1969).

Using a pulse-labelling technique with [5-³H]orotic acid, Cohen and

Ruddon (1970) have shown that although phenobarbital treatment leads to an increased level of cytoplasmic 28S and 18S ribosomal RNA in rat liver, no increase occurs in the synthesis of nuclear ribosomal RNA or of its precursors. The suggestion of these authors that phenobarbital may enhance the processing and transport of ribosomal RNA to the cytoplasm, or reduce its rate of degradation, rather than to stimulate its rate of synthesis in the nucleus has been confirmed by Smith *et al.* (1972). These workers, by a similar technique have shown that phenobarbital results in increased post-transcriptional stabilization of the rat hepatic nuclear ribosomal-RNA precursor (45S-RNA), rather than to enhanced synthesis. Furthermore, different mechanisms may exist for the effect of single and multiple doses of phenobarbital, for although a single dose increases the 18S and 28S RNA some 50% within 16 hours, these were unaltered by 4 days of treatment. These findings do not rule out the possibility that phenobarbital increases transcription, since as messenger RNA accounts for such a small percentage of nuclear RNA no increases in labelling of nuclear RNA would be observed even if phenobarbital did markedly stimulate the synthesis of specific messenger RNA.

The Role of Translation

Alternatively, the increased synthesis of enzymic protein could be the result of stabilization of messenger-RNA (Matsumura and Omura, 1973), and result of an increase in the translation process since phenobarbital pretreatment of rats results in inhibition of ribonuclease activity which parallels an increase in the microsomal demethylation of aminopyrine (Stolman and Loh, 1970; Louis-Ferdinand and Fuller, 1970).

More detailed studies have shown that phenobarbital decreases the activities of ribonucleases I, II and III in nuclear, mitochondrial, microsomal and supernatant fractions of rat liver and also decreases the activity of 5'-nucleotidase (5'-UMP-ase) (Seifert and Vácha, 1970). The inhibition of microsomal ribonuclease by phenobarbital is dose-dependent, is independent of the adrenal-pituitary axis and is probably due to the presence of a cytoplasmic inhibitor, which is not phenobarbital itself as this has no effect on ribonuclease *in vitro* (Louis-Ferdinand and Fuller, 1970; 1972). In contrast, methylcholanthrene did not result in inhibition of microsomal ribonuclease (Louis-Ferdinand and Fuller, 1972; Mycek, 1971).

Phenobarbital greatly enhances the stability of rat liver polyribosomes, possibly the result of the lowered ribonuclease activity (Cohen and Ruddon, 1971). A single dose of phenobarbital increased protein synthesis on membrane-bound polyribosomes more than two-fold over 3 to 24 hr after pretreatment, whereas synthesis on free polyribosomes was increased to only a minor extent (30%) at 5 hr after

the barbiturate treatment. Similarly, by an immunospecific precipitation method using an antiserum to rat liver $NADPH_2$-cytochrome
c reductase, it has been shown that phenobarbital increases the rate
of synthesis of $NADPH_2$-cytochrome c reductase by membrane-
bound polyribosomes within 3–5 hr of drug treatment, but no increase
was observed in the synthesis of this enzyme by free polyribosomes
(Glazer and Sartorelli, 1972).

Microsomal Protein Turnover

The proteins of the hepatic endoplasmic reticulum are in a dynamic
state, with a mean half-life of two days, but rates of degradation of
individual protein components of the endoplasmic reticulum are
heterogeneous. Although phenobarbital produces a marked proliferation of the liver endoplasmic reticulum and a two- to four-fold
increase of cytochrome P-450 and $NADPH_2$-cytochrome c reductase,
only a slight increase in cytochrome b_5 results (Arias, Doyle and
Schinke, 1969). However, other workers have shown that although
a single dose of phenobarbital to rats increases the hepatic $NADPH_2$-
cytochrome c reductase, but not cytochrome b_5, repeated dosage
produces a large increase in $NADPH_2$-cytochrome c reductase,
together with a moderate increase of cytochrome b_5, which result
from reduction in the rates of degradation of these enzymes (Kuriyama
et al., 1969). Nevertheless, Holtzman (1969) found that chronic phenobarbital pretreatment had no effect on the rate of degradation of
total microsomal protein and suggested that the increase of this
protein is due, almost entirely, to increased synthesis.

A reduced rate of degradation of microsomal protein might be
expected to be reflected in some change in the stability of the hepatic
lysosomes or to some decrease in lysosomal enzyme activity. Mulder
(1971) has shown that phenobarbital had no effect on the total activity
of three lysosomal enzymes, acid phosphatase and cathepsins B and C,
although because of the dilution effect of the proliferated endoplasmic
reticulum the specific activity of these enzymes was slightly reduced
(80–85%). However, because the lysosomal enzymes were measured
after treatment with the detergent, Triton X-100, it is not possible to
eliminate a contribution to the enzyme induction from an enhanced
stabilization of lysosomes.

The rate of synthesis of certain other proteins of the endoplasmic
reticulum may even be decreased by phenobarbital treatment, and a
mechanism for this coupled induction/repression of protein synthesis
involving a cascade regulation of genetic expression has been advanced
by Venkatesan, Arcos and Argus (1971).

Although low levels of cytochrome P-450, $NADPH_2$-cytochrome c
reductase and certain drug-metabolizing enzyme activities were found
to be present in the hepatocyte nuclear membrane of the normal

rat, these were not increased by treatment with phenobarbital, indicating that distinctly different regulatory mechanisms are operative for the nuclear and microsomal membranes (Kasper, 1971).

Phospholipid Turnover

In addition to the increase in protein synthesis, treatment with phenobarbital also results in an increase in microsomal phospholipid which Orrenius and his colleagues (1965) attributed to increased synthesis as at 3–24 hr after phenobarbital treatment of fasted male rats increased incorporation of ^{32}P into microsomal phospholipid occurred. However, Holtzman and Gillette (1966) subsequently found no increase in phospholipid until 24 hr after dosage with phenobarbital, and they alleged that this was due to inhibition of phospholipid degradation and not to increased synthesis.

These views have been supported by studies involving the simultaneous labelling of phospholipids with ^{14}C-fatty acids and ^{32}P-phosphate (Infante et al., 1971) and the finding of Stein and Stein (1969) that microsomal phospholipase activity is depressed in rat liver after phenobarbital treatment. Studies of the enzymes of phospholipid biosynthesis have revealed that in vivo methylation of microsomal phosphatidylethanolamine into phosphatidylcholine was increased three-fold by phenobarbital treatment, suggesting that a substantial synthesis of lecithins rather than of total phospholipid, accompanies the phenobarbital-stimulated proliferation of the endoplasmic reticulum (Young, Powell and McMillan, 1971). It may therefore be of some significance that a number of good inducers of the hepatic drug-metabolizing enzymes, such as 4-methylcoumarin and nicotine, have been shown to act as biological methyl-donors in phosphatidylcholine synthesis and it has been suggested that the transference of a methyl group from these inducing agents may play an active, but non-specific, role in the enzyme induction (Feuer et al., 1973).

Turnover of Microsomal Haemoproteins

It is now evident that induction of the microsomal enzymes requires the biosynthesis of both protein and haem, and Granick (1966) has postulated that chemicals stimulate drug metabolism by increasing haem synthesis and derepress 5-aminolaevulinate synthetase (EC 2.3.1.37) by competing with haem for a binding site on an aporepressor. Administration of either phenobarbital or benzo[a]pyrene to rats results in a rapid and marked induction of hepatic 5-aminolaevulinate synthetase, the enzyme which catalyses the rate-determining reaction for haem biosynthesis, and this is followed sequentially by increases in hepatic microsomal haem and cytochrome P-450, and in the stimulation of certain microsomal drug-metabolizing enzymes (Baron and Tephly, 1970). Microsomal cytochrome b_5 is not similarly changed. 3-Amino-1,2,4-triazole, an inhibitor of aminolaevulinate dehydratase (EC 4.2.1.24), the catalyst for the second step in haem synthesis, inhibits the phenobarbital-induced increase in hepatic haem synthesis but does not inhibit the increase in 5-aminolaevulinate synthetase (Tephly, Hasegawa and Baron, 1971). Similarly, haemopexin, a serum β-globulin which binds and transports haem to the liver, is induced together with cytochrome P-450, by treatment of rabbits with 3-methylcholanthrene or allylisopropylacetamide (Ross and Muller-Eberhard, 1970). That the increase of cytochrome P-450 produced in rat liver by phenobarbital treatment does involve increased haem synthesis was shown by an increased incorporation of 5-amino[4-^{14}C]-laevulinate; furthermore the half-lives of cytochromes P-450 and b_5 were unchanged in the steady-state induced animals as the rate of haem catabolism was increased to balance the increased rate of synthesis (Greim et al., 1970).

A number of different barbiturates have been shown to increase the level of 5-aminolaevulinate synthetase activity, the mechanism of which may be dependent on the decreased concentration of haem, a repressor of aminolaevulinate synthetase, thereby withdrawing the normal feedback control as the haem is removed for the induced synthesis of cytochrome P-450 (Moore et al., 1970). Secobarbital, allobarbital and other barbiturates containing the allyl group are potent porphyrogenic agents and have been shown to result in the destruction of cytochrome P-450, probably mediated by a metabolite in which the allyl moiety has undergone biotransformation (Levin et al., 1972c). Barbiturates without an allyl group do not have this porphyrogenic effect but allylisopropylacetamide and carbon tetrachloride have been shown also to lead to destruction of cytochrome P-450 (Levin, Jacobson and Kuntzman, 1972a).

De Matteis and Gibbs (1972), however, have shown that the stimulation of the drug-metabolizing enzymes, and the marked increase in the activity of 5-aminolaevulinate synthetase are separate, independent events likely to result from different actions of the drugs. Phenylbutazone increases microsomal cytochrome P-450 with no effect on aminolaevulinate synthetase, whereas 3,5-diethoxycarbonyl-1,4-dihydrocollidine showed the converse behaviour. However, Rifkind et al. (1973) examined a number of drugs, in a chick embryo liver system, for their effects on the stimulation of 5-aminolaevulinate synthetase and cytochrome P-450 induction and they suggest that in this system these two activities are related.

8.6. THE DIFFERENT CLASSES OF INDUCING AGENT

It has been suggested that the different types of inducer of the microsomal mixed-function oxygenases may be classified according to their effects on the various components of the enzyme system (Gillette, 1971). In this classification, phenobarbital and many drugs are associated with increases in both cytochrome P-450 and $NADPH_2$-cytochrome c reductase; methylcholanthrene and polycyclic hydrocarbons produce an increase of a modified form of cytochrome P-450, namely cytochrome P-446 or 448, but do not increase $NADPH_2$-cytochrome c reductase nor the rate of reduction of cytochrome P-450; whereas spironolactone and other steroids increase $NADPH_2$-cytochrome c reductase and the rate of reduction of cytochrome P-450 but do not affect the levels of cytochrome P-450. Although this classification is undoubtedly a gross oversimplification it is nevertheless a useful guide to broad principles, which might also include considerations of the differences in characteristics of the induced enzymes and the differences in their mechanisms of induction.

Different Forms of Cytochrome P-450

Phenobarbital induces an increase in hepatic microsomal cytochrome P-450 which shows the characteristic spectral absorption maxima at 450 nm in the reduced state with carbon monoxide as ligand, and at 430 and 455 nm with ethyl isocyanide. In contrast, methylcholanthrene induces an hepatic microsomal haemoprotein with maxima at 448 nm with carbon monoxide as ligand, and at 455 but not 430 nm with ethyl isocyanide (Alvares et al., 1967; Alvares and Mannering, 1967). Considerable evidence has accumulated to show that cytochrome P-448 is not merely a stable complex of cytochrome P-450 with 3-methylcholanthrene or one of its metabolites, as is known to occur with safrole and piperonyl butoxide (Parke and Rahman, 1971; Gray and Parke, 1973; Philpot and Hodgson, 1971; Franklin, 1972). Unequivocal evidence of the absence of a stable complex is provided by the observation that purified, soluble cytochrome P-448 obtained from rats treated with ^{3}H-labelled 3-methylcholanthrene contained only a trace of methylcholanthrene (0.04 mol/mol haemoprotein) (Fujita and Mannering, 1971) and that the cytochrome P-420 derived from this, while differing from normal cytochrome P-420 in its spectral extinction coefficient, electrophoretic migration and drug-binding characteristics, contains no radioactivity whatsoever (Shoeman, Vane and Mannering, 1973).

The many differences seen in the interaction of the microsomal enzymes with various substrates might indicate that several different species of cytochrome P-450 exist. For example, with the substrate 7-ethoxycoumarin, microsomes from untreated rats show no O-dealkylating activity, possibly related to the inability of rats to hydroxylate coumarin in the 7-position, whereas microsomes from phenobarbital-pretreated rats are able to de-ethylate 7-ethoxycoumarin but exhibit an anomalous substrate-binding spectrum (410 nm max., 428 nm min.); and microsomes from 3-methylcholanthrene-treated

rats also de-ethylate 7-ethoxycoumarin but show the normal characteristic type I binding spectrum (386 nm max. and 422 nm min.) (Ullrich,
Frommer and Weber, 1973). Similar evidence for a multiplicity of
cytochrome P-450 species has also been observed in the hydroxylation
of N-methylphenobarbital (Bohn, Ullrich and Staudinger, 1971) and
of (+)- and (−)-hexobarbital (Degkwitz et al., 1969). By diethylamino-
ethylcellulose chromatography of rat-liver microsomes treated with
subtilisin and deoxycholate, three spectrally-distinguishable forms of
cytochrome P-450 have now been isolated (Comai and Gaylor, 1973).
Furthermore, the relative amounts of the three forms are altered by
pretreatment with inducing agents, one form (probably identical with
cytochrome P-448) being increased after pretreatment with 3-methyl-
cholanthrene, another being increased after phenobarbital, and the
third after ethanol. Different rates of incorporation of ^{3}H- or ^{14}C-
labelled δ-aminolaevulinic acid into the haem moieties of the three
forms argues against their being the same cytochrome complexed
with different tightly-bound endogenous substrates (Comai and Gaylor,
1973; Conney et al., 1973), but is compatible with there being a family
of haemoproteins differing in the nature of the apoenzyme and their
enzymic characteristics yet very similar in spectral behaviour.

Solubilized preparations of purified cytochrome P-450 and P-448
prepared by the sonication/sodium cholate method of Lu and Levin
(1972), when reconstituted by addition of $NADPH_2$-dependent
reductase and phospholipid, catalyse the hydroxylation of a variety
of substrates but exhibit different substrate specificities (see Table 8.4)
(Jacobson et al., 1972). Both haemoproteins show a Soret peak at
418–419 nm in the oxidized absolute spectra, exhibit typical type I

TABLE 8.4

Differences in Substrate Specificity of Liver Microsomal Cytochromes from
Rats Pretreated with Phenobarbital and 3-Methylcholanthrene

Substrate	Relative increase in specific enzyme activities*	
	Phenobarbital-treated	Methylcholanthrene-treated
Pentobarbital	1000	60
Benzphetamine	250	50
Ethylmorphine	100	100
Aniline	120	170
Benzo[a]pyrene	70	390
Testosterone:		
6β-hydroxylation	100	100
7α-hydroxylation	100	200
16α-hydroxylation	550	50

* The activities of cytochrome P-450 from untreated (control) rats being 100
for all substrates.
(From Lu et al., 1973).

(benzphetamine) and type II (aniline) binding spectra, but differ in their binding of hexobarbital (another type I substrate) (Levin *et al.*, 1972b). From similar data it is evident that the cytochrome P-450 present in the liver microsomes of rats pretreated with phenobarbital is not identical with that of untreated animals (see Table 8.4) (Lu *et al.*, 1973).

Furthermore, from a study of the biosynthesis of the different cytochromes P-450 and of the ratios (430 nm to 450 nm maxima) of the two ethyl isocyanide complexes it has been shown that cytochrome P-448 is formed after treatment of rats with methylcholanthrene even when synthesis of new haem and protein are inhibited (Imai and Siekevitz, 1971). It thus appears that during induction by polycyclic hydrocarbons, conversion of cytochrome P-450 to P-448 occurs and may be quite separate from the synthesis of new haemoprotein.

Alvares and his colleagues have also shown that whereas after treatment with phenobarbital the rate of increase in N-demethylase activity parallels the increase in cytochrome P-450, after 3-methylcholanthrene treatment the increases in N-demethylation and in the 455/430 nm ratio of the ethyl isocyanide ligand complex occur after only 2–6 hr and long before the increase in total haemoprotein becomes manifest (Alvares, Parli and Mannering, 1973c). This suggests that the spectrally different cytochromes produced by phenobarbital and methylcholanthrene, are two physically distinct forms of a single molecular species of haemoprotein, and possibly result from differences in their immediate environment in the microsomal lipoprotein membrane (Imai and Siekevitz, 1971). The changes observed in the major species of cytochrome present after induction could thus be the combined result of enhanced cytochrome synthesis and modification of the lipophilicity of the endoplasmic reticulum by the inducing agent.

Differences in Enzyme Characteristics

The induced increases in hepatic microsomal enzyme activity are not always directly related to increases in the cytochromes P-450, and the reductases may often be the rate-limiting factors. From time-course studies of induction by phenobarbital and 3-methylcholanthrene it has been shown that the changes in O- and N-demethylase activities correlate closely with concentration of cytochrome P-450, whereas aniline hydroxylase correlates more closely with changes in the levels of $NADPH_2$-cytochrome P-450 reductase (Flynn, Lynch and Zannoni, 1972). Addition of certain type I substrates, such as hexobarbital and ethylmorphine, to microsomal preparations activate $NADPH_2$-cytochrome P-450 reductase by an allosteric effect. Holtzmann and Rumack (1973) have shown that there are separate activation and catalytic sites for ethylmorphine on the cytochrome P-450 complex, the presence of substrate being a regulator of the flux of electrons to cytochrome P-450, with the increased flux through the microsomal chain being tightly coupled to metabolism of the substrate.

Studies of binding spectra of hepatic microsomal cytochrome P-450 with hexobarbital (type 1) and aniline (type 2) show that whereas the hexobarbital binding spectrum was increased by phenobarbital pretreatment it was reduced by methylcholanthrene, but the aniline reaction spectra was increased by both. These changes in the binding spectra were reflected in the *in vitro* oxidations of aniline and hexobarbital, for after methylcholanthrene pretreatment the microsomes gradually increased their ability to oxidize aniline and lost their ability to oxidize hexobarbital, whereas both activities increased after phenobarbital (Kato and Takanaka, 1969; Shoeman, Chapman and Mannering 1969). Similarly, Sladek and Mannering (1969) have shown that whereas phenobarbital pretreatment of rats increases the N-demethylation of both ethylmorphine and 3-methyl-4-methylaminoazobenzene, treatment with 3-methylcholanthrene or benzo[a]pyrene enhances the N-demethylation of only the latter substrate and has no effect on ethylmorphine metabolism. However, some of the changes in microsomal binding spectra that appear to result from enzyme induction have been shown to be associated with the presence of residual inducing agent in the microsomes (Kutt, Waters and Fouts, 1971).

In the hydroxylation of benzo[a]pyrene by rat liver microsomes, methylcholanthrene-pretreatment reduced the K_m value, indicating that the induced cytochrome has a greater affinity than the normal liver enzyme for the substrate (Alvares, Schilling and Kuntzman, 1968). On the other hand, phenobarbital resulted in an increase of normal enzyme since for the oxidation of benzo[a]pyrene there was an increase of V_{max} with no decrease of K_m. Similar increases of V_{max} with no change of K_m have also been observed for the O-demethylation of p-nitroanisole (Netter and Seidel, 1964) and the N-demethylation of ethylmorphine (Rubin, Tephly and Mannering, 1964) by hepatic microsomal preparations from phenobarbital-stimulated animals. Treatment of young rats and guinea-pigs with 3-methylcholanthrene increases the hepatic microsomal o-aminophenol glucuronyltransferase activity, microsomal preparations from both species showing an increase in V_{max} together with a decrease in K_m of the enzyme for rats but an increase in K_m for the guinea-pigs. The liver glucuronyltransferase was not similarly increased by pretreatment with phenobarbital, nikethamide or chlorcyclizine (Howland and Burkhalter, 1971), and is probably masked by latency of this enzyme.

Differences in Mechanisms of Induction

From studies in various strains of mice it has been shown that induction of the enzymes benzo[a]pyrene hydroxylase, p-nitroanisole-

O-demethylase, 7-ethoxycoumarin de-ethylase and 3-methyl-4-methyl-aminoazobenzene N-demethylase are all associated with the same genetic locus, termed the Ah locus, or at least are in very close proximity. On the other hand, the expression of aminopyrine N-demethylase, benzphetamine N-demethylase, NADPH$_2$-cytochrome c reductase and NADPH$_2$-cytochrome P-450 reductase activities are not associated with this locus (Nebert, Considine and Owens, 1973). The formation of cytochrome P-448, as distinct from cytochrome P-450, after treatment with 3-methylcholanthrene, was observed only in hepatic microsomes of those strains of mice which are genetically responsive to polycyclic inducers and exhibit marked induction of benzo[a]pyrene hydroxylase. Phenobarbital induces cytochrome P-450 and benzo[a]-pyrene hydroxylase to the same extents in both polycyclic hydrocarbon-inducible and non-inducible strains of mice (see Table 8.5) (Gielen, Goujon and Nebert, 1972).

Injection of young rats with 3-methylcholanthrene is followed by marked activation of the hepatic nuclear RNA-polymerase system resulting from an increased chromatin template efficiency (Bresnick and Mossé, 1969). This activation of the genome resulted in the production of not only a quantitative increase in RNA but also of qualitative changes in the types of RNA (Bresnick and Mossé, 1969; Yee and Bresnick, 1971). Further studies have shown that 3-methyl-cholanthrene induces a qualitative change in the composition of the chromatin proteins and also alters the rate of incorporation of ^{14}C-lysine into chromatin proteins (Yee and Bresnick, 1971). It is reasonably well established that histones are involved in the regulation of gene activity and as benzo[a]pyrene and several other carcinogens have been shown to interact with the non-basic regions of arginine-rich rat liver histones (Sluyser, 1968) this interaction might well be involved in the initiation of genomal activation by polycyclic hydrocarbon inducing agents. 3-Methylcholanthrene also results in reduced activities of rat liver ribosomal ribonuclease, increased polysome stabilization (Pousada and Lechner, 1972), increased incorporation of amino acids and δ-aminolaevulinic acid into microsomal haemo-proteins, and a reduced rate of their degradation (Black et al., 1971; Lanclos and Bresnick, 1973).

Some of the major differences in the mechanisms and effects of induction by phenobarbital and polycyclic hydrocarbons are summarized in Table 8.6.

8.7. INDUCTION OF HEPATIC MICROSOMAL ENZYMES BY STEROIDS

Steroid hormones of all kinds, including corticosteroids, androgens, estrogens and progestogens, together with various synthetic steroids,

TABLE 8.5

Genetic Variation in Induction of Hepatic Microsomal CO-binding Cytochrome and Benzo[a]pyrene Hydroxylase by Administration of Phenobarbital and 3-Methylcholanthrene to Mice

Strain of mouse	Phenotype	Benzo[a]pyrene hydroxylase (units/mg microsomal protein)			CO-Binding cytochrome (picomol/mg microsomal protein)		
		Control	Methylcholanthrene	Phenobarbital	Control	Methylcholanthrene	Phenobarbital
C57BL/6N	CC	630	2700	1160	810	1380	1580
DBA/2N	DD	570	550	1120	860	790	1550
NZW/BCN	WW	770	760	1270	950	850	1950
ZZB/BCN	BB	590	560	1270	850	800	1680
CC × DD	CD	590	2500	1320	700	1140	1510
CC × BB	CB	620	3020	1200	720	1130	1660
DD × BB	DB	580	590	1090	830	800	1430

(From Gielen et al., 1972).

TABLE 8.6

Differences Between Phenobarbital and Polycyclic Hydrocarbon Inducers

Characteristic	Phenobarbital	Polycyclic hydrocarbon	Reference
Liver enlargement	Marked hepatic hypertrophy	No marked hypertrophy	Remmer and Merker (1963)
Enzyme system components	Cytochrome P-450 increased; $NADPH_2$-cyt.c reductase increased	Cytochrome P-448 increased. No major increase of $NADPH_2$-cyt.c reductase	Gillette (1971)
Nuclear chromatin	—	Increased template efficiency and activation of nuclear RNA-polymerase	Bresnick and Mossé (1969)
Protein synthesis	Increased; inhibition of microsomal ribonuclease	Increased; inhibition of ribosomal ribonuclease	Pousada and Lechner (1972); Louis-Ferdinand and Fuller (1972)
Phospholipid synthesis	Marked increase in microsomal phospholipid; UDP-glucuronyltransferase increase is latent	No marked increase in phospholipid; UDP-glucuronyltransferase increase is expressed	Howland and Burkhalter (1971)
Substrate binding	Increase type I and type II bindings	Increases only type II binding	Shoeman et al. (1969)
Ethyl isocyanide ligand spectra with reduced cytochrome	Ratio of maxima at 430 and 455 nm unchanged	Preferential increase of maxima at 455 nm	Sladek and Mannering (1966)
Substrate specificity	Increases N-demethylation of ethylmorphine and 3-methyl-4-methylaminobenzene	Increases N-demethylation of 3-methyl-4-aminoazobenzene but not ethylmorphine	Sladek and Mannering (1969)
	Increases hydroxylation of hexobarbital and phenobarbital	Does not increase hydroxylation of hexobarbital or pentobarbital	Kato and Takanaka (1969)
	Increase in 4-hydroxylation but not 2-hydroxylation of biphenyl	Increase in 2- and 4-hydroxylation of biphenyl	Creaven and Parke (1966)

have been shown to affect the metabolism of drugs by stimulating or inhibiting the activities of the drug-metabolizing enzymes of the liver (Conney, 1967). As drugs and steroids appear to be alternative substrates for the hepatic microsomal mixed-function oxygenase system, they would be expected to exhibit competitive inhibition of this enzyme system *in vitro* and to induce the enzymes after an appropriate time interval, *in vivo*.

Classic inhibitors of drug metabolism, such as SKF 525-A and carbon monoxide, also inhibit the microsomal synthesis of cholesterol (Kato, Vassanelli and Chiesara, 1963), and the microsomal hydroxylation of cholesterol and steroid hormones such as testosterone (Astrup, Kjeldsen and Wanstrup, 1967). Phenobarbital, a classic inducer of drug metabolism, similarly induces cholesterol biosynthesis (Mitoma *et al.*, 1968), cholesterol breakdown (Wada *et al.*, 1967), hydroxylation of the cholic acids (Einarsson and Johansson, 1969), the 6β-hydroxylation of cortisol (Conney *et al.*, 1967) and the hydroxylation of testosterone (Conney and Klutch, 1963), estrogens (Levin, Welch and Conney, 1968) and progesterone (Conney, 1967). In contrast, methylcholanthrene has little or no effect on the hydroxylation of the steroid hormones (Conney, 1967).

Corticosteroids

Adrenalectomy decreases the rates of metabolism of hexobarbital and many other drugs in the male rat (Conney, 1967) but these activities may be restored by administration of prednisolone, cortisol or phenobarbital (Furner and Stitzel, 1968). Removal of the adrenals decreases hepatic $NADPH_2$-cytochrome c reductase and cytochrome P-450 reductase more than it decreases cytochrome P-450 (Castro *et al.*, 1970). Activity of these enzymes is restored by corticosterone, in adrenalectomized rats, and by both corticosterone and ACTH in hypophysectomized animals which suggests that the pituitary-adrenal system exerts a regulatory function on the hepatic drug-metabolizing enzymes. This was substantiated by Radzialowski and Bousquet (1968) who showed that in the rat these enzymes exhibit a circadian rhythm which matches the circadian rhythm of the plasma corticosterone concentration.

Daily administration of various corticosteroids to normal rats for seven days showed no induction of the hepatic microsomal drug-metabolizing enzymes by corticosterone, a modest induction of microsomal protein but no increase in enzymic activity by prednisolone, and an enhancement of microsomal hydroxylase activity with no induction of cytochrome P-450 or microsomal protein from betamethasone; presumably betamethasone exerts this effect by stimulating the reduction of cytochrome P-450 (Tredger, Chakraborty and Parke, 1973).

At the transcription level cortisol is bound to the F3 histones of the liver, stimulates their acetylation, and enhances hepatic enzyme synthesis (Sluyser, 1969). At the level of translation, the synthetic steroid, triamcinolone diacetate, results in a decrease in rat liver ribonuclease activity paralleled by increases in liver ribonucleic acid and enzyme activity (Sarkar, 1969). Steroid hormones also regulate the pattern of hepatic enzyme synthesis by inducing the "smooth" membranes of liver endoplasmic reticulum to react with polysomes to form "rough" membranes (James, Rabin and Williams, 1969). Cortisol and dexamethasone have been shown to enhance the activity of 5-aminolaevulinate synthetase in isolated perfused rat liver similar to the action of allylisopropylacetamide; testosterone and etiocholanolone do not have this effect (Bock, Krauss and Fröhling, 1971).

Sex Hormones

The well-known differences between male and female rats in the pharmacological response to drugs is due to the enhanced level of hepatic microsomal drug-metabolizing enzymes present in the male. This is the result of an inductive effect of the male hormones, for the stimulating effect appears only at puberty and may be abolished by castration. Testosterone administered to female rats increases the rates of metabolism of many drugs and, conversely, estrogens administered to male animals decreases drug metabolism (Conney, 1967). Various anabolic 19-nortestosterone derivatives, such as norethandrolone and 19-nortestosterone, also increase the metabolism of drugs, such as hexobarbital and zoxazolamine in mice and rats (Novick, Stohler and Swagzdis, 1966). This sex difference in drug-metabolizing enzyme activities is probably not so marked in other species.

Catatoxic Steroids

Certain synthetic steroids have been shown to inhibit the action of various drugs and toxic compounds, e.g. digitoxin and picrotoxin, by increasing their catabolic detoxication—the so-called "catatoxic" effect. Administration of the diuretic spironolactone to male mice increased the microsomal metabolism of hexobarbital, aniline and ethylmorphine (Gerald and Feller, 1970) and also increased the hepatic contents of cytochrome P-450 and other components of the microsomal electron transport chain (Feller and Gerald, 1971). Administration of spironolactone to rats (200 mg/kg/day for 4 days) also results in induction of the drug-metabolizing enzymes but, unlike phenobarbital and methylcholanthrene, does not significantly increase cytochrome P-450 or microsomal protein in this species (Stripp et al., 1971). Enzyme induction by spironolactone is sex-dependent, and although NADPH$_2$-cytochrome c reductase, ethylmorphine N-demethyl-ase, hexobarbital oxidase and benzo[a]pyrene hydroxylase were increased several-fold in female rats, only the first two activities were induced in males (Stripp et al., 1971). This capricious induction by spironolactone and other steroids (methyltestosterone, pregnenolone-16α-carbonitrile) in male rats is attributed to the interaction of opposing effects, namely, a direct induction of hepatic drug metabolism and an indirect inhibition due to the anti-androgenic effect of these steroids resulting in reduction of testicular cytochrome P-450 and inhibition of testosterone biosynthesis (Stripp, et al., 1973). Ethylestrenol, like spironolactone, enhances NADPH$_2$-cytochrome c reductase, and benzo[a]pyrene hydroxylase in the rat, but like phenobarbital, also induces microsomal protein and cytochrome P-450 (Solymoss et al., 1971). Thus, although steroids, like drugs, induce the drug-metaboliz-ing enzymes of the liver, there are both qualitative and quantitative differences in the pattern of enzyme induction. This enhanced enzymic activity involves the synthesis of new enzyme protein, since the stimul-atory effects of spironolactone or ethylestrenol were inhibited by actinomycin D, puromycin aminonucleoside or cycloheximide, inhibi-tors of RNA- and protein-synthesis (Solymoss et al., 1970). Spironolac-tone was found to be only a weak inducer of the drug-metabolizing enzymes in man (Taylor et al., 1972).

Over 1,300 steroids have been screened by Selye and his colleagues (1970) for catatoxic and enzyme-inducing activity and the most potent compound discovered was pregnenolone-16α-carbonitrile (PCN). This substance has been shown to be far more potent than pheno-barbital in inducing hepatic microsomal hydroxylation in rats (Talcott and Stohs, 1973), and produces a marked increase in liver weight and extensive proliferation of the smooth endoplasmic reticulum (Tuchweber et al., 1972). It increases cytochrome P-450, NADPH$_2$-cytochtome c reductase and several drug-metabolizing enzyme activities (Lu et al., 1972), and markedly decreases the toxicity to rats of digitoxigenin by increasing its hydroxylation to polar metabolites (Talcott and Stohs, 1973). Although PCN is like phenobarbital in inducing cytochrome P-450, and not cytochrome P-448 as does 3-methylcholanthrene, the three inducing agents differ in their effects on different enzymes, with PCN, phenobarbital and methylcholanthrene having their greatest effects on ethylmorphine N-demethylation, benzphetamine N-demethylation and benzo[a]pyrene hydroxylation, respectively (see Table 8.7).

Oral Contraceptive Steroids

Juchau and Fouts (1966) showed that one hour after administration to rats of the oral contraceptive steroid, norethynodrel, the metabolism of both hexobarbitone (a type I substrate) and zoxazolamine (a type

TABLE 8.7

Effect of Various Inducing Agents on Rat Hepatic Microsomal Enzymes

Inducing agent	Microsomal protein	Cytochrome P-450	NADPH₂-cytochrome c reductase	Ethylmorphine N-demethylase	Benzphetamine N-demethylase	Benzo[a]pyrene hydroxylase
			Percentage of control activity			
None	100	100	100	100	100	100
Phenobarbital	110	230	190	460	730	400
PCN	105	180	230	1040	350	640
Methylcholanthrene	95	180	110	230	90	1380

(From Lu *et al.*, 1972).

II substrate) *in vitro* were inhibited, whereas twenty-four hours later the metabolism of both compounds was increased. Similarly, acute dosage of norethynodrel and ethinylestradiol, but not medroxyprogesterone nor mestranol, inhibited *in vivo* metabolism of pentobarbital in rats, but after chronic treatment (30 days) metabolism was significantly increased (Jori, Bianchetti and Prestini, 1969).

Continuous parenteral treatment of rats (18 days) with the contraceptive steroids, mestranol plus ethynodiol diacetate, resulted in a significant increase in the hydroxylation of biphenyl (30%), which was not seen after the same doses of mestranol or ethynodiol acetate alone; but neither cytochrome P-450 nor glucuronyl transferase was similarly enhanced by this combined treatment (Neale, 1970). This is similar to the observations of Jori, and his co-workers (1969) who showed that pretreatment of rats for 30 days with mestranol, norethynodrel, medroxyprogesterone acetate or ethynyl estradiol, alone and in combination, frequently led to induction of drug metabolism as measured by 4-nitroanisole *O*-demethylation, aminopyrine *N*-demethylation and aniline hydroxylation. Chronic oral treatment for 120 days with ethynodiol diacetate, norethynodrel or chlormadinone acetate showed only minor changes (Neale, 1970). In short, it would appear that many contraceptive steroids are substrates for the hepatic microsomal drug-metabolizing enzymes, giving rise to modest inhibition of drug metabolism *in vitro*, and variable extents of induction *in vivo*, but on prolonged treatment the enzyme activities finally return to normal values. The long-term use of these oral-contraceptive steroids thus appears unlikely to result in any major changes in the levels of the hepatic drug-metabolizing enzymes, and hence is unlikely to result in any marked changes in the activity of drugs or the toxicity of environmental chemicals.

8.8. NATURAL ANUTRIENTS AS HEPATIC ENZYME INDUCERS

It has been postulated that the normal physiological function of the drug-metabolizing enzymes is to facilitate the elimination from the body of foreign compounds or xenobiotics naturally present in the diet. As the induction of these enzymes appears to be readily effected by unnatural substrates such as drugs, and polycyclic hydrocarbons, especially when these are only slowly metabolized (Ioannides and Parke, 1973), it is not unreasonable to suppose that natural xenobiotics present in the diet (anutrients) may also effect the induction of these enzymes. The substitution of a purified diet (casein, starch, corn oil and salt-mix plus vitamin supplement) free from xenobiotics for the natural Purina Chow fed to rats, resulted in substantial reductions in the levels of benzo[a]pyrene hydroxylase activity present in lung

TABLE

The Effects on the Hepatic Enzymes of Pretreatment

						Enzyme	
Species	Inducing agent	Cytochrome P-450 (O.D. units/g)	% change	Microsomal protein (mg/g)	% change	Biphenyl-4-hydroxylase (μmol/g/hr)	% change
Rat	None	2.4		30		4.4	
	β-Ionone	3.8	+60	33	n.s.	6.8	+50
	Safrole	4.6	+90	36	+20	5.8	+30
Rabbit	None	2.4		28		4.3	
	β-Ionone	3.6	+50	32	n.s.	6.5	+50
	Safrole	5.0	+110	36	+30	2.0	−50
Mouse	None	2.5		—		7.8	
	β-Ionone	3.0	+20	—		8.0	n.s.
	Safrole	3.6	+40	—		6.4	−20

Animals were treated with β-ionone (150 mg/kg) or safrole (125 mg/kg) intraperitoneally each
n.s. indicates that no significant change was observed.
(From Parke and Rahman, 1969).

and intestinal mucosa of these animals (Wattenberg, 1972). Similar
studies conducted with mice, hamsters and rabbits have shown that
the levels of benzo[a]pyrene hydroxylase in the two tissues was related
to the amount of Purina Chow diet consumed. From these findings
it was concluded that this natural diet contains an inducing agent(s)
for this enzyme, probably present in the alfalfa meal, a constituent
of this diet (Wattenberg, 1972). Many other vegetables such as spinach,
turnip, celery and dill, but especially members of the Brassicacae
family, have been shown to have enzyme-inducing activity, and this
may vary with the particular variety of vegetable and also with the
different parts of the plant itself (Wattenberg, 1972).

   The administration of eucalyptol (1,8-cineole) to rats and human
subjects, even by aerosol (daily for 4–10 consecutive days), markedly
decreases plasma levels, and hence increases rates of metabolism
of amphetamine, zoxazolamine, pentobarbital and aminopyrine
given 24 hours after the last aerosol treatment. Furthermore, the
livers from rats treated with eucalyptol showed an increased rate of
metabolism of aminopyrine, aniline and p-nitroanisol in vitro (Jori
et al., 1970). Similarly, the hepatic drug-metabolizing enzymes of
mice and rats have been shown to be induced by the terpenes present
in cedarwood bedding (Vessel, 1967).

   The study of a wide range of terpenoids and other naturally
occurring xenobiotics, administered orally to rats for three days,
mostly showed only modest ability to induce the hepatic drug-
metabolizing enzymes, although β-ionone was found to be a potent
inducer of these enzymes, especially when administered for several

8.8

of Rats, Rabbits and Mice with β-Ionone or Safrole

parameter							
Biphenyl-2-hydroxylase		p-Nitrobenzoate reductase		UDP-Glucuronyl transferase		Inducing agent	Species
($\mu mol/g/hr$)	% change	($\mu mol/g/hr$)	% change	($\mu mol/g/hr$)	% change		
0.12		1.3		49		None	Rat
0.13	n.s.	2.3	+80	74	+60	β-Ionone	
0.77	+650	2.6	+100	80	+60	Safrole	
0.13		1.4		54		None	Rabbit
0.14	n.s.	2.1	+50	105	+90	β-Ionone	
0.04	−70	2.0	+40	110	+100	Safrole	
0.49		—		—		None	Mouse
0.52	n.s.	—		—		β-Ionone	
0.65	+30	—		—		Safrole	

day for 3 days; enzyme activities are expressed as $\mu mol/g$ liver/hr or as % change from control.

days (see Table 8.8) (Parke and Rahman, 1969). Similarly, the terpenoid, linalool, has been shown to increase the cytochrome P-450 content and UDP-glucuronyltransferase activity of rat liver when administered for several days (Parke, Rahman and Walker, 1974).

Although coumarin itself is not a potent inducer of these enzymes, 4-methylcoumarin and various derivatives of this compound, administered orally to rats at 1 mmol/kg/day for 7 days, gave rise to marked induction of hepatic coumarin 3-hydroxylase (Feuer, 1970a) and a number of other liver microsomal mono-oxygenases (see Table 8.9) (Feuer, 1970b). This difference in induction-potential between coumarin and 4-methylcoumarin is not due to differences in lipid solubility, nor would it seem is it due to differences in rates of metabolism of these two compounds; the suggestion of the author is that it may be attributed to the lower electron density of 4-methylcoumarin in position

TABLE 8.9

The Effect of Coumarin and 4-Methylcoumarin on Some Rat Hepatic Microsomal Mono-oxygenases

Microsomal enzyme activities (nmol substrate metabolized/mg protein/h)	Inducing agent		
	None	Coumarin	4-Methyl-coumarin
Hexobarbital oxidase	5.2	4.8	10.6
Codeine demethylase	11.2	7.9	21
Nitroanisole demethylase	11.8	7.9	20
Coumarin-3-hydroxylase	4.3	4.8	18

(From Feuer, 1970b).

4 which may bring about a better interaction with the induction effector.

Induction of hepatic microsomal acetanilide hydroxylase, p-chloromethylaniline N-demethylase, and o-nitroanisole O-demethylase by oral and intraperitoneal pretreatment with caffeine has been shown to occur in the rat (Mitoma et al., 1969), and although no increase in liver weight or in hepatic cytochrome could be demonstrated directly, a significant increase in the 455 nm peak of the ethyl isocyanide difference-reduced spectra was observed. This suggests that caffeine, like 3-methylcholanthrene, results in the formation of a modified form of cytochrome P-450, but unlike methylcholanthrene, caffeine also results in an increase in the rate of NADPH-cytochrome P-450 reductase activity (Lombrozo and Mitoma, 1970).

A number of naturally occurring flavones have been shown to induce microsomal drug metabolism, both in the liver and in various extrahepatic tissues (Wattenberg, 1972). The mechanism of induction by these compounds has been investigated using β-naphthoflavone, a highly potent inducing agent, although it may be questioned whether the findings for this synthetic analogue may apply equally to all natural flavones (Cantrell and Bresnick, 1971). The marked increase in activity of benzo[a]pyrene hydroxylase (aryl hydrocarbon hydroxylase) in both liver and extrahepatic tissues of rat, the potent reduction in zoxazolamine-induced paralysis and the absence of any effect on hexobarbital sleeping time, suggest that β-naphthoflavone, like 3-methylcholanthrene and in contrast to phenobarbital, produces a type II—rather than a type I—substrate induction. The similarity to methylcholanthrene induction was further substantiated by the 448 nm CO-difference spectrum of the haemoprotein induced by β-naphthoflavone and by the similarities of the "absolute" spectra and the aniline-binding spectra of the oxidized haemoproteins produced by these two inducing agents (Cantrell and Bresnick, 1971).

Pyrethrum, at high dosage to rats (200 mg/kg/day orally for 23 days) results in hepatic enlargement, increased microsomal enzyme activities, including hexobarbital oxidation, p-nitroanisole demethylation and hydrolysis of O-ethyl-O-(4-nitrophenyl)phenylphosphonothioate(EPN), but not aminopyrine demethylation, and increased NADPH$_2$-cytochrome c reductase activity and cytochrome P-450 concentration (Springfield, Carlson and De Foe, 1973). Piperonyl butoxide, a methylenedioxyphenyl derivative and a widely used synergist for pyrethroid insecticides, acts by inhibiting insect microsomal enzyme activity and the detoxication of insecticides. Piperonyl butoxide similarly inhibits the hepatic microsomal enzyme activity in several mammalian species and this has been shown to involve the binding of a metabolite, probably a hydroquinone derivative, to the microsomal enzymes (Friedman et al., 1972).

A logical consequence of this binding would be the induction of these enzymes, and Kamienski and Murphy (1971) have noted biphasic effects of piperonyl butoxide and other methylenedioxyphenyl synergists, on the actions of hexobarbital and organophosphate insecticides in mice, indicating marked inhibition of the microsomal drug-metabolizing enzymes at 1 hr after dosage of piperonyl butoxide, followed by enhanced enzymic activity at 48 hr after dosage. The bimodal effect of piperonyl butoxide on the 2- and 4-hydroxylations of biphenyl that has been observed with mouse liver microsomes (Jaffe et al., 1969) is probably similar to that observed with safrole and isosafrole, naturally occurring methylenedioxyphenyl compounds (Parke and Rahman, 1970; 1971) which produce inhibition of the 4-hydroxylation of biphenyl but a non-inductive stimulation of the 2-hydroxylation of biphenyl for the first 6 hr after dosage, followed by induction of both these enzyme activities at 24 hr after dosage. Like piperonyl butoxide, safrole and isosafrole also give rise to the formation of a stable complex of cytochrome P-450 with the methylenedioxyphenyl metabolite, which is probably instrumental in the induction of cytochrome P-450 and the various drug-metabolizing enzyme activities (Parke and Rahman, 1971; Lake and Parke, 1972a; 1972b).

Induction of the hepatic microsomal hydroxylating enzymes has also been observed with several analogues of piperonyl butoxide,

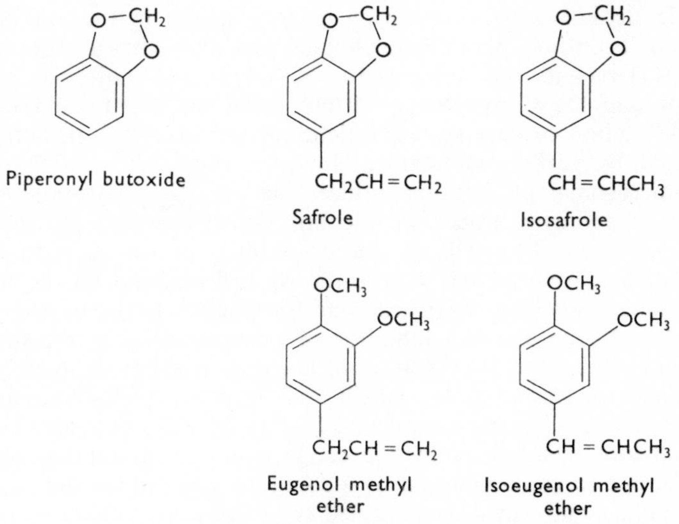

Fig. 8.3. Piperonyl butoxide and some structurally related inducing agents.

including safrole and isosafrole, which contained the methylenedioxyphenyl grouping, and eugenol methyl ether and isoeugenol methyl
ether, which contained the dimethoxyphenyl moiety; the response
to the methylenedioxyphenyl structures was greater than that to the
dimethoxyphenyl structures (see Fig. 8.3) (Wagstaff and Short, 1971).
Pretreatment of rats and hamsters with safrole and isosafrole also
induce the ring hydroxylation (3-, 5- and 7-hydroxy acetamidofluorenes),
and to a lesser extent the N-hydroxylation, of acetamidofluorene but
the inductive effects were much less marked than those obtained
after treatment with 3-methylcholanthrene (Lotlikar and Wasserman,
1972).

Ethanol

Much attention has recently been focused on the microsomal
ethanol-oxidizing system (MEOS) of mammalian liver, the induction
of this by drugs, and the induction of the microsomal drug-metabolizing enzyme system by alcohol (Rubin, Hutterer and Lieber, 1968).
Ethanol is, in small amounts, a natural component of many foodstuffs,
a product of intestinal fermentation of dietary carbohydrates, and is
also possibly an intermediate of mammalian carbohydrate metabolism.
Under these conditions it is probably wholly metabolized by liver
alcohol dehydrogenase and aldehyde oxidase, but possibly when
ingested in large quantities other mechanisms for its oxidative metabolism may come into play.

The microsomal ethanol-oxidizing system is distinguished from
hepatic alcohol dehydrogenase by its localization in the microsomal
fraction, its optimal pH of 7 (alcohol dehydrogenase has a pH optimum
of 10–11) and its requirements for $NADPH_2$ and molecular oxygen
(Lieber and De Carli, 1970a; 1970b). Like the hepatic microsomal
mixed-function oxidase system which metabolizes drugs, its activity is
inhibited by carbon monoxide. However, repeated washing of the
liver-microsomal preparation reduces the MEOS activity and simultaneously removes traces of alcohol dehydrogenase and catalase
activity. When the washings, or crystalline alcohol dehydrogenase
and catalase, were added back to the washed microsomes the MEOS
activity was fully restored (Carter and Isselbacher, 1971). Furthermore,
SKF 525-A, the classical inhibitor of microsomal drug metabolism,
does not inhibit the MEOS *in vitro*, nor does it affect the metabolism
of ethanol *in vivo* (Tephly, Tinelli and Watkins, 1969; Khanna and
Kalant, 1970; Khanna, Kalant and Lin, 1970). All this would suggest
that ethanol metabolism by the crude microsomal fraction of liver
may be due to contamination by alcohol dehydrogenase and catalase,
or that it is metabolized by a coupled system of microsomal and soluble
enzymes. Even if a system for the oxidation of ethanol does exist
in the endoplasmic reticulum it does not appear to be completely

identical to the microsomal enzyme system which oxidizes drugs and other xenobiotics.

It is well known that simultaneous administration of alcohol may potentiate the action of barbiturates and other drugs, that alcoholics are resistant to barbiturates and that barbiturate addicts are similarly resistant to intoxication by alcohol. The mechanisms of this synergism and cross-tolerance of alcohol and barbiturates is not completely understood but it has been suggested that the binding of ethanol to cytochrome P-450 may competitively inhibit the metabolism of barbiturates, although this does not appear to give rise to a subsequent induction of this enzyme (Ioannides and Parke, 1973). Administration of ethanol and a variety of drugs, including phenobarbital and 3-methylcholanthrene, have been reported to produce an increase in MEOS activity, and in the rate of drug metabolism both *in vitro* and *in vivo* (Lieber and De Carli, 1970a; 1970b; Rubin *et al.*, 1970). Furthermore, it has been shown that chronic administration to rats of ethanol, together with an adequate diet, results in proliferation of the hepatic smooth endoplasmic reticulum and an increase in microsomal protein, and it has also been claimed that this produces a 40–80% increase in the hepatic microsomal cytochrome P-450 (Rubin *et al.*, 1970).

This is in contrast to the findings of other workers who have shown that chronic administration of alcohol to rats on an adequate diet results in a moderate increase in the rate of ethanol metabolism *in vivo*, but to no increase in the MEOS or the microsomal metabolism of drugs (Khanna, Kalant and Lin, 1972; Kalant, Khanna and Marshman, 1970), and that inhibitors and inducers of the hepatic microsomal drug-metabolizing enzymes did not affect ethanol metabolism *in vivo* (Khanna and Kalant, 1970; Tephly *et al.*, 1969; Mezey, 1971). Furthermore, although treatment of rats daily for one week with phenobarbital or chlorcyclizine increased the MEOS activity it did not increase ethanol metabolism (Khanna *et al.*, 1972). The changes produced in MEOS activity are therefore unlikely to be related to change in ethanol metabolism and probably reflect the increase in smooth endoplasmic reticulum.

One possible explanation of these apparently conflicting results may be found in the kinetics of hepatic ethanol metabolism and the concentrations that occur within the hepatocyte. When the intake of ethanol is moderate the concentrations within the hepatocyte would be such that liver alcohol dehydrogenase could effectively metabolize this and thus prevent accumulation. On the other hand, when the intake of alcohol is excessive the concentrations within the hepatocyte may accrue at a faster rate than can be metabolized by alcohol dehydrogenase, so that the concentration of ethanol would exceed a certain critical level, pass into the endoplasmic reticulum and interact with cytochrome P-450. Ethanol has been shown to interact with cytochrome P-450 *in vitro* to give a modified type II binding spectrum (Rubin *et al.*, 1971). The ethanol thus competes with other substrates for cytochrome P-450, inhibits the reduction of cytochrome P-450 and as a consequence could lead to induction of the microsomal enzyme system and proliferation of the endoplasmic reticulum, even if the ethanol were not metabolized by the microsomal mixed-function oxidase. The finding that induction of the microsomal enzyme system by drug pretreatment does not result in increased metabolism of ethanol *in vivo* suggests that, although ethanol interacts with cytochrome P-450 and (in the absence of alcohol dehydrogenase) is oxidized by microsomal preparations *in vitro* (MEOS activity), this has little physiological significance. For in the *in vivo* situation, metabolism is usually effected at an adequate rate by alcohol dehydrogenase.

Thus in the chronic ethanol administration experiments of Ioannides and Parke (1973) alcohol administered to the rats at a dose level of 6–8 g/kg body weight per day for six weeks resulted in no increases in cytochrome P-450, cytochrome b_5, NADPH$_2$-cytochrome c reductase or ethylmorphine N-demethylase activity, slight but significant increases in aniline hydroxylase and p-nitrobenzoate reductase, but a 100% increase in biphenyl-4-hydroxylase (see Table 8.10). This last increase in enzyme activity has been attributed to enzyme activation by a solvent effect on the membrane rather than to enzyme induction. In contrast, three days' pretreatment with phenobarbital leads to significant increases in all these enzymic parameters.

Other workers, using higher doses of ethanol, namely 36% of the total dietary calories (approx. equivalent to 12 g ethanol/kg, Khanna *et al.*, 1972), daily for 5 weeks found significant increases in microsomal protein, cytochrome P-450, NADPH$_2$-cytochrome c reductase, aminopyrine demethylase and ethylmorphine demethylase, which was confined largely to the smooth microsomal fraction (see Table 8.11) (Joly *et al.*, 1973). Tobon and Mezey (1971), who found significant increases of microsomal protein, cytochrome P-450 and aniline hydroxylase activity, were feeding their rats about 14 g ethanol/kg a day, and in the experiments of Lieber and De Carli (1970b), which showed a marked increase in MEOS activity after administration of ethanol, administration was at a dose level of 16.5 g/kg/day.

Chronic treatment of rats with ethanol also results in a substantial increase in NADPH$_2$-dependent hydrogen peroxide formation, which is consistent with the view that microsomal ethanol oxidation may be due to peroxidation via catalase and hydrogen peroxide (Thurman, 1973). It has further been shown that treatment of mice with phenobarbital, but not ethanol, leads to an increase in hepatic acetaldehyde dehydrogenase activity, which may account for the tolerance to alcohol shown by persons taking barbiturates (Redmond and Cohen,

TABLE 8.10

The Effects of Chronic Administration of Ethanol on Rat Hepatic Microsomal Drug Metabolism

Microsomal enzyme	Treatment of animals			
	None	Ethanol	Phenobarbitone	Ethanol plus phenobarbitone
Microsomal protein	35	40	39	46
Cytochrome P-450	0.040	0.034	0.084	0.091
Cytochrome b_5	0.042	0.036	0.050	0.054
$NADPH_2$-cytochrome c reductase	0.33	0.32	0.63	0.51
Biphenyl-4-hydroxylase	0.7	1.5	2.7	2.8
Ethylmorphine N-demethylase	20	18	28	26
Aniline hydroxylase	1.4	1.9	2.3	3.0
p-Nitrobenzoate reductase	6.3	7.1	7.9	7.9

(From Ioannides and Parke, 1973).

Rats were given ethanol in the drinking water equivalent to 6–8 g/kg body wt daily for six weeks, or phenobarbital at 75 mg/kg for 3 days only, prior to killing, or both treatments.

Cytochrome P-450 and cytochrome b_5 are expressed as O.D. units/mg protein; $NADPH_2$-cytochrome c reductase as O.D. units/mg protein/min; biphenyl 4-hydroxylase, ethylmorphine N-demethylase, aniline hydroxylase and p-nitrobenzoate reductase are expressed μmol substrate metabolized/g liver/hr; microsomal protein is expressed as mg/g liver.

TABLE 8.11

Effect of High-Dose Chronic Ethanol Feeding on the Drug-Metabolizing Enzymes in the Rough and Smooth Microsomes of Rat Liver

	Rough microsomes		Smooth microsomes	
	control	ethanol	control	ethanol
Microsomal protein (mg/g liver)	17	15	13	16
Cytochrome P-450 (nmol/mg protein)	0.5	0.6	0.4	0.6
NADPH$_2$-cytochrome P-450 reductase (O.D./sec/10^{-3} mg protein)	3.0	3.2	2.7	4.0
NADPH$_2$-cytochrome c reductase (nmol cyt.c reductase/min/mg protein)	59	77	63	95
Aminopyrine demethylase (nmol HCHO/min/mg protein)	3.2	4.3	2.8	5.4
Ethylmorphine demethylase (nmol HCHO/min/mg protein)	1.6	2.2	1.5	3.1

(From Joly *et al.*, 1973).

1971). This has been demonstrated in the rat, where the effect is under genetic control being inherited as an autosomal dominant characteristic (Deitrich, 1971).

8.9. EXTRAHEPATIC ENZYME INDUCTION

Although induction of the microsomal drug-metabolizing enzymes has been studied extensively in hepatic tissue of many species it is only during the last few years that attention has also been focused on other tissues. For even though drug-metabolizing enzyme activity is generally highest in the liver, these enzymes have also been reported in several other mammalian tissues including kidney, intestine, lung, skin, brain, testes, placenta and adrenals. In many circumstances, the metabolism of drugs in extrahepatic tissues may play a significant role in determining the pharmacological and toxicological activity of a drug. It is therefore of more than academic interest to determine the extent to which these enzymes are induced in organs and tissues other than the liver.

The extents of induction vary widely from tissue to tissue (Lake *et al.*, 1973) as do the effects of various inducing agents. UDP-Glucuronyltransferase (4-methylumbelliferone) activity is increased by phenobarbital, 3-methylcholanthrene or cinchophen in rat liver, by methylcholanthrene and cinchophen, but not phenobarbital, in rat lung, and by cinchophen only in rat kidney (Aitio, 1973).

Kidney

Induction of the drug-metabolizing enzymes in this organ shows certain species differences, at least between rat and rabbit. Pretreatment of rabbits with phenobarbital leads to similar percentage increases in both kidney and liver of microsomal cytochrome P-450, cytochrome b_5, and several mixed-function oxidases (p-C-hydroxylation, N-oxidation and dealkylation of N-methylaniline, and N-hydroxylation of 4-aminobiphenyl), although the normal levels of these enzymes in the kidney is only some 10–30% of those found in liver. In contrast, although phenobarbital increases these cytochromes and microsomal hydroxylating enzyme activities in rat liver, no induction occurred in rat kidney (Uehleke and Greim, 1968). In further studies of the mono-oxygenase system of rat kidney it has been shown that the CO-complex of the reduced CO-binding haemoprotein of kidney cortex microsomes has an absorption maximum at 454 nm instead of at 450 nm (the characteristic wavelength for the hepatic cytochrome), and that there are also differences from rat liver microsomal mono-oxygenase in substrate specificity. Furthermore, although phenobarbital pretreatment has no inductive effect on the amount of cytochrome P-454 or on the ω-oxidation of fatty acids by rat kidney cortex microsomes, pretreatment with the fatty acid, lauric acid, markedly increases the cytochrome content and moderately increases the ω-oxidation (Jakobsson, Thor and Orrenius, 1970). Other workers have shown that induction of rat kidney microsomal UDP-glucuronyltransferase occurs following intragastric administration of 3-methylcholanthrene (Aitio, Vainio and Hänninen, 1972), and induction of benzo[a]pyrene hydroxylase (aryl hydrocarbon hydroxylase) is produced by 3-methylcholanthrene or isosafrole (Lake et al., 1973). Nevertheless, Feuer et al. (1971) were unable to demonstrate any increase in the activities of aniline hydroxylase, hexobarbital oxidase, aminopyrine N-demethylase or coumarin 3-hydroxylase of rat kidney following pretreatment with phenobarbital. Induction of the drug-metabolizing enzymes of the kidney is therefore somewhat capricious and, unlike the situation in the liver, seems to vary with species and to be highly dependent on the inducing agent and the particular hydroxylating enzyme under study.

Lung

The microsomal mono-oxygenase most widely studied for inducibility in lung tissue is benzo[a]pyrene hydroxylase (aryl hydrocarbon hydroxylase). Welch and his colleagues have shown that in rat lung this enzyme activity was enhanced 12-fold by exposure to cigarette smoke for several days, and although this enzymic activity was similarly increased in other tissues, induction was most pronounced in the lung (Welch, Loh and Conney, 1971) (see Table 8.12). These

9

workers have also shown that the de-ethylation of phenacetin is similarly enhanced in rat lung by exposure to cigarette smoke, although the extent of this induction was only a fraction of that observed with benzo[a]pyrene hydroxylase (see Table 8.12) (Welch, Cavallito and Loh, 1972). These two microsomal mono-oxygenase activities, although widely differing in both substrate and enzymic transformation, have

TABLE 8.12

Effect of Cigarette Smoke on Benzpyrene Hydroxylase and Phenacetin De-ethylase in Various Rat Tissues

Tissue	Benzo[a]pyrene hydroxylation (μg 3-hydroxybenzpyrene/g/hr)			Phenacetin de-ethylation (μg p-acetamidophenol/g/hr)		
	control	cigarette smoke	% increase	control	cigarette smoke	% increase
Liver	41	91	120	55	65	20
Intestine	9.4	20	120	1.2	2.4	100
Lung	0.6	7.6	1200	7	11	60
Placenta*	0.1	0.7	500			

* Benzpyrene hydroxylation was studied in pregnant rats; the other tissues were maternal.
(From Welch et al., 1971; 1972).

been associated because unlike most other mono-oxygenase activities they are both induced in liver and extrahepatic tissues of rat by treatment with carcinogenic polycyclic hydrocarbons such as 3-methylcholanthrene, or by cigarette smoking. Furthermore, in a study of phenacetin metabolism in humans it was found that smokers had significantly lower plasma levels of phenacetin than non-smokers, indicating that smoking may have stimulated the metabolism of the drug, by induction of the enzymes concerned (Pantuck, Kuntzman and Conney, 1972).

Benzo[a]pyrene hydroxylase of rat and mouse lung has also been shown to be induced by a number of flavonoid compounds and especially by the synthetic analogue, β-naphthoflavone (Wattenberg, 1972). Feeding the experimental animal a balanced purified diet (casein, starch, corn-oil, salts) led to almost complete loss of the basal level of benzo[a]pyrene hydroxylase in both small intestine and lung, indicating that the levels of this enzyme normally found in these tissues is probably the result of induction brought about by xenobiotics present in the natural diet. Pretreatment with 3-methylcholanthrene, isosafrole, or piperonyl butoxide, but not phenobarbital, also increases the benzo[a]pyrene hydroxylase of rat lung, and methylcholanthrene but no other inducing agent increases the activity of another aryl hydrocarbon hydroxylase, biphenyl 4-hydroxylase, in rat lung (see Table 8.13) (Lake *et al.*, 1973).

Intestine

As this tissue is usually one of the first with which the many foreign compounds of the diet and orally administered drugs come into contact, it is reasonable to suppose that the drug-metabolizing enzymes present in this tissue may exhibit marked induction. This has been confirmed for benzo[a]pyrene hydroxylase which is induced by poly-cyclic hydrocarbons such as 3-methylcholanthrene and benzpyrene (Gelboin and Blackburn, 1964), by flavonoids and other naturally occurring xenobiotics (Wattenberg, 1972), isosafrole and piperonyl butoxide (see Table 8.13) (Lake *et al.*, 1973), and by cigarette smoking (see Table 8.12). Cigarette smoke induces the enzyme phenacetin de-ethylase even more than it does benzo[a]pyrene hydroxylase (Welch *et al.*, 1972). The dealkylation of another substrate, 7-ethoxycoumarin, by enzymes of mouse small intestine was also markedly induced (up to 10-fold), as was also the intestinal cytochrome P-450, by pretreatment with phenobarbital (Lehrmann, Ullrich and Rummel, 1973). This induction of 7-ethoxycoumarin de-ethylation has also been observed after incubation of mouse mucosal cell homogenate with pheno-barbital (5×10^{-4} M) *in vitro*. As noticed in the induction of enzymes in cell culture, there is a lag phase of about 30 min, after which there is a two- to three-fold increase in de-ethylation activity and a two-fold increase in the incorporation of ^{14}C-leucine into protein, both of these being inhibited by addition of puromycin or actinomycin D (Scharf and Ullrich, 1973).

Several other mono-oxygenases, such as biphenyl 4-hydroxylase, chloromethylaniline *N*-demethylase (Lake *et al.*, 1973), aminopyrine *N*-demethylase, aniline hydroxylase and hexobarbital oxidase (Feuer *et al.*, 1971) were found not to be inducible in rat small intestine after pretreatment of the animals with phenobarbital, methylcholanthrene and other known inducing agents of the hepatic drug-metabolizing enzymes. This suggests that induction of microsomal mono-oxygenases in the small intestine is very selective despite the obvious contact of the enzymes of this tissue with a wide variety of xenobiotics, and it thus differs markedly from the pattern of enzyme induction found in the liver.

The polycyclic hydrocarbons, 3,4-benzpyrene and 3-methylcholan-threne, given intragastrically induce also the UDP-glucuronyltrans-ferase of rat duodenal mucosa (Aitio *et al.*, 1972). Similarly, the triterpenoid drug, carbenoxolone, which when administered orally is absorbed largely via the stomach and is excreted in the bile as glucuronide conjugate(s), gives rise to marked induction of UDP-glucuronyltransferase of rat gasteric mucosa (Shillingford, Lindup and

TABLE 8.13

Induction of Some Drug-metabolizing Enzymes of Extrahepatic Tissues of Rat

Enzyme	Inducing agent	Liver	Enzyme activities after pretreatment with inducing agents (ratio of induced level/control)			
			Intestinal mucosa	Kidney	Lung	
Benzo[a]pyrene hydroxylase	Phenobarbital	1.1	1.1	1.2	1.1	
	Methylcholanthrene	5.2	20	20	5.0	
	Isosafrole	2.4	7.7	4.7	2.6	
Biphenyl 4-hydroxylase	Phenobarbital	2.5	0.9	1.0	1.1	
	Methylcholanthrene	1.6	1.1	1.3	3.2	
	Isosafrole	3.0	2.1	2.4	1.0	
Biphenyl 2-hydroxylase	Phenobarbital	2.3	1.0			
	Methylcholanthrene	6.2	1.0			
	Isosafrole	5.4	0.9			
Aniline 4-hydroxylase	Phenobarbital	1.6		1.3	1.1	
	Methylcholanthrene	0.7		0.6	0.7	
	Isosafrole	0.7		0.9	0.8	
Chloromethylaniline N-demethylase	Phenobarbital	1.3	1.0	1.0	1.0	
	Methylcholanthrene	1.6	1.0	1.0	1.1	
	Isosafrole	1.9	0.9	1.1	0.5	

(From Lake et al., 1973).

Parke, 1973). This drug, which is used in the treatment of gastric and duodenal ulcer, leads to the induction of several other gastric mucosal enzymes concerned in the biosynthesis of gastric mucus glycoproteins (unpublished observations). Several anti-inflammatory drugs, such as phenylbutazone, indomethacin, and various steroids, all of which tend to result in gastric irritation and ulcer formation, have been shown to exert the opposite effect and to inhibit the synthesis of glycoproteins.

Other tissues

Benzo[a]pyrene hydroxylase is also markedly induced in rat skin by painting with 3-methylcholanthrene (Schlede and Conney, 1970) and in human foreskin by treating with benzo[a]anthracene *in vitro* (Alvares *et al.*, 1973b). Cultures of skin therefore offer another *in vitro* system for examining the inducibility of this aromatic hydrocarbon hydroxylase and may be of value in determining individual differences in the capacities of humans to metabolize environmental carcinogens.

The testes of rat exhibit levels of biphenyl 4-hydroxylase comparable with those of the kidney, lung and intestine, but the levels of benzo[a]-pyrene hydroxylase are lower in the testes than in other extrahepatic tissues. No induction of either of these enzymes of the testes could be effected using phenobarbital, methylcholanthrene or isosafrole (Lake *et al.*, 1973).

Pregnancy

Following the induction of benzo[a]pyrene hydroxylase with poly-cyclic hydrocarbons or by exposure to cigarette smoke, the most marked increases are found in lung and placenta (see Table 8.12) (Welch *et al.*, 1971). Treatment of pregnant rats with a variety of polycyclic aromatic hydrocarbons known to be present in cigarette smoke also increased the activity of placental benzo[a]pyrene hydroxy-lase (see Table 8.14) and one of the most potent of these inducers produced an almost linear dose response with doses up to 1 mg/kg bodyweight (Welch *et al.*, 1969). Cigarette smoking also markedly increased the activities of benzo[a]pyrene hydroxylase and 3-methyl-4-monomethylaminoazobenzene demethylase present in human placentae, the induction of these enzymes probably being due to systemic absorp-tion of the polycyclic hydrocarbons present in cigarette smoke and their subsequent transport to the placenta. In the placentae of non-smoking mothers these enzyme activities are almost undetectable (see Table 8.15).

The rat placenta, like other extrahepatic tissues, does not show any induction of the drug-metabolizing enzymes after treatment with phenobarbital, and although treatment with 3-methylcholanthrene enhances the activity of benzo[a]pyrene hydroxylase, many other

TABLE 8.14

Induction of Benzpyrene Hydroxylase of Rat Placenta by Some Polycyclic Hydrocarbons

	Benzpyrene hydroxylase activity (μg 3–hydroxybenzpyrene/g/hr)	ratio of control
None	0.2	1.0
1,2-Benzanthracene	4.0	20
1,2,5,6-Dibenzanthracene	3.6	18
3,4-Benzpyrene	3.5	17.5
Chrysene	3.3	16.5
3,4-Benzofluorene	1.9	9.5
Anthracene	1.4	7
Pyrene	1.2	6
Phenanthrene	0.7	3.5

(From Welch et al., 1969).

mono-oxygenase activities such as biphenyl 4-hydroxylase, 4-chloromethylaniline N-demethylase, together with UDP-glucuronyltransferase and nitrobenzoic acid reductase, remain unaffected (Lake et al., 1973). Similarly, in rabbit placenta, benzo[a]pyrene hydroxylase is induced by 3-methylcholanthrene but not by phenobarbital, and although no other mono-oxygenase was induced, glucuronyltransferase was significantly increased by phenobarbital (Lake et al., 1973). This species difference in the induction of glucuronyltransferase suggests that the inability of phenobarbital to induce other placental enzymes is not due to poor bioavailability to the placenta. It would seem that the placenta, like other extrahepatic tissues, exhibits a high degree of selectivity in the induction of the drug-metabolizing enzymes, both as regards the enzyme, the inducing agent, and the species.

Genetic differences

The criteria governing induction of the drug-metabolizing enzymes in extrahepatic tissues remain obscure. Although bioavailability of the inducing agent to the tissue, and its persistence at the enzyme site, would appear to be necessary prerequisites, other factors are undoubtedly involved. Tissue-specific responses to the different classes of inducing agent are seen to exist, since although polycyclic hydrocarbons induce benzo[a]pyrene hydroxylase in the liver, lung, kidney, gastrointestinal tract and placenta, phenobarbital stimulated only the hepatic enzyme (Nebert and Gelboin, 1969). Species differences between rat and rabbit have been found to occur, as in the induction of the microsomal mono-oxygenases of the kidney (Uehleke and

TABLE 8.15

Effects of Cigarette Smoking and Drugs on Human Placental Mono-oxygenase Activities

No. of cigarettes smoked	Drugs administered	Benzpyrene hydroxylase (μg-3-hydroxybenzpyrene/g/hr)	3-Methyl-4-monomethylaminoazo-benzene-N-demethylase (μg 3-methyl AB/g/hr)
0	None	<0.1	<1.0
0	Phenobarbital (30 mg)	<0.1	<1.0
0	Secobarbital (200 mg)	<0.1	<1.0
10	None	0.5	1.3
10	None	5.0	5.9
20	None	0.6	1.4
20	Meperidine (100 mg) + secobarbital (200 mg)	4.3	5.2
20	None	17	8.6
20	Meperidine (50 mg)	15	20.6
40	None	23	12.2

(From Welch et al., 1969).

TABLE 8.16

Aryl Hydrocarbon Hydroxylase Activity and its Induction in Tissues of Different Strains of Mouse

| Strain* | Specific enzyme activity (units/mg protein) | | | | | | | |
| | Liver | | Lung | | Small intestine | | Kidney | |
	Control	Induced†	Control	Induced†	Control	Induced†	Control	Induced†
Swiss	2000–4800	7700–10800	6–45	45–320	18–460	75–770	<6	<6–100
C-57GK	2400–2900	10500–15500	44–96	410–510	33–48	140–200	<6–9	78–250
C3H/HEN	390–2000	9600–11800	6–120	39–350	33–48	140–200	<6	72–160
A/HEN	1300	6900	78	332	120	190	<6	5–6
AKR/N	2900–3500	3400–3600	<6–72	35–260	9	<6–9	<6	12–15
DBA	850–940	800–890	68–81	180–200	<6	<6	<6	<6

(From Nebert and Gelboin, 1969).

* Female animals.

† 3-Methylcholanthrene (100 mg/kg) was administered intraperitoneally.

Greim, 1968) or in the induction of placental UDP-glucuronyltransferase by phenobarbital (Lake *et al.*, 1973), and it would seem that genetic differences are of considerable importance in the determination of extrahepatic enzymic induction.

The genetic differences seen in the induction of hepatic benzo[a]pyrene hydroxylase by 3-methylcholanthrene in different strains of mice are paralleled by a similar pattern in the extrahepatic tissues such as kidney and intestine, although in the lung all strains show induction of this enzyme to some extent (see Table 8.16) (Nebert and Gelboin, 1969). Using the more potent enzyme inducer, benzo[a]anthracene, it has been shown that benzo[a]pyrene hydroxylase activity may be induced in a variety of extrahepatic tissues of all strains of mice, even those strains in which the hepatic benzo[a]pyrene hydroxylase is not inducible (Wiebel, Leutz and Gelboin, 1973). This more recent information indicates that the inducibility of benzo[a]pyrene hydroxylase varies from tissue to tissue in a given strain of mouse and is not a genetically determined all-or-none phenomenon.

The induction of the microsomal enzyme, epoxide hydrase, in various inbred strains of mice, by phenobarbital exhibits a genetic expression which is not identical with the genetic expression of benzo[a]pyrene hydroxylase, and hence although these enzymes are closely related both functionally and in their sub-cellular location, they are not organized within the same operon. The genetic differences in the activities and inducibilities of these two enzymes, catalysing sequential metabolic reactions of polycylic hydrocarbons, may be of considerable consequence in the ultimate determination of cytotoxicity and carcinogenicity (Oesch *et al.*, 1973).

8.10. ENZYME INDUCTION IN CELL CULTURE

Studies of the mechanisms of enzyme induction in cell culture have the advantages of a homogeneous simplified system free from the complexities of hormonal regulation and inducer distribution kinetics which the intact mammalian organism incurs. A carefully detailed study of the induction of benzo[a]pyrene hydroxylase (aryl hydrocarbon hydroxylase), a cytochrome P-450 enzyme, has been undertaken in cultures of hamster foetal cells (Nebert and Gelboin, 1968a; 1968b), using a highly sensitive spectrophotofluorimetric method capable of detecting 10^{-12} mol hydroxylated benzo[a]pyrene metabolites per ml (Nebert and Gelboin, 1968a). The benzo[a]pyrene hydroxylase enzyme system is inducible in hamster foetal cell cultures of liver, intestine, lung, limb and brain and is greatest in cells from older foetuses. Induction has been similarly demonstrated in foetal cell cultures of rat and mouse, but to only a lesser extent with tissues from guinea-pig and chick. Additions of inducing agents such as benzo[a]anthracene and other polycyclic hydrocarbons, after a lag period of 35 min, increase the amount of hydroxylase activity linearly for 16–32 hr, after which time the enzyme levels are 15- to 40-fold greater than those of the control cells. Non-carcinogenic polycyclic hydrocarbons such as 8-methylbenzo[a]anthracene, pyrene and chrysene, were as effective as carcinogenic hydrocarbons as inducing agents of this enzyme in the cell culture system. On the other hand the drugs phenobarbital and chlorcyclizine, and the steroids cortisol and dexamethasone did not induce this enzyme in heterogeneous cell cultures from whole hamster foetus. Cycloheximide, actinomycin D or puromycin, added to the cell culture simultaneously with the inducing agent, benzo[a]anthracene, completely inhibit induction of the aryl hydroxylase, but addition of actinomycin D to previously induced cells produces a greater increase of hydroxylase activity than is obtained with the inducing agent alone, indicating that induction by benzo[a]anthracene is due to increased genetic transcription into messenger-RNA (Nebert and Gelboin, 1968b). The benzo[a]pyrene hydroxylase of rat liver cells can also be induced by the temporary inhibition of protein synthesis with cycloheximide or puromycin, which suggests that a post-transcriptional regulation of this enzyme is occurring, mediated by labile protein(s) which inhibit the expression of the hydroxylase-specific RNA (Whitlock and Gelboin, 1973).

Despite the stimulation of the benzo[a]pyrene hydroxylase some 20- to 40-fold by benzo[a]anthracene, the microsomal CO-binding cytochrome (cytochrome P-450) content of the treated hamster foetal cells was increased only two-fold. The enhanced enzymic activity was, however, related to the appearance of a new microsomal CO-binding cytochrome with an absorption maximum at 446 nm, suggesting that the benzo[a]pyrene hydroxylase activity is more closely related to this new cytochrome than it is to the normal cytochrome P-450 (Nebert, 1969).

The elegance of this cell culture technique becomes apparent when elucidating the time sequence of events of enzyme induction. In continuation of earlier studies of benzo[a]pyrene hydroxylase induction, Nebert has observed that the inducing agent (benzo[a]anthracene) enters the cells by passive diffusion, reaches equilibrium after about 30 min exposure, is physically bound to the nuclear fraction and to other subcellular components, and to a lesser extent ($<1\%$ total intracellular inducer) is covalently bound as a metabolite (Nebert and Bausserman, 1970a). Subsequently, polar metabolites of the benzo[a]-anthracene appear in the cell culture medium with the intracellular concentration of inducing agent decreasing, and the extracellular concentration of its metabolites increasing, as induced enzyme activity enhances this metabolism (Nebert and Bausserman, 1970b).

Detailed studies of the molecular mechanisms of the induction of benzo[a]pyrene hydroxylase by benzo[a]anthracene have revealed that it involves at least two phases: (a) an initial phase (transcription) which concerns the synthesis of induction-specific RNA and is independent of translation (cycloheximide-insensitive), and (b) a second phase (translation) which involves protein synthesis related to an induction-specific RNA and is transcription-independent (actinomycin D-insensitive) (Nebert and Gelboin, 1970). The resultant enzyme synthesis appears to produce a modified cytochrome P-450, the reduced CO-complex of which exhibits maximum absorption at 446 nm. This new cytochrome is unlikely to be an adduct of cytochrome P-450 with benzo[a]anthracene or some metabolite of this, or to be the result of the effects of the inducing agent on the membrane environment or conformation of the enzyme, since the concentration of cytochrome P-446 is independent of the intracellular concentration of benzo[a]anthracene and appears to be dependent on *de novo* protein synthesis (Nebert, 1970).

Although phenobarbital produced no induction of benzo[a]pyrene hydroxylase in cultures of heterogeneous cell populations of whole hamster foetus (Nebert and Gelboin, 1968a), it does induce this enzyme in cultures of foetal rat hepatocytes (Nebert and Gielen, 1971). Phenobarbital and polycyclic hydrocarbons represent two distinct classes of microsomal enzyme inducers in that their inductive effects are additive when added simultaneously to rat foetal hepatocyte cultures (Gielen and Nebert, 1971); yet the mechanisms of both are mediated through transcription plus translation, although nucleolar synthesis of ribosomal RNA may be of greater importance in induction by phenobarbital than that by benzo[a]anthracene (Nebert and Gielen, 1971). Furthermore, although both types of inducer have their primary effects at the level of transcription there is also a lesser effect at the post-translational level, which reduces the normal rate of enzyme degradation (Nebert and Gielen, 1971).

Benzo[a]pyrene hydroxylase has also been induced in cultures of guinea-pag peritoneal macrophages treated with benzo[a]anthracene (Bast *et al.*, 1973), and in human pulmonary alveolar macrophages by smoking (Cantrell *et al.*, 1973). Induction of the enzyme in human lymphocytes by 3-methylcholanthrene (Busbee, Shaw and Cantrell, 1972) revealed a biphasic distribution pattern of high and low inducers, suggesting that the capacity of induction of benzo[a]pyrene hydroxylase in man is genetically determined (Kellermann, Cantrell and Shaw, 1973). The full significance of this observation remains to be elucidated but a number of individuals with a family history of cancer all showed high levels of this enzyme induction.

Genetic differences in the induction of aryl hydrocarbon hydroxylase have also been observed in foetal cell cultures of various inbred strains of mice (Nebert and Bausserman, 1970c). The maximal inducible level of this enzyme in cells of the C57BL/6N strain is some five times greater

than that from the DBA/2N strain, despite the absence of any differences
in the rates of cellular uptake and the intracellular binding of the
inducing agent (benzo[a]pyrene), or in the rates of degradation of the
induced hydroxylase activity. However, the DBA/2N cells show a lower
extent of formation of the CO-binding cytochrome (see Table 8.5,
page 229), a smaller shift of the absorption maximum (448–449 nm
instead of 446 nm), and a decreased expression of an induction-specific
RNA, which the authors consider may be related to decreased extent of
benzo[a]pyrene hydroxylase induction (Nebert and Bausserman, 1970c).
The presence or absence of this enzyme-inducing ability in the mouse is
expressed as a simple autosomal dominant trait, the possession of this
ability being associated with a greater tendency to skin tumorigenesis in
these animals (Nebert, Goujon and Gielen, 1972). A study of benzo[a]-
pyrene hydroxylase in a number of hybrid cell strains, derived from
parent cells differing widely in both basal and induced enzyme activities,
suggests that closely coupled mechanisms of the regulation of both
basal enzyme activity and its inducibility may exist (Wiebel, Gelboin
and Coon, 1972).

The magnitude of induction of benzo[a]pyrene hydroxylase activity
in vivo in various tissues of inbred strains of mice has been shown to
correlate with the extent of induction in cultures of foetal cells derived
from the corresponding strains. As this enzyme results in the formation
of highly reactive and therefore toxic metabolites of polycyclic hydrocarbons, possibly arene oxides, the toxicity of these hydrocarbons may
be greater in those strains in which the enzyme is highly inducible. It
has therefore been suggested that cell culture may prove valuable in the
evaluation of potential toxicity of drugs, carcinogens and other chemicals
by providing an *in vitro* system which reflects the genetically regulated
response of the adult individual *in vivo*. Thus by the use of foetal cells,
fibroblasts and leucocytes it may be possible to predict potential
carcinogenesis, teratogenesis and other toxicity of a new chemical, and
even human pharmacogenetic differences (Nebert, 1973).

8.11. EFFECTS OF ENZYME INDUCTION ON TOXICITY

Induction of the drug-metabolizing enzymes of the hepatic microsomes
or of other tissues may result in marked changes in the toxicity of drugs
and foreign compounds. If the foreign compound is *detoxicated* by
these enzymes the enhancement of enzyme activity leads to more rapid
detoxication and a lowered toxicity. However, if the converse is the
case, that is, metabolism leads to the formation of a more toxic
metabolite, enzyme induction may consequently increase toxicity. The
ability of enzyme-inducing agents such as phenobarbital, methylcholanthrene and DDT to alter the toxicity of exogenous chemicals is
now extensively documented (see Table 8.17).

TABLE 8.17

Effects of enzyme induction on toxicity

Toxic chemical	Inducing agent	Effect	Reference
Reduced toxicity:			
3'-Methyl-4-dimethyl-aminoazobenzene	3-Methylcholanthrene	Loss of hepatocarcinogenicity	Richardson, Stier and Borsos-Nachtnebel (1952)
Strychnine	Thiopental, phenaglycodol	Increased hydroxylation reduced acute toxicity	Kato (1961)
Phenylbutazone	Phenylbutazone (continuous administration)	Decreased gastrointestinal ulceration	Welch, Harrison and Burns (1967)
Warfarin	Phenobarbital, DDT, chlordane	Decreased haemorrhagic effects, decreased acute toxicity	Ikeda, Conney and Burns (1968)
Dimethylnitrosamine	Pregnenolone-16α-carbonitrile	Loss of acute hepatotoxicity	Solymoss (1972)
Increased toxicity:			
Schradan	Thiopental, phenaglycodol	Increased oxidation, increased acute toxicity	Kato (1961)
2-Naphthylamine	Phenobarbital	Increased *N*-hydroxylation to the proximate bladder carcinogen	Uehleke and Brill (1968)
Monocrotaline (pyrrolizidine alkaloid)	Phenobarbital	Increased hepatic necrosis and lung lesions	Allen, Chesney and Frazee (1972)
Cyclophosphamide	Phenobarbital	Increased cytotoxicity	Bus, Short and Gibson (1973)

Recent studies have shown that changes in the rates of metabolism may have more complex effects on toxicity. For example the centrilobular hepatic necrosis produced by bromobenzene is markedly increased in the rat following stimulation of metabolism by phenobarbital but is decreased following pretreatment with 3-methylcholanthrene. The 3-methylcholanthrene, unlike phenobarbital, does not markedly accelerate the metabolism of bromobenzene but shifts the pattern of metabolism away from the more toxic pathways, increasing the amounts of bromophenyldihydrodiol and 2-bromophenol formed but decreasing the amount of 4-bromophenol and bromophenylmercapturic acid (Zampaglione *et al.*, 1973). The acute toxicity of the pyrrolizidine alkaloid, monocrotaline, is enhanced by pretreatment with phenobarbital which stimulates the metabolic formation of the cytotoxic pyrroles, but inhibitors such as chloramphenicol which reduce the rate of metabolism and the acute effects also permit the development of more chronic lesions (Allen, Chesney and Frazee, 1972). Pregnenolone-16α-carbonitrile, a potent inducer of the drug-metabolizing enzymes, greatly reduces the acute toxicity of dimethylnitrosamine but does not appear to accelerate its overall rate of metabolism. The mechanism of this protection remains unknown but presumably toxic metabolites are more rapidly detoxicated or non-toxic pathways of metabolism are accelerated (Solymoss, 1972).

Hepatomegaly and Hepatotoxicity

Although administration of many diverse xenobiotics leads to induction of the hepatic microsomal enzymes, to hypertrophy of the endoplasmic reticulum and hepatomegaly, these drug-induced changes in the liver need not necessarily implicate toxicity *per se*; indeed since Goldberg's hypothesis (Goldberg, 1966) that such phenomena should be regarded as an adaptive functional response to increased work load, there has been a general tendency to regard many of these drug-induced liver changes as "physiological" rather than "pathological." Differentiation between toxic and non-toxic responses can often be made by morphological examination, particularly when histochemical and electron microscope studies are employed. These can become even more meaningful when coupled with sequential studies of the hepatic enzymes and their characteristics over a period of prolonged administration, and thus to correlate the cellular and subcellular sequence of events. The primary locus of action of most xenobiotics, toxic or non-toxic, appears to be the endoplasmic reticulum but it would seem that the subsequent events, often occurring in other organelles of the hepatocyte or even in other tissues, determine whether the consequence is toxic or not. In a series of studies conducted by Platt and Cockrill (1969a; 1969b; 1969c) the effects on the liver of some 30 drugs and other xenobiotics have been classified into 11 different patterns of response from which it may be seen that hepatotoxicity and hepatomegaly may be differentiated by biochemical criteria. In a more recent comparative correlation of the sequential morphological and biochemical changes produced in liver by a number of toxic and non-toxic enzyme-inducing agents, including butylated hydroxytoluene, safrole, Ponceaux MX and acetamido-fluorene, an attempt has been made to identify some of the molecular processes which initiate the toxic phenomena (Gray *et al.*, 1972). Marked differences in the progressive changes in hepatic enzyme activities and in the nature of cytochrome P-450 have been demon-strated which appear to explain some of the molecular events preceding nodule formation, hepatic necrosis and hepatocarcinogenicity (Gray *et al.*, 1972; Gray and Parke, 1973).

Hepatotoxicity of Carbon Tetrachloride

The toxicity of CCl_4 is markedly increased by pretreatment of the dosed animals with phenobarbital (Garner and McLean, 1969). Pretreatment of dogs with phenobarbital decreases the lethal dose of CCl_4 by as much as 200-fold (Farber *et al.*, 1970). The extensive liver necrosis resulting from ingestion of CCl_4 has been attributed to metabolic dechlorination of the chlorinated hydrocarbon, possibly by homolytic cleavage to the free radical $CCl_3 \cdot$, (Slater and Sawyer, 1971), which then gives rise to a peroxidative decomposition of the structural lipids of mitochondrial and other intracellular membranes (Recknagel, 1967). Alternatively, the binding of CCl_4 to cytochrome P-450 could block the electron transport chain to xenobiotic hydroxyl-ation and activate a shunt pathway to lipid peroxidation (Suarez, 1972). The increased toxicity of CCl_4 induced by phenobarbital has been ascribed to the enhanced rate of dechlorination.

In contrast, administration of 3-methylcholanthrene does not similarly increase the hepatotoxicity of CCl_4 but exerts a slight protective effect (Suarez *et al.*, 1972). This has been attributed to the smaller CCl_4-induced loss of cytochrome P-450 and the smaller increase in NADPH-cytochrome c reductase resulting from methylcholanthrene pretreat-ment (see Table 8.18), which thus create a change of balance in the microsomal electron transport chain favouring utilization of electrons at the terminal acceptor, cytochrome P-450, in preference to the lipid peroxidation shunt (Suarez *et al.*, 1972).

Hepatotoxicity of Methylenedioxyaryl Compounds

Safrole (4-allyl-1,2-methylenedioxybenzene), a naturally occurring plant constituent, has been shown to produce liver tumours in rats and mice, and although no longer used as a food additive it still remains in most countries a minor constituent of diet because of its presence in spices and other plant foodstuffs. Safrole has been shown to be an effective enzyme-inducing agent, more closely resembling the

TABLE 18.18

Comparative Effects of Pretreatment of Rats with Phenobarbital or 3-Methylcholanthrene Prior to Exposure to Carbon Tetrachloride

Pretreatment	Treatment	NADP-Cytochrome c reductase (nmol reduced/mg protein/min)	Cytochrome P-450 ($E_{450-500}$/mg protein × 10^4)	Serum glutamate pyruvate transaminase (SGPT values)	Acute hepato-toxicity
None	None	95	130	25	—
None	CCl_4	85	105	50	+ +
Phenobarbital	None	175	270	35	—
Phenobarbital	CCl_4	165	105	115	+ + + +
Methylcholanthrene	None	120	325	25	—
Methylcholanthrene	CCl_4	95	195	20	—

(From Suarez et al., 1972).

carcinogenic polycyclic hydrocarbons like methylcholanthrene than agents like phenobarbital. Cytochromes P-450 and b_5, NADPH$_2$-cytochrome c reductase, biphenyl 4-hydroxylase and many other enzyme activities of rat hepatic microsomal preparations are increased (Parke and Rahman, 1970) and like other carcinogenic agents that are enzyme inducers, marked increases also occur in the 2-hydroxylation of biphenyl and in the hydroxylation of benzo[a]pyrene in extrahepatic tissues (Lake and Parke, 1972b). On repeated dosage of rats with safrole or isosafrole, a new redox difference spectral absorption peak at 455 nm was observed in the liver microsomal preparation, which appears to be due to the tenacious binding of a metabolite of safrole or isosafrole to cytochrome P-450 (Parke and Rahman, 1971; Lake and Parke, 1972a; 1972b). This binding of a metabolite profoundly affects the enzymic properties of cytochrome P-450, and removal may only be effected by treating the microsomes with safrole and certain other type I substrates (Gray and Parke, 1973). The binding of the safrole or its metabolite to cytochrome P-450 thus appears to initiate the induction of new microsomal enzymes and haemoprotein, the latter binding further safrole metabolite with concomitant loss of enzymic activity. Thus, despite the enzyme induction that occurs, there is no increase, and indeed there may even be a loss, of enzymic activity with consequent appearance of hepatotoxicity and hepatocarcinogenicity (Gray et al., 1972). It has recently been concluded that a metabolite of safrole namely 1'-hydroxysafrole, is the proximate carcinogen (Borchert et al., 1973a; 1973b).

Similarly, effects have also been observed with other methylenedioxyaryl compounds, such as piperonyl butoxide, which after an initial inhibitory period (Freidman et al., 1972) leads to the induction of hepatic microsomal drug-metabolizing enzyme activity (Kamienski and Murphy, 1971) and the appearance of the characteristic 455 nm cytochrome peak (Philpot and Hodgson, 1971). This new cytochrome absorption band has also been obtained by incubation of rat liver microsomes with safrole or piperonyl butoxide and a NADPH$_2$-generating system in vitro (Franklin, 1972), thus giving further evidence that it is a metabolite of the methylenedioxyaryl compound which binds to the cytochrome. However, Ullrich has suggested that this 450 nm haemoprotein is a carbanion complex of reduced cytochrome P-450 and has shown the formation of a similar ligand complex with the carbanion of fluorene (Ullrich and Schnabel, 1973). It would seem that cytochrome P-450 in the presence of O_2 and NADPH$_2$ has a dual role of carbanion formation and hydroxylation, and from a comparison of microsomal preparations from control, phenobarbital-treated and 3-methylcholanthrene-treated rats it is evident that the extent of carbanion formation and binding is greater after phenobarbital induction (Ullrich and Schnabel, 1973).

The recent observation that trichlorofluoromethane gives rise to a similar spectral change in cytochrome P-450 *in vitro* (Cox, King and Parke, 1973) as does carbon tetrachloride (Reiner and Uehleke, 1971), may explain the potentiation by piperonyl butoxide of the toxicity of the freons (fluorohalogenohydrocarbons) (Epstein *et al.*, 1967). A similar type of complex of cytochrome P-450 with an oxygenated metabolite and having a characteristic maximum at 455 nm has also been demonstrated with the classical inhibitor of drug metabolism, SKF 525-A (Schenkman, Wilson and Cinti, 1972), which may explain its non-competitive mode of inhibition.

Dimethylbenzo[a]anthracene-induced adrenal necrosis

7,12-Dimethylbenzo[a]anthracene (DMBA) administered to Sprague–Dawley rats produces necrosis of the adrenal cortex which has been shown to be mediated by hepatic microsomal oxygenation of DMBA to 7-hydroxymethyl-12-methylbenz[a]anthracene (Boyland, Sims and Huggins, 1965). Pretreatment of rats with the hepatotoxins, carbon tetrachloride or *p*-dimethylaminoazobenzene (butter yellow), or partial hepatectomy, at short intervals before the administration of DMBA, prevent this adrenocorticolytic action (Wheatley *et al.*, 1972). The short time intervals involved between pretreatment and DMBA administration (0–12 hr for CCl_4, 3–48 hr for butter yellow, and 1–3 days for partial hepatectomy) suggests that the protective effect is due to inhibition or impairment of the hepatic microsomal oxygenase, although the authors suggest that enzyme induction may be involved in the pretreatment with butter yellow (Wheatley *et al.*, 1972).

Pretreatment of rats with the steroids spironolactone, ethylestrenol, pregnenolone-16α-carbonitrile and dexamethasone acetate for two days before parenteral administration of DMBA also gave complete protection against adrenal necrosis and, furthermore, enhanced the metabolism of DMBA two- to six-fold (Somogyi *et al.*, 1971). These results suggest that the enhanced metabolism may represent a more rapid conversion of the adrenotoxic metabolite of DMBA (7-hydroxy-methyl-12-methylbenzanthracene) to non-toxic compounds, or that alternative and less-toxic metabolic pathways of DMBA, such as aromatic hydroxylation, are stimulated by these steroids. In this context it has recently been shown that hepatic microsomes contain a tightly coupled enzyme system of mono-oxygenase and epoxide hydrase, catalysing the pathway from arene oxide to dihydrodiol, and that this detoxicating system may be induced by 3-methylcholanthrene or phenobarbital (Oesch, 1973).

8.12. ENZYME INDUCTION AND CARCINOGENESIS

The microsomal aryl hydrocarbon hydroxylase, 3,4-benzo[a]pyrene hydroxylase, oxygenates both carcinogenic and non-carcinogenic polycyclic hydrocarbons, and is found in most tissues of mouse, rat, hamster and monkey. It is a highly inducible enzyme, the activity being increased up to 300-fold in certain tissues, by polycyclic hydrocarbons, drugs and other xenobiotics. Although this enzyme may function largely as a detoxication system it may also result in increased toxicity and may activate pro-carcinogens. Its involvement in carcinogenesis initiated by polycyclic hydrocarbons is evidence by the following observations: (a) covalent binding of benzo[a]pyrene to DNA is catalysed by this microsomal hydroxylase, (b) inhibition of the hydroxylase in mouse skin homogenates by 7,8-benzoflavone parallels the inhibition of 7,12-dimethylbenz[a]anthracene-skin tumorigenesis, and (c) increased levels of the hydroxylase parallel increased tumorigenesis (Gelboin, Kinoshita and Wiebel, 1972). On the other hand, pretreatment of rats with the inducing agent, DDT, significantly reduces the incidence of dimethylbenzanthracene-induced mammary tumours, and this was attributed to a stimulation of the overall metabolism of dimethylbenzanthracene, rather than to stimulation of carcinogen activation, thus resulting in reduced amounts of the carcinogen reaching the mammary tissue (Okey, 1972).

The induction of benzo[a]pyrene hydroxylase is genetically mediated and in mice occurs only in certain inbred strains. Kouri, Ratrie and Whitmore (1973) have shown that strains of mice exhibiting inducibility of this enzyme system were more susceptible than other strains to the formation of skin tumours following subcutaneous administration of 3-methylcholanthrene. However, genetic inducibility of benzo[a]pyrene hydroxylase could not be correlated to skin tumorigenesis resulting from topical application of 7,12-dimethylbenz[a]anthracene or benzo[a]pyrene, or to tumours resulting from intraperitoneal administration of large doses of benzo[a]pyrene (Benedict, Considine and Nebert, 1973). The inducibility of benzo[a]pyrene hydroxylase activity in human lymphocytes also shows genetic variation, with high, medium and low response to polycyclic hydrocarbon inducing agents. In a recent study of 50 patients with bronchial carcinoma a marked correlation was observed between this disease and higher levels of inducibility of this enzyme (Kellerman, Shaw and Luyten-Kellerman, 1973).

The effects of enzyme-inducing agents in carcinogenesis may also involve an enhancement of DNA synthesis and the promotion of hyperplasia which may then facilitate rapid development of clones of malignantly transformed cells before DNA-repair can be effected. Inducing agents may thus potentiate carcinogenesis initiated by oncogenic viruses and chemical carcinogens (Roe and Rowson, 1968). Furthermore, it has been suggested that metabolites of dimethylbenzanthracene may also activate oncogenic viruses or subviral particles (Benedict et al., 1973). Malignant transformation of cells by oncogenic viruses and carcinogenic chemicals appears to require the stimulation of DNA synthesis, and Marquardt and Heidelberger (1972) have shown that in hamster embryo cells the K-region epoxide of benz[a]anthracene, possibly the proximate carcinogen, and, to a lesser extent the K-region *cis*-dihydrodiol, both stimulate DNA synthesis and produce malignant transformation.

The possible potentiation of carcinogenesis by enzyme-inducing agents that are not themselves carcinogens may be illustrated by the following studies of Peraino, Fry and Staffeldt (1971). The simultaneous feeding to rats of phenobarbital and 2-acetamidofluorene reduced the hepatocarcinogenic effect of the acetamidofluorene and also decreased the effect of the carcinogen on hepatic cell proliferation. In contrast, the sequential feeding of acetamidofluorene then phenobarbital significantly increases the incidence of hepatomas over that resulting from the same initial period of treatment with carcinogen alone (Peraino *et al.*, 1971). Phenobarbital administered alone is known to increase DNA-synthesis and the proliferation of hepatocytes but the effect is relatively short-lived (Argyris and Magnus, 1968). This stimulation of mitosis by phenobarbital may thus potentiate the carcinogenic activity of previously administered acetamidofluorene by facilitating the development of cells, malignantly transformed by the acetamidofluorene metabolites, into established clones. Conversely, enzyme induction resulting from simultaneous administration of phenobarbital would be likely to enhance the detoxication of the carcinogen and minimize its oncogenic effects. Butylated hydroxytoluene (BHT), another inducing agent of the hepatic microsomal enzymes, was also found to reduce the carcinogenicity of 2-acetamidofluorene and of its more toxic metabolite *N*-hydroxy-2-acetamidofluorene (Ulland *et al.*, 1973). This protective action of BHT has been shown to result from an increased rate of detoxication of the carcinogens and, in particular, from an increased excretion of glucuronide conjugates which may result from the induction of the glucuronyl transferases (Grantham, Weisburger and Weisburger, 1973). Pretreatment with enzyme inducers, such as phenobarbital, chlordane and β-naphthoflavone, also reduces the number of pulmonary tumours produced in mice following administration of urethane (Yamamoto, Weisburger and Weisburger, 1971). The effects of enzyme inducers on chemical carcinogenesis may thus be either potentiation or inhibition, dependent on the nature of the carcinogen, the enzymic pathways for activation and detoxication, and the relative times of administration of the enzyme inducer and the carcinogen.

REFERENCES

Aitio, A. (1973). *Life Sci.*, **13**, 1705
Aitio, A., Vainio, H. and Hänninen, O. (1972). *FEBS Lett.*, **24**, 237

Allen, J. R., Chesney, C. F. and Frazee, W. J. (1972), *Toxicol. Appl. Pharmacol.*, **23**, 470

Alvares, A. P. and Mannering, G. J. (1967). *Fed. Proc.*, **26**, 462

Alvares, A. P., Schilling, G. R., Levin, W. and Kuntzman, R. (1967). *Biochem. Biophys. Res. Commun.*, **29**, 521

Alvares, A. P., Schilling, G. R. and Kuntzman, R. (1968). *Biochem. Biophys. Res. Commun.*, **30**, 588

Alvares, A. P., Schilling, G. R. and Levin, W. (1970). *J. Pharmacol. Exp. Ther.*, **175**, 4

Alvares, A. P., Bickers, D. R. and Kappas, A. (1973a). *Proc. Nat. Acad. Sci. U.S.A.*, **70**, 1321

Alvares, A. P., Kappas, A., Levin, W. and Conney, A. H. (1973b). *Clin. Pharmacol. Ther.*, **14**, 30

Alvares, A. P., Parli, C. J. and Mannering, G. J. (1973c). *Biochem. Pharmacol.*, **22**, 1037

Argyris, T. S. and Magnus, D. R. (1968). *Develop. Biol.*, **17**, 187

Arias, I. M., Doyle, D. and Schinke, R. T. (1969). *J. Biol. Chem.*, **244**, 3303

Astrup, P., Kjeldsen, K. and Wanstrup, J. (1967). *J. Atheroscler. Res.*, **7**, 343

Baron, J. and Tephly, T. R. (1970). *Archs. Biochem. Biophys.*, **139**, 410

Bast, R. C., Jr., Shears, B. W., Rapp, H. J. and Gelboin, H. V. (1973). *J. Natl. Cancer Inst.*, **51**, 675

Benedict, W. F., Considine, N. and Nebert, D. W. (1973). *Mol. Pharmacol.*, **9**, 266

Bickers, D. R., Harber, L. C., Kappas, A. and Alvares, A. P. (1972). *Res. Comm. Path. Pharmacol.*, **3**, 505

Black, O., Cantrell, E. T., Buccino, R. J., Bresnick, E. (1971). *Biochem. Pharmacol.*, **20**, 2989

Bock, K. W., Krauss, E. and Fröhling, W. (1971). *Europ. J. Biochem.*, **23**, 366

Bohn, W., Ullrich, V. and Staudinger, Hj. (1971). *Naunyn-Schmiedeberg's Arch. Pharmacol.*, **270**, 41

Borchert, P., Wislocki, P. G., Miller, J. A. and Miller, E. C. (1973a). *Cancer Res.*, **33**, 575

Borchert, P., Miller, J. A., Miller, E. C. and Shires, T. K. (1973b). *Cancer Res.*, **33**, 590

Boyland, E., Sims, P. and Huggins, C. (1965). *Nature (London)*, **207**, 816

Bresnick, E. and Mossé, H. (1969). *Mol. Pharmacol.*, **5**, 219

Breyer, U. (1972). *Naunyn-Schmiedeberg's Arch. Pharmacol.*, **272**, 277

Brown, B. R. (1971). *Anaesthesiology*, **35**, 241

Bus, J. S., Short, R. D. and Gibson, J. E. (1973). *J. Pharmacol. Exp. Ther.*, **184**, 749

Busbee, D. L., Shaw, C. R. and Cantrell, E. T. (1972). *Science (Washington)*, **178**, 315

Cantrell, E. and Bresnick, E. (1971). *Life Sci.*, **10**, 1195

Cantrell, E. T., Warr, G. A., Busbee, D. L. and Martin, R. R. (1973). *J. Clin. Investig.*, **52**, 1881

Capdevila, J., Morello, A., Perry, A. S. and Agosin, M. (1973). *Biochemistry*, **12**, 1445

Carter, E. A. and Isselbacher, M. D. (1971). *Ann. N.Y. Acad. Sci.*, **179**, 282

Castro, J. A., Greene, F. E., Gigon, P., Sasame, H. and Gillette, J. R. (1970). *Biochem. Pharmacol.*, **19**, 2461

Cohen, A. M. and Ruddon, R. W. (1970). *Mol. Pharmacol.*, **6**, 540

Cohen, A. M. and Ruddon, R. W. (1971). *Mol. Pharmacol.*, **7**, 484

Comai, K. and Gaylor, J. L. (1973). *J. Biol. Chem.*, **248**, 4947

Conney, A. H. (1967). *Pharmacol. Rev.*, **19**, 317

Conney, A. H. and Gilman, A. G. (1963). *J. Biol. Chem.*, **238**, 3682

Conney, A. H. and Klutch, A. (1963). *J. Biol. Chem.*, **238**, 1611

Conney, A. H., Lu, A. Y. H., Levin, W., Somogyi, A., West, S., Jacobson, M., Ryan, D. and Kuntzman, R. (1973). *Drug Metabolism and Disposition*, **1**, 199

Conney, A. H., Welch, R. M., Kuntzman, R. and Burns, J. J. (1967). *Clin. Pharmacol. Ther.*, **8**, 2

Coon, M. J., Strobel, H. W. and Boyer, R. F. (1973). *Drug Metabolism and Disposition*, **1**, 92

Cox, P. J., King, L. J. and Parke, D. V. (1973). *Biochem. J.* **130**, 87P

Cram, R. L., Juchau, M. R. and Fouts, J. R. (1965). *Proc. Soc. Exptl. Biol. Med.*, **118**, 872

Creaven, P. J. and Parke, D. V. (1966). *Biochem. Pharmacol.*, **15**, 7

Daly, J., Jerina, D., Farnsworth, J. and Guroff, G. (1969). *Archs. Biochem. Biophys.*, **131**, 238

Degkwitz, E., Ullrich, V., Staudinger, Hj. and Rummel, W. (1969). *Hoppe-Seyler's Z. Physiol. Chem.*, **350**, 547

De Matteis, F. and Gibbs, A. (1972). *Biochem. J.*, **126**, 1149

Deitrich, R. A. (1971). *Science (Washington)*, **173**, 334

Einarsson, K. and Johnasson, G. (1969). *FEBS Lett.*, **4**, 177

Epstein, S. S., Andrea, J., Clapp, P. and Mackintosh, D. (1967). *Toxicol. Appl. Pharmacol.*, **11**, 442

Estabrook, R. W., Matsubara, T., Mason, J. I., Werringloer, J. and Baron, J. (1973). *Drug Metabolism and Disposition*, **1**, 98

Farber, T. M., Heider, A., Peters, E. L., Ritter, D. L., Disraely, M. and Van Loon, E. J. (1970). *Toxicol. Appl. Pharmacol.*, **17**, 286

Feller, D. R. and Gerald, M. C. (1971). *Biochem. Pharmacol.*, **20**, 1991

Feuer, G. (1970a). *Chem.-Biol. Interactions*, **2**, 203

Feuer, G. (1970b). *Can. J. Physiol. Pharmacol.*, **48**, 232

Feuer, G. and Granda, V. (1970). *Toxicol. Appl. Pharmacol.*, **16**, 626

Feuer, G., Sosa-Lucero, J. C., Lumb, G. and Moddel, G. (1971). *Toxicol. Appl. Pharmacol.*, **19**, 579

Feuer, G , Miller, D. R., Cooper, S. D., de la Inglesia, F. A. and Lumb, G. (1973). *Int. J. Clin. Pharmacol. Ther. Toxicol.*, **7**, 13

Flynn, E. J., Lynch, M. and Zannoni, V. G. (1972). *Biochem. Pharmacol.*, **21**, 2577

Franklin, M. R. (1972). *Xenobiotica*, **2**, 517

Friedman, M. A., Greene, E. J., Csillag, R. and Epstein, S. S. (1972). *Toxicol. Appl. Pharmacol.*, **21**, 419

Fujita, T. and Mannering, G. J. (1971). *Chem.-Biol. Interactions*, **3**, 264

Furner, R. L. and Stitzel, R. E. (1968). *Biochem. Pharmacol.*, **17**, 121

Garner, R. C. and McLean, A. E. M. (1969). *Biochem. Pharmacol.*, **18**, 645

Gelboin, H. V. and Blackburn, N. R. (1964). *Cancer Res.*, **24**, 356

Gelboin, H. V., Wortham, J. S. and Wilson, R. G. (1967). *Nature (London)*, **214**, 281

Gelboin, H. V., Kinoshita, N. and Wiebel, F. J. (1972). *Fed. Proc.*, **31**, 1298

Gerald, M. C. and Feller, D. R. (1970). *Biochem. Pharmacol.*, **19**, 2529

Gielen, J. E. and Nebert, D. W. (1971). *Science (Washington)*, **172**, 167

Gielen, J. E., Goujon, F. M. and Nebert, D. W. (1972). *J. Biol. Chem.*, **247**, 1125

Gillette, J. R. (1971). *Metabolism*, **20**, 215

Gillette, J. R., Kamm, J. J. and Sasame, H. A. (1968). *Mol. Pharmacol.*, **4**, 541

Glazer, R. I. and Sartorelli, A. C. (1972). *Mol. Pharmacol.*, **8**, 701

Goldberg, L. (1966). *Proc. Eur. Soc. Study Drug Toxicity*, **7**, 171

Granick, S. (1966). *J. Biol. Chem.*, **241**, 1359

Grantham, P. H., Weisburger, J. H. and Weisburger, E. K. (1973). *Fd. Cosmet. Toxicol.*, **11**, 209

Gray, T. J. B., Grasso, P., Crampton, R. F. and Parke, D. V. (1972). *Biochem. J.*, **130**, 91P

Gray, T. J. B. and Parke, D. V. (1973). *9th Int. Congr. Abstr.*, 7d3, p. 341

Greig, J. B. (1972). *Biochem. Pharmacol.*, **21**, 3196
Greim, H., Schenkman, J. B., Klotzbücher, M. and Remmer, H. (1970). *Biochim. Biophys. Acta*, **201**, 20
Holtzman, J. L. (1969). *Biochim. Pharmacol.*, **18**, 2573
Holtzmann, J. L. and Gillette, J. R. (1966). *Biochim. Biophys. Res. Commun.*, **24**, 63
Holtzman, J. L. and Rumack, B. H. (1973). *Biochemistry*, **12**, 2309
Howland, R. D. and Burkhalter, A. (1971). *Biochem. Pharmacol.*, **20**, 1463
Ikeda, M., Conney, A. H. and Burns, J. J. (1968). *J. Pharmacol. Exp. Ther.*, **162**, 338
Imai, Y. and Siekevitz, P. (1971). *Arch. Biochem. Biophys.*, **144**, 143
Infante, R., Petit, D., Polonovski, J. and Caroli, J. (1971). *Experienta (Basel)*, **27**, 640
Ioannides, C. (1973). *Ph. D. Thesis*. University of Surrey, U.K.
Ioannides, C. and Parke, D. V. (1973). *Biochem. Soc. Trans.*, **1**, 716
Irving, C. C. (1962). *Cancer Res.*, **22**, 867
Jacobson, M., Lu, A. Y. H., Sernatinger, E., West, S. and Kuntzman, R. (1972). *Chem.-Biol. Interactions*, **5**, 183
Jaffe, H., Fujii, K., Guerin, H., Sengupta, M. and Epstein, S. S. (1969). *Biochem. Pharmacol.*, **18**, 1045
Jakobsson, T., Thor, H. and Orrenius, S. (1970). *Biochem. Biophys. Res. Commun.*, **39**, 1073
James, D. W. Rabin, B. R. and Williams, D. J. (1969). *Nature (London)*, **224**, 371
Jansen, P. L. M. and Henderson, P. Th. (1972). *Biochem. Pharmacol.*, **21**, 2457
Japundzić, M., Knezević, B., Djordjević, V. and Japundzić, I. (1967). *Exp. Cell Res.*, **48**, 163
Japundzić, I., Cupić, Z., Japundzić, M. and Knezević, B. (1969). *Experientia, (Basel)*, **25**, 599
Joly, J. G., Ishi, H., Teschke, R., Hasumura, Y. and Lieber, C. S. (1973). *Biochem. Pharmacol.*, **22**, 1532
Jori, A., Bianchetti, A. and Prestini, P. E. (1969). *Eur. J. Pharmacol.*, **7**, 196
Jori, A., Bianchetti, A., Prestini, P. E. and Garattini, S. (1970). *Europ. J. Pharmacol.*, **9**, 362
Juchau, M. R., Cram, R. L., Plaa, G. L. and Fouts, J. R. (1965). *Biochem. Pharmacol.*, **14**, 473
Juchau, M. R. and Fouts, J. R. (1966). *Biochem. Pharmacol.*, **15**, 89
Kalant, H., Khanna, J. M. and Marshman, J. (1970). *J. Pharmacol. Exp. Ther.*, **175**, 318
Kamienski, F. X. and Murphy, S. D. (1971). *Toxicol. Appl. Pharmacol.*, **18**, 883
Kasper, C. B. (1971). *J. Biol. Chem.*, **246**, 577
Kato, R. (1961). *Arzneimittel-Forsch.*, **11**, 797
Kato, R., Vassanelli, P. and Chiesara, E. (1963). *Biochem. Pharmacol.*, **12**, 349
Kato, R., Loeb, L. and Gelboin, H. V. (1965). *Biochem. Pharmacol.*, **14**, 1164
Kato, R. and Takanaka, A. (1968). *J. Biochem., Tokyo*, **63**, 406
Kato, R. and Takanaka, A. (1969). *Jap. J. Pharmacol.*, **19**, 171
Kellerman, G., Cantrell, E. and Shaw, C. R. (1973). *Cancer Res.*, **33**, 1654
Kellerman, G., Shaw, C. R. and Luyten-Kellerman, M. (1973). *N. Engl. J. Med.*, **289**, 934
Khanna, J. M. and Kalant, H. (1970). *Biochem. Pharmacol.*, **19**, 2033
Khanna, J. M., Kalant, H. and Lin, G. (1970). *Biochem. Pharmacol.*, **19**, 2493
Khanna, J. M., Kalant, H. and Lin, G. (1972). *Biochem. Pharmacol.*, **21**, 2215
Kouri, R. E., Ratrie, H. and Whitmore, C. E. (1973). *J. Natl. Cancer Inst.*, **51**, 197
Kuntzman, R., Lawrence, D. and Conney, A. H. (1965). *Mol. Pharmacol.*, **1**, 163
Kuriyama, Y., Omura, T.. Siekevitz, P. and Palade, G. E. (1969). *J. Biol. Chem.*, **244**, 2017
Kutt, H., Waters, L. and Fouts, J. R. (1971). *J. Pharmacol. Exp. Ther.*, **179**, 101

Lake, B. G. and Parke, D. V. (1972a). *Biochem. J.*, **127**, 23P
Lake, B. G. and Parke, D. V. (1972b). *Biochem. J.*, **130**, 86P
Lake, B. G., Hopkins, R., Chakraborty, J., Bridges, J. W. and Parke, D. V. (1973) *Drug Metabolism Disposition*, **1**, 342
Lanclos, K. D. and Bresnick, E. (1973). *Drug Metabolism and Disposition*, **1**, 239
Lange, G. (1967). *Naunyn-Schmiedeberg's Arch. Exp. Path. Pharmacol.*, **257**, 230
Lehrmann, Ch., Ullrich, V. and Rummel, W. (1973), *Naunyn-Schmiedeberg's Arch. Pharmacol.*, **276**, 89
Levin, W., Welch, R. M. and Conney, A. H. (1968). *J. Pharmacol. Exp. Ther.*, **159**, 36
Levin, W., Jacobson, M. and Kuntzman, R. (1972a). *Arch. Biochem. Biophys.*, **148**, 262
Levin, W., Lu, A. Y. H., Ryan, D., West, S., Kuntzman, R. and Conney, A. H. (1972b). *Arch. Biochem. Biophys.*, **153**, 543
Levin, W., Sernatinger, E., Jacobson, M. and Kuntzman, R. (1972c). *Science, (Washington)*, **176**, 1341
Levin, W., Ryan, D., West, S. and Lu, A. Y. H. (1973). *Drug Metabolism and Disposition*, **1**, 602
Lieber, C. S. and De Carli, L. M. (1970a). *J. Biol. Chem.*, **245**, 2505
Lieber, C. S. and De Carli, L. M. (1970b). *Life Sci.*, **9**, 267
Litterst, C. L. and van Loon, E. J. (1972). *Proc. Soc. Exp. Biol. Med.*, **141**, 765
Lombrozo, L. and Mitoma, C. (1970). *Biochem. Pharmacol.*, **19**, 2317
Lotlikar, P. D. and Wasserman, M. B. (1972). *Biochem. J.*, **129**, 937
Louis-Ferdinand, R. T. and Fuller, G. C. (1970). *Biochem. Biophys. Res. Commun.*, **38**, 811
Louis-Ferdinand, R. T. and Fuller, G. C. (1972). *Toxicol. Appl. Pharmacol.*, **23**, 492
Lu, A. Y. H. and Levin, W. (1972). *Biochem. Biophys. Res. Commun.*, **46**, 1334
Lu, A. Y. H., Somogyi, A., West, S., Kuntzman, R. and Conney, A. H. (1972). *Arch. Biochem. Biophys.*, **152**, 457
Lu, A. Y. H., Levin, W., West, S., Jacobson, M., Ryan, D., Kuntzman, R. and Conney, A. H. (1973). *Ann. N.Y. Acad. Sci.*, **212**, 156
McMahon, R. E. and Sullivan, H. R. (1966). *Life Sci.*, **5**, 921
Marquardt, H. and Heidelberger, C. (1972). *Chem.-Biol. Interactions*, **5**, 69
Matsumura, S. and Omura, T. (1973). *Drug Metabolism and Disposition*, **1**, 248
Maylin, G. A., Cooper, M. J. and Anders, M. W. (1973). *J. Med. Chem.*, **16**, 606
Mezey, E. (1971). *Biochem. Pharmacol.*, **20**, 508
Mitoma, C., Yasuda, D., Tagg, J. S., Neubauer, S. E., Calderoni, F. J. and Tanabe, N. (1968). *Biochem. Pharmacol.*, **17**, 1377
Mitoma, C., Lombrozo, L., Le Valley, S. E. and Dehn, F. (1969). *Arch. Biochem. Biophys.*, **134**, 434
Moore, M. R., Battistini, V., Beattie, A. D. and Goldberg, A. (1970), *Biochem. Pharmacol.*, **19**, 751
Mowat, A. P. and Arias, I. M. (1970). *Biochim. Biophys. Acta*, **212**, 175
Mulder, G. J. (1970). *Biochem. J.*, **117**, 319
Mulder, G. J. (1971). *Biochem. Pharmacol.*, **20**, 1328
Mycek, M. J. (1971). *Biochem. Pharmacol.*, **20**, 325
Neale, M. G. (1970). Ph.D. Thesis, University of Surrey
Nebert, D. W. (1969). *Biochem. Biophys. Res. Commun.*, **36**, 885
Nebert, D. W. (1970). *J. Biol. Chem.*, **245**, 519
Nebert, D. W. (1973). *Clin. Pharmacol. Ther.*, **14**, 693
Nebert, D. W. and Bausserman, L. L. (1970a). *Mol. Pharmacol.*, **6**, 293
Nebert, D. W. and Bausserman, L. L. (1970b). *Mol. Pharmacol.*, **6**, 304
Nebert, D. W. and Bausserman, L. L. (1970c). *J. Biol. Chem.*, **245**, 6373
Nebert, D. W. and Gelboin, H. V. (1968a). *J. Biol. Chem.*, **243**, 6242

Nebert, D. W. and Gelboin, H. V. (1968b). *J. Biol. Chem.*, **243**, 6250
Nebert, D. W. and Gelboin, H. V. (1969). *Arch. Biochem. Biophys.*, **134**, 76
Nebert, D. W. and Gelboin, H. V. (1970). *J. Biol. Chem.*, **245**, 160
Nebert, D. W. and Gielen, J. E. (1971). *J. Biol. Chem.*, **246**, 5199
Nebert, D. W., Goujon, F. M. and Gielen, J. E. (1972). *Nature New Biology*, **236**, 107
Nebert, D. W., Considine, N. and Owens, I. S. (1973). *Arch. Biochem. Biophys.*, **157**, 148
Netter, K. J. and Seidel, G. (1964). *J. Pharmacol. Exp. Ther.*, **146**, 61
Norback, D. H. and Allen, J. R. (1972). *Proc. Soc. Exp. Biol. Med.*, **139**, 1127
Novick, W. J., Jr., Stohler, C. M. and Swagzdis, J. (1966). *J. Pharmacol. Exp. Ther.*, **151**, 139
Oesch, F. (1973). *Xenobiotica*, **3**, 5, 305
Oesch, F., Morris, N., Daly, J. W., Gielen, J. E. and Nebert, D. W. (1973). *Mol. Pharmacol*, **9**, 692
Ohnhaus, E. E., Thorgeirsson, S. S., Davies, D. S. and Breckenridge, A. (1971). *Biochem. Pharmacol.*, **20**, 2561
Okey, A. B. (1972). *Life Sci.*, **11**, 833
O'Malley, K., Browning, M., Stevenson, I. and Turnbull, M. J. (1973). *Eur. J. Clin. Pharmacol.*, **6**, 102
Orrenius, S., Ericson, J. E. and Ernster, L. (1965). *J. Cell Biol.*, **25**, 627
Owen, N. V., Griffing, W. J., Hoffman, D. G., Gibson, W. R. and Anderson, R. C. (1971). *Toxicol. Appl. Pharmacol.*, **18**, 720
Page, J. G. and Vesell, E. S. (1969). *Proc. Soc. Exptl. Biol. Med.*, **131**, 256
Pandhi, P. N. and Baum, H. (1970). *Life Sci.*, **9**, 87
Pantuck, E. J., Kuntzman, R. and Conney, A. H. (1972). *Science (Washington)*, **175**, 1248
Parke, D. V. (1968). *The Biochemistry of Foreign Compounds*. International Series of Monographs in Pure and Applied Biology, Pergamon Press; Oxford
Parke, D. V. (1972). *Biochem. J.*, **130**, 53P
Parke, D. V. and Rahman, H. (1969). *Biochem. J.*, **113**, 12P
Parke, D. V. and Rahman, H. (1970). *Biochem. J.*, **119**, 53P
Parke, D. V. and Rahman, H. (1971). *Biochem. J.*, **123**, 9P
Parke, D. V., Rahman, K. Q. and Walker, R. (1974). *Biochem. Soc. Trans.*, **2**, 615
Periano, C., Fry, R. J. M. and Staffeldt, E. (1971). *Cancer Res.*, **31**, 1506
Philpot, R. M. and Hodgson, E. (1971). *Chem.-Biol. Interactions*, **4**, 185
Platt, D. S. and Cockrill, B. L. (1969a). *Biochem. Pharmacol.*, **18**, 429
Platt, D. S. and Cockrill, B. L. (1969b). *Biochem. Pharmacol.*, **18**, 445
Platt, D. S. and Cockrill, B. L. (1969c). *Biochem. Pharmacol.*, **18**, 459
Poland, A. and Glover, E. (1973). *Science (Washington)*, **179**, 476
Pousada, C. R. and Lechner, M. C. (1972). *Biochem. Pharmacol.*, **21**, 2563
Radzialowski, F. M. and Bousquet, W. F. (1968). *J. Pharmacol. Exp. Ther.*, **163**, 229
Recknagel, R. O. (1967). *Pharmacol. Rev.*, **19**, 145
Redmond, G. and Cohen, G. (1971). *Science (Washington)*, **171**, 387
Reiner, O. and Uehleke, H. (1971). *Hoppe-Seyler's Z. Physiol. Chem.*, **352**, 1048
Remmer, H. and Merker, H. J. (1963). *Science, (Washington)*, **142**, 1657
Remmer, H., Schenkman, J., Estabrook, R. W., Sasame, H., Gillette, J., Narasimhulu, S., Cooper, D. Y. and Rosenthal, O. (1966). *Mol. Pharmacol.*, **2**, 187
Richardson, H. L., Stier, A. R. and Borsos-Nachtnebel, E. (1952). *Cancer Res.*, **12**, 356
Rifkind, A. B., Gillette, P. G., Song, C. S. and Kappas, A. (1973). *J. Pharmacol. Exp. Ther.*, **185**, 214
Roe, F. J. C. and Rowson, K. E. K. (1968). *Int. Rev. Exper. Path.*, **6**, 181
Ross, J. D. and Muller-Eberhard, V. (1970). *J. Lab. Clin. Med.*, **75**, 694

Rubin, A., Tephly, T. R. and Mannering, G. J. (1964). *Biochem. Pharmacol.*, **13**, 1007

Rubin, E., Hutterer, F. and Lieber, C. S. (1968). *Science (Washington)*, **159**, 1469

Rubin, E., Bacchin, P., Gang, H. and Lieber, C. S. (1970). *Lab. Invest.*, **22**, 569

Rubin, E., Lieber, C. S., Alvares, A. P., Levin, W. and Kuntzman, R. (1971). *Biochem. Pharmacol.*, **20**, 229

Ruddon, R. W. and Rainey, C. H. (1970). *Biochem. Biophys. Res. Commun.*, **40**, 152

Sarkar, N. S. (1969). *FEBS Lett.*, **4**, 37

Scharf, R. and Ullrich, V. (1973). *Naunyn-Schmiedeberg's Arch. Pharmacol.*, **278**, 329

Schenkman, J. B., Wilson, B. J. and Cinti, D. L. (1972). *Biochem. Pharmacol.*, **21**, 2373

Schlede, E. and Conney, A. H. (1970). *Life Sci.*, **9**, 1295

Schulte-Hermann, R., Schlicht, I., Koransky, W., Leberl, C., Eulenstedt, C. and Zimeck, M. (1972). *Naunyn-Schmiedeberg's Arch. Pharmacol.*, **273**, 109

Seifert, J. and Vácha, J. (1970). *Chem.-Biol. Interactions*, **2**, 297

Selye, H. (1970). *Rev. Can. Biol.*, **29**, 49

Shillingford, J. S., Lindup, W. E. and Parke, D. V. (1973). *Biochem. Soc. Trans.*, **1**, 966

Shoeman, D. W., Chapman, M. D. and Mannering, G. J. (1969). *Mol. Pharmacol.*, **5**, 412

Shoeman, D. W., Vane, F. H. and Mannering, G. J. (1973), *Drug Metabolism and Disposition*, **1**, 40

Sladek, N. E. and Mannering, G. J. (1966). *Biochem. Biophys. Res. Commun.*, **24**, 668

Sladek, N. E. and Mannering, G. J. (1969). *Mol. Pharmacol.*, **5**, 186

Slater, T. F. and Sawyer, B. C. (1971). *Biochem. J.*, **123**, 805

Sluyser, M. (1968). *Biochim. Biophys. Acta*, **154**, 606

Sluyser, N. (1969). *Biochim. Biophys. Acta*, **182**, 235

Smith, S. J., Hill, R. N., Gleeson, R. A. and Vesell, E. S. (1972). *Mol. Pharmacol.*, **8**, 691

Solymoss, B. (1972). *Nature New Biology*, **237**, 61

Solymoss, B., Krajny, M., Varga, S. and Werringloer, J. (1970). *J. Pharmacol. Exp. Ther.*, **174**, 473

Solymoss, B., Toth, S., Varga, S., Werringloer, J. and Zsigmond, G. (1971). *Can. J. Physiol. Pharmacol.*, **49**, 841

Somogyi, A., Kovacs, K., Solymoss, B., Kuntzman, R. and Conney, A. H. (1971). *Life Sci.*, **10**, 1261

Springfield, A. C., Carlson, G. P. and De Feo, J. J. (1973). *Toxicol. Appl. Pharmacol.*, **24**, 298

Stein, Y. and Stein, O. (1969). *Israel J. Med. Sci.*, **5**, 985

Stolman, S. and Loh, H. H. (1970). *J. Pharm. Pharmacol.*, **22**, 713

Stripp, B., Hamrick, M. E., Zampaglione, N. G. and Gillette, J. R. (1971). *J. Pharmacol. Exp. Ther.*, **176**, 766

Stripp, B., Menard, R. H., Zampaglione, N. G., Hamrick, M. E. and Gillette, J. R. (1973). *Drug. Metabolism and Disposition*, **1**, 216

Suarez, K. A., Carlson, G. P., Guller, G. C. and Fausto, N. (1972). *Toxicol. Appl. Pharmacol.*, **23**, 171

Talcott, R. E. and Stohs, S. J. (1973). *Res. Comm. Chem. Path. Pharmacol.*, **5**, 663

Taylor, S. A., Rawlins, M. D. and Smith, S. E. (1972). *J. Pharm. Pharmacol.*, **24**, 578

Tephly, T. R., Tinelli, F. and Watkins, W. D. (1969). *Science (Washington)*, **166**, 627

Tephly, T. R., Hasegawa, E. and Baron, J. (1971). *Metabolism*, **20**, 200

Thurman, R. C. (1973). *Mol. Pharmacol.*, **9**, 670

Tobon, F. and Mezey, E. (1971). *J. Lab. Clin. Med.*, **77**, 110
Tredger, J. M., Chakraborty, J. and Parke, D. V. (1973). *Biochem. Soc. Trans.*, **1**, 1000
Tuchweber, B., Solymoss, B., Khandekar, J. D., Kovacs, K. and Garg, B. D. (1972). *Exp. Molec. Pathol.*, **17**, 281
Uehleke, H. and Brill, E. (1968). *Biochem. Pharmacol.*, **17**, 1459
Uehleke, H. and Greim, H. (1968). *Naunyn-Schmiedeberg's Arch. Pharmacol. Path.*, **259**, 199
Ulland, B. M., Weisburger, J. H., Yamamoto, R. S. and Weisburger, E. K. (1973). *Fd. Cosmet. Toxicol.*, **11**, 199
Ullrich, V., Frommer, U. and Weber, P. (1973). *Hoppe-Seyler's Z. Physiol. Chem.*, **354**, 514
Ullrich, V. and Schnabel, K. H. (1973). *Arch. Biochem. Biophys.*, **159**, 240
Venkatesan, N., Arcos, J. C. and Argus, M. F. (1971). *J. Theor. Biol.*, **33**, 517
Vessel, E. S. (1967). *Science (Washington)*, **157**, 1057
Wada, F., Hirata, K., Shibota, H., Higashi, K. and Sakamoto, Y. (1967). *J. Biochem. Tokyo*, **62**, 134
Wagstaff, D. J. and Short, C. R. (1971). *Toxicol. Appl. Pharmacol.*, **19**, 54
Wattenberg, L. W. (1972). *Toxicol. Appl. Pharmacol.*, **23**, 741
Welch, R. M., Harrison, Y. E. and Burns, J. J. (1967). *Toxicol. Appl. Pharmacol.*, **10**, 340
Welch, R. M., Harrison, Y. E., Gommi, B. W., Poppers, P. J., Finster, M. and Conney, A. H. (1969). *Clin. Pharmacol. Ther.*, **10**, 100
Welch, R. M., Loh, A. and Conney, A. H. (1971). *Life Sci.*, **10**, 215
Welch, R. M., Cavallito, J. and Loh, A. (1972). *Toxicol. Appl. Pharmacol.*, **23**, 749
Wheatley, D. N., Gerrard, M. E., Kernohan, I. R. and Currie, A. R. (1972). *Br. J. Cancer*, **26**, 99
Whitlock, J. P., Jr. and Gelboin, H. V. (1973). *J. Biol. Chem.*, **248**, 6114
Wiebel, F. J., Gelboin, H. V. and Coon, H. G. (1972). *Proc. Natl. Acad. Sci.*, **69**, 3580
Wiebel, F. J., Leutz, J. C. and Gelboin, H. V. (1973). *Arch. Biochim. Biophys.*, **154**, 292
Wisnes, A. (1969). *Biochim. Biophys. Acta*, **191**, 279
Wisnes, A. (1971). *Biochem. Pharmacol.*, **20**, 1249
Wisnes, A. (1972). *Biochim. Biophys. Acta*, **284**, 394
Wold, J. W. and Steele, W. J. (1969). *Fed. Proc.*, **28**, 48
Yamamoto, R. S., Weisburger, J. H. and Weisburger, E. K. (1971). *Cancer Res.*, **31**, 483
Yee, M. and Bresnick, E. (1971). *Mol. Pharmacol.*, **7**, 191
Young, D. L., Powell, G. and McMillan, W. O. (1971). *J. Lipid Res.*, **12**, 1
Zakim, D., Goldenberg, J. and Vessey, D. A. (1973). *Biochim. Biophys. Acta*, **297**, 497
Zampaglione, N. G., Jollow, D. J., Mitchell, J. R., Stripp, B., Hamrick, M. and Gillette, J. R. (1973). *J. Pharmacol. Exp. Ther.*, **187**, 218

Chapter 9

Clinical Implications of Enzyme Induction

Alasdair Breckenridge
Department of Clinical Pharmacology,
Royal Postgraduate Medical School,
London W.12.

9.1. INTRODUCTION

The implications of enzyme induction extend beyond alterations in rates of hepatic microsomal drug oxidation. In the last few years it has been shown in man that administration of inducing agents may be associated with increased turnover of endogenous substrates such as cholesterol, vitamin D and cortisol, with alterations in liver blood flow and bile flow, and even with the amelioration of disease processes such as Gilbert's syndrome and perhaps Cushing's syndrome.

Not only are enzymes of the hepatic endoplasmic reticulum stimulated by administration of inducing agents, but enzyme activity within white blood cells, placenta, skin and gut may also be augmented. However, the functional counterparts of these changes have not as yet been clearly delineated. Thus this chapter deals predominantly with the effects of inducing agents on the enzymes of the hepatic endoplasmic reticulum. The purists may rightly contest the widespread use of the term "enzyme induction" in this chapter. In the few instances in man where stimulation of drug metabolism has been demonstrated, it has never been unequivocally shown to be due to enzyme induction. To demonstrate induction unequivocally one would have to prove:

1. Increased enzyme activity within the tissues due to increased enzyme enzyme protein synthesis and not to decreased protein catabolism, or merely to a general increase in cell size and protein content.

2. An increased rate of production of drug metabolites *in vivo* when the inducing agent was given, with no change in the apparent volume of distribution of either the parent drug or the metabolites formed from the parent drug.

In view of this, the term "stimulation" may be preferred to "induction," but the latter term is retained for the sake of uniformity in the full knowledge that it may well be inaccurate.

9.2. ASSESSMENT OF AN AGENT AS A POSSIBLE INDUCING AGENT IN MAN

Two principal methods of assessing whether a drug is an inducing agent in man have been used. The first is to monitor changes in an endogenous substance after administration of the agent under study. The substances most widely used for this are urine 6β-hydroxycortisol, urine D-glucaric acid and the enzyme plasma γ-glutamyl transpeptidase (γGT). It should be stressed that it is *changes* in these parameters when the subject is given an agent which are important; an isolated reading of the parameter is in itself of little value in the individual patient.

The second method is to use changes in exogenous substrates (usually drugs) produced by administration of the agent under study.

Changes in Endogenous Substances

6β-Hydroxycortisol

6β-Hydroxycortisol (Fig. 9.1) is a polar metabolite of cortisol which is found unconjugated in urine and which normally forms between 2% and 6% of the total urinary 17-oxogenic steroid output (Frantz *et al.*, 1961). Current evidence favours the endoplasmic reticulum of the liver as the source of 6β-hydroxycortisol, although it has been suggested that in some patients with adrenal tumours, the adrenal cortex

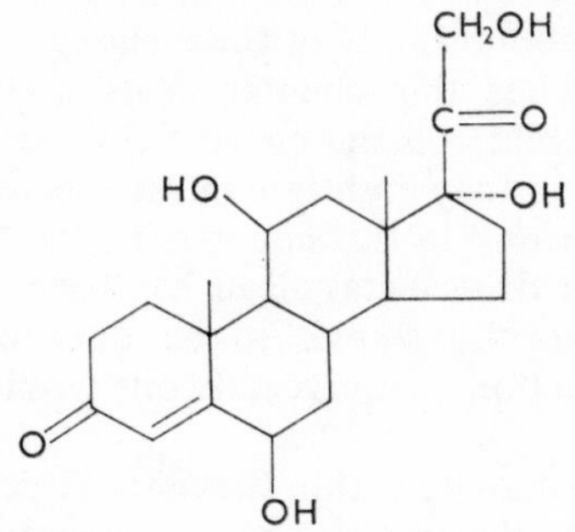

Fig. 9.1. 6β-Hydroxycortisol.

itself may secrete 6β-hydroxycortisol (Thrasher *et al.*, 1969). Administration of certain drugs increases the ratio of 6β-hydroxycortisol to the total 17-oxogenic steroid output, presumably by increasing the proportion of cortisol oxidized to this metabolite within the liver.

To utilize the ratio as an index of enzyme induction, one must compare 6β-hydroxycortisol excretion prior to administration of the agent under study, with values obtained after a suitable interval of treatment. It has been shown in man that diphenylhydantoin (Werk *et al.*, 1964), phenobarbitone (Burstein and Klaiber, 1965), antipyrine, dichloralphenazone, (Breckenridge *et al.*, 1971) phenylbutazone (Kuntzman *et al.*, 1966) o,p′-DDD (Bledsoe *et al.*, 1964), and DDT (Poland *et al.*, 1970) cause an increase in urine 6β-hydroxycortisol excretion in man.

However, it has also been shown that chlordiazepoxide (30 mg/day) may also cause an increase in 6β-hydroxycortisol excretion, and at this dose there were no changes in the rate of drug oxidation (Orme *et al.*, 1972). Figure 9.2 shows increases in urine 6β-hydroxycortisol excretion in a group of patients on long-term treatment with warfarin, who were given a variety of hypnotic and sedative drugs including chlordiazepoxide; these changes are correlated with alteration in plasma warfarin concentration. One cannot use changes in urine 6β-hydroxycortisol excretion as an index of changes in rates of drug metabolism in the

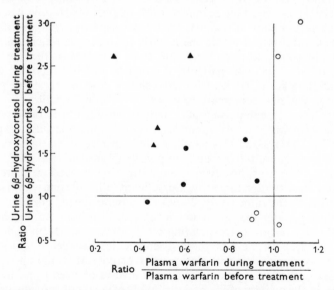

Fig. 9.2. Relation between changes in steady state plasma warfarin concentrations and urine 6β-hydroxycortisol excretion in 15 patients given barbiturates (●); benzodiazepines (○); dichloralphenazone and antipyrine (▲).

individual patient, although as a screening procedure in large groups of patients exposed to an agent it may be a valid method of assessing stimulation of the hepatic microsomal enzymes. For example, subjects exposed to DDT in their work have been shown to excrete more 6β-hydroxycortisol in urine and to have shorter plasma half-lives of phenylbutazone than non-exposed subjects; there was, however, no correlation between the 6β-hydroxycortisol excretion and the phenyl-butazone half-life (Poland et al., 1970).

Changes in plasma γ-glutamyl transpeptidase (γGT)

γ-Glutamyl transpeptidase is a ubiquitous enzyme found in highest concentrations within the kidney, liver, pancreas and testis (Orlowski, 1963). Its function is largely undetermined, but it may be concerned with the intracellular transfer of γ-glutamyl residues and has nothing to do with drug metabolism. The enzyme would appear to be stable over several months in frozen plasma and shows relatively little day-to-day variation, so that increases can be easily detected. Plasma concentrations of γGT are increased in patients with acute hepatocellular damage and in some patients with prostatic tumours. γ-Glutamyl transpeptidase concentrations are only sporadically increased in the plasma of patients with renal disease, e.g. untreated nephrotic syndrome and some patients with renal tumours (Orlowski, 1963). Rosalki et al. (1971) showed that administration of phenobarbitone caused an increase in plasma γGT activity, the hypothesis being that an increased enzyme concentration within the hepatic endoplasmic reticulum spilled into the plasma. Although it remains to be proven that the liver is the source of γGT in these patients, this hypothesis is supported by the work of Ideo et al., (1972) who have shown that in rats given pheno-barbitone there is an increase in hepatic γGT concentration.

A significant rise in plasma γGT activity was seen in 13 of 14 patients on long-term warfarin given amylobarbitone, quinalbarbitone or anti-pyrine for 30 days and this paralleled falls in plasma warfarin concentrations both quantitatively and temporally (Whitfield et al., 1973). One patient in this series, however, given antipyrine showed no increase in γGT activity although there was a significant fall in plasma warfarin concentration. In the same study, 2 out of 4 patients given benzodiazepines (chlordiazepoxide, diazepam or nitrazepam) showed an increase in γGT activity but no corresponding fall in plasma warfarin concentration (Fig. 9.3). In view of these discrepancies between alterations in rates of drug oxidation and plasma γGT activity, it would appear that increases in the latter parameter cannot always be used as an index of changes in the activity of liver microsomal enzymes concerned with drug oxidation. In this respect conclusions about changes seen in γGT activity are very like those reached concerning changes in urine 6β-hydroxycortisol excretion produced by inducing agents.

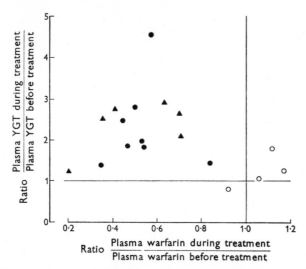

Fig. 9.3. Relation between changes in steady state plasma warfarin concentrations and plasma γ-glutamyl transpeptidase concentration in 18 patients given barbiturates (●); benzodiazepines (○); dichloralphenazone and antipyrine (▲). (From, Whitfield *et al.*, 1973, reproduced by kind permission of the editor of the *British Medical Journal*).

Urine excretion of D-*Glucaric acid*

It was originally shown by Aarts (1965) that treatment of patients with phenobarbitone or aminopyrine caused an increased urine excretion of D-glucaric acid. D-Glucaric acid is an end-product of carbohydrate metabolism produced via the glucuronic acid pathway (Fig. 9.4). The rate-limiting step of this pathway is unknown, but phenobarbitone has been shown to increase the activity of all three enzymes shown (Kawada *et al.*, 1961). In many mammals, D-glucuronic acid is converted mainly to ascorbic acid, and thus it is only in those mammals incapable of synthesizing ascorbic acid (i.e. primates and guinea-pigs) that increases in D-glucaric excretion are a reliable index of increased enzyme activity. In guinea-pigs given phenobarbitone, a good correlation between the daily excretion of D-glucaric acid and total liver *content* of cytochrome P-450 has been found (Hunter *et al.*, 1973) (whether or not there is a correlation with P-450 *activity* is uncertain; this is more relevant). Increased D-glucaric acid excreton has been shown in epileptics given barbiturates and diphenylhydantoin; and a positive correlation between the dose of inducing agent and the increase in D-glucaric acid excretion has also been shown (Hunter *et al.*, 1971a). It has further been found that factory workers exposed to insecticides

10

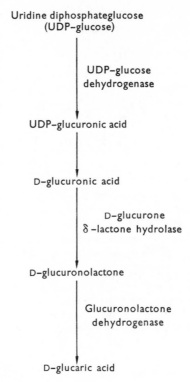

Fig. 9.4. Pathway of formation of D-glucaric acid.

(aldrin, dieldrin, endin) which are known to increase rates of drug oxidation in man, have a significantly higher D-glucaric acid excretion than control subjects (Hunter *et al.*, 1972).

Patients on oral contraceptives have also been shown to excrete more D-glucaric acid than the control population (Mowat, 1968) and it is of interest that oral contraceptives have not been shown to increase rates of drug oxidation in man.

It should be stressed, however, that, like plasma γGT, the enzymes concerned with the synthesis of D-glucaric acid are not those concerned with hepatic microsomal drug oxidation. There is no correlation between D-glucaric acid excretion and basal rates of drug oxidation, nor any correlation, for that matter, between γGT, urine excretion of 6β-hydroxycortisol and D-glucaric excretion (Smith and Rawlins, 1974).

It would thus appear that these indirect indices may be valuable in group studies to screen for increases in hepatic microsomal enzyme activity, but to use them in individual subject as markers for alterations in rates of drug oxidation is not justified.

Changes in Rates of Drug Metabolism

Evidence for classifying a drug as an inducing agent in man is frequently accrued from measurements of changes in rates of drug metabolism produced by administration of the putative agent. Three parameters have been studied in man.

Change in rate of metabolite formation

The best evidence for a change in the rate of drug metabolism is a change in the rate of metabolite formation when the putative agent is given. Breckenridge et al. (1971) performed such a study in which the urine excretion of the metabolites of warfarin were measured after single doses of ^{14}C-warfarin, before and after administration of antipyrine, 600 mg daily for 30 days, to two patients. Since virtually no unchanged warfarin is excreted in urine, urine radioactivity can be taken as a measurement of metabolite excretion. In two patients studied, the percentage of dose eliminated in urine over 24 and 48 hours increased after antipyrine administration; concomitantly there was a shortening of the plasma half-life of warfarin.

Pantuck et al. (1974) examined the plasma concentrations of phenacetin and its metabolite p-acetamidolphenol (paracetamol) in smokers and non-smokers, and found that the ratio of plasma p-acetamidophenol to phenacetin was increased several-fold in smokers. This suggested that cigarette smoking had stimulated the rate of metabolism of phenacetin.

Phenobarbitone which may increase the rate of metabolism of diphenylhydantoin, will also increase the rate of excretion of 5-hydroxyphenyl-5-phenylhydantoin in urine (Morselli et al., 1971).

Change in the plasma half-life of a marker drug

Figure 9.5 (taken from Reigelman et al., 1971) shows the theoretical changes which should be seen in the elimination of drug before and after administration of an enzyme inducing agent. In the upper curve (the intravenous data), the early steep phase represents distribution from the blood into tissues; the later less steep slope represents elimination by metabolism. If the rate of metabolism is increased by enzyme induction, a steeper curve at a lower level is seen. After oral dosing, curves like those seen in the lower diagram would be obtained. After induction, the peak blood concentration is reached earlier, the peak is lower and the subsequent slope is steeper due to increased metabolism. If the agent alters either the apparent volume of distribution or the absorption of the marker drug, a change in the plasma half-life may be

10A

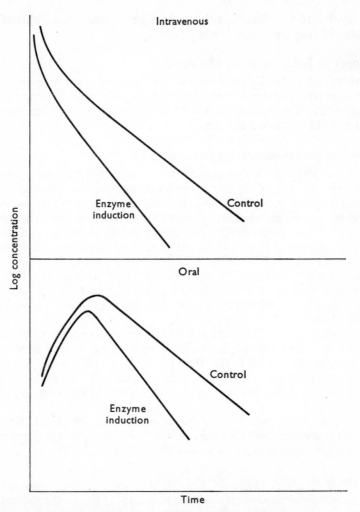

Fig. 9.5 Theoretical changes produced by enzyme induction in the plasma
half-life of 2 drug given intravenously and orally

obtained, but there may be no change in the rate of metabolism. On
occasions, the inducing agent may cause a change in metabolism *and*
absorption of the marker drug, as was shown by Aggeler and O'Reilly
(1969) with heptabarbital and bishydroxycoumarin. This complex
interaction was solved only by giving the coumarin by oral then intravenous routes, and measuring the amount of unchanged drug in faeces
under the several conditions.

Changes in steady state plasma concentration

The steady state plasma concentration (C_{ss}) of a drug is determined by several factors (Wagner et al., 1965):

$$C_{ss} = \frac{F.D}{V_{\mathrm{D}}.K_{\mathrm{el}}.\Delta_t}$$

where F is the fraction of the dose (D) absorbed, V_{D} is the apparent volume of distribution, K_{el} is the elimination rate constant, and Δ_t is the dosing interval.

Thus a second drug may alter C_{ss} by changing either F, V_D or K_{el}. A fall in C_{ss} does not imply enzyme induction.

9.3. FACTORS CONTROLLING ENZYME INDUCTION IN MAN

The factors which determine whether a drug will stimulate drug oxidation in man are poorly understood. As Conney (1967) noted: "The characteristic pharmacological actions of these compounds on the organism are extremely diverse, and there is no apparent relationship between either their actions or structure and their ability to induce enzymes. It is of interest that most of the inducers are soluble in lipid at a physiological pH'. Lipid solubility, however, is not the most relevant physical property of inducing agents. In a series of studies on the inducing potency of six barbiturates, relative lipid solubility (as measured by heptane:water partition coefficient) was not shown to be a good predictor of their relative potency as an inducer; e.g., phenobarbitone was many times less lipid soluble than thiopentone but was a much more potent inducing agent in the rat (Breckenridge et al., 1974). Nor is the inducing ability of barbiturates in rats inversely correlated with lipid solubility (Pelkonen and Karki, 1973).

One factor which is important in determining the degree of induction is the dose of inducing agent. Enzyme induction as applied to drug oxidation is frequently thought of as an "all or nothing" response, whereas, in fact, it is graded like many other pharmacological effects. In a recent study, quinalbarbitone, 100 mg nightly, was given for 30 nights to 6 patients stabilized on long-term warfarin; the percentage fall in steady state plasma warfarin concentration was found to range from 5% to 70%. By increasing the dose of quinalbarbitone to 200 mg nightly (2 patients) and 300 mg quinalbarbitone (1 patient) in separate studies and measuring the percentage fall in plasma warfarin concentrations, a dose response curve for enzyme induction in man was constructed (Breckenridge et al., 1973b).

The role of genetic factors in controlling enzyme induction in man was studied by Vesell and Page (1969) in a series of twins. These authors showed that phenobarbitone induced a more pronounced reduction of the antipyrine plasma half-life in subjects whose initial antipyrine half-life was long. They found no correlation between the plasma concentration of phenobarbitone and the relative shortening of the antipyrine plasma half-life. This latter finding was confirmed by Breckenridge *et al.* (1973b) who showed no correlation between the percentage fall in steady state plasma warfarin concentration and plasma quinalbarbitone concentration in patients given quinalbarbitone. The demonstration of enzyme induction in organ and tissue culture, however, emphasize that factors other than those operating *in vivo* exert an important control over the whole process (Wattenberg *et al.*, 1968).

9.4. DRUGS SHOWN TO BE INDUCING AGENTS IN MAN

Many textbooks of pharmacology contain lists of drugs which are 'inducing agents'. Great care must be taken in interpreting such data. Firstly, the inter-individual variability of induction (mentioned above) must be considered. Secondly, critical examination of many of these lists reveals that frequently data have been derived from animal studies and extrapolated to man. If a drug is found *not* to increase rates of drug oxidation in, say, the rat when given at doses 10–20 times those studied in man, it appears to be unlikely that an inducing effect will be seen in man; but if enzyme induction is observed in the rat under these circumstances, this does not necessarily mean the same effect will be observed in man when the drug is given in smaller doses. Thirdly, the evidence for induction in man has frequently been a decrease in the pharmacological response of the marker drug. There are many possible factors, both dynamic and kinetic, other than enzyme induction, to account for such a change.

Types of Inducing Agent

Based on animal studies, inducing agents have been divided into several sub-groups (Conney, 1967). The first type of inducing agent exemplified by phenobarbitone is characterized by the ability to stimulate the metabolism of a large number of substrates by a number of pathways in liver microsomal preparations. These include oxidation and reduction reactions and glucuronide formation. The second type, typified by 3-methylcholanthrene and 3,4-benzpyrene, stimulates a more limited group of reactions. These inducing agents also differ in their time course and intensity of induction and in the production of either cytochrome P-450 or P-448 within the hepatic endoplasmic reticulum. One further sub-group, steroids which under certain conditions will induce liver microsomal enzyme activity, has been suggested. Steroids which increase rates of drug oxidation usually possess androgenic, anabolic, progestational or glucocorticoid activity. Spironolactone, the aldosterone antagonist, appears to be a slightly unusual inducer of drug metabolism since, with the exception of slight progestational activity, it is reported to lack the above steroid properties (Hertz and Tullner, 1968).

The relevance of these sub-groups in clinical medicine is uncertain. Most therapeutic agents which act as inducing agents belong to the first sub-group; some environmental contaminants and food additives which may be inducing agents in man, belong to the second.

Barbiturates

These are probably the best documented and most widely studied inducing agents in man. This effect has been shown in man for many barbiturates, including barbitone (Dayton *et al.*, 1961), amylobarbitone (Robinson and Sylwester, 1970), aprobarbitone (Johansson, 1968), butobarbitone (Anlintz *et al.*, 1968), heptabarbitone (Aggeler and O'Reilly, 1969), phenobarbitone (Cucinell *et al.*, 1965), quinalbarbitone (Breckenridge *et al.*, 1973b) and vinbarbitone (Johannson, 1968).

Interactions with oral anticoagulants

This has been a fruitful area for clinical pharmacological research, since not only can changes in plasma concentration of the oral anti-coagulant be measured, but also changes in its pharmacological effect. The basis for most of the interactions between barbiturates and oral anticoagulants is enzyme induction, but it is worth noting that hepta-barbitone decreases the effect of bishydroxycoumarin by two mechanisms: increasing its rate of oxidation and decreasing its absorption as noted above (Aggeler and O'Reilly, 1969). Bishydroxycoumarin is, of course, relatively poorly absorbed and no such interaction has been found with warfarin, which is well absorbed.

Several general points about barbiturate–coumarin interactions are important. First, there is marked interindividual variation in the degree of induction as discussed above. In view of this, it is obviously difficult to state what is a significant period of barbiturate administration to cause enzyme induction. If a change in plasma coumarin concentrations is to be seen, six days of continuous barbiturate administration is sufficient to show it. The maximum effect is usually seen after 14–21 days of barbiturate administration (Breckenridge and Orme, 1971). In one study, 2 days of phenobarbitone at 210 mg/day caused a significant shortening of the plasma half-life of warfarin but in only 13 of 21 subjects (Corn, 1966).

When a fall in plasma coumarin concentration has been observed, stopping the barbiturate results in a return to previous anticoagulant concentrations, and this is usually complete by 18–24 days, although under certain conditions may take longer (Fig. 9.6). There would appear

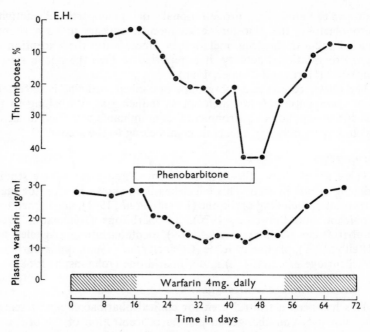

Fig. 9.6. Change in steady state plasma warfarin concentration and thrombo-
test produced by phenobarbitone (120 mg/day for 30 days).

to be relatively less interindividual variation in the time course of the
effect, however, than in the magnitude of the effect.

Although under certain clinical circumstances it is unavoidable to
administer barbiturates and coumarins together, this practice should be
discouraged as far as possible. Even with careful monitoring, optimal
anticoagulant control may be difficult to achieve in these patients
(Robinson and McDonald, 1966).

Interactions with diphenylhydantoin

Addition of phenobarbitone may cause a rise, a fall, or no change in
the steady state plasma concentration of diphenylhydantoin (Morselli
et al., 1971). In few patients would these changes appear to be of great
clinical significance. There is still considerable debate as to whether the
addition of phenobarbitone to diphenylhydantoin represents good
therapeutic practice; undoubtedly on many occasions these two drugs
are given when one drug (diphenylhydantoin) appropriately prescribed
would be sufficient. However, when the side effects of diphenylhydantoin become limiting, it is reasonable to add phenobarbitone to the
therapeutic regimen, especially as it has been suggested that the mode of

action of the two anticonvulsants is different and may be additive (Calne, 1973). It has been shown in rats that brain concentrations of diphenylhydantoin are higher in those treated with phenobarbitone and diphenylhydantoin than when diphenylhydantoin is given alone (Rizzo et al., 1972). The mechanism of this augmentation is not clear, but in a study in man where brain, plasma and CSF concentrations of both agents were measured, the brain/plasma concentrations of diphenyl- hydantoin were similar whether or not phenobarbitone was given (Vajda et al., 1974).

Interactions with other drugs

(1) Digitoxin is metabolized by hepatic microsomal enzymes; its principal metabolic product is digoxin. It has been demonstrated in man that phenobarbitone administration results in increased conver- sion (Jelliffe and Blankenhorn, 1966). This is probably of little clinical importance, since digoxin too is active as an inotropic and antiarrhyth- mic agent.

(2) Phenobarbitone administration has been reported to cause a shortening of the plasma half-life of the antibiotic doxycycline. This has been attributed to enzyme induction but there was no evidence whether the absorption or apparent volume of distribution of doxycline was altered by phenobarbitone (Neuvonen and Pentilla, 1974).

(3) In patients with liver disease exposed to various barbiturates, the plasma half-life of phenylbutazone was shorter than in the control population (Levi et al. 1968). The duration of barbiturate administra- tion varied from 2 days upwards and no evidence for enzyme induction other than the change in half-life was given.

(4) Phenobarbitone has also been shown to increase the rate of elimination of desmethylimipramine (Hammer and Sjöqvist, 1967), nitroglycerin (Bogaert et al., 1971) and chlorpromazine (Forrest et al., 1969).

The clinical relevance of all these interactions is uncertain.

Other Hypnotic and Sedative Drugs

(1) *Glutethimide* administration has been known to shorten the plasma half-life of warfarin (Corn, 1966; McDonald et al., 1969) and ethylbiscoumacetate (van Dam and Gribnau-Overkamp, 1967). Glutethimide has also been shown to increase the rate of elimination of dipyrone, which is a derivative of aminopyrine (Remmer, 1962). It will also stimulate its own metabolism (Schmid et al., 1964).

(2) *Dichloralphenazone.* As its name suggests this is a complex of 2 parts of chloral hydrate with 1 part of phenazone (antipyrine). Its administration causes an increased elimination of warfarin metabolites, a shortening of the plasma half-life of warfarin and a fall in plasma warfarin steady state concentrations (Breckenridge et al., 1971).

Increases in urine 6β-hydroxycortisol excretion and plasma γ-glutamyl-transpeptidase have also been shown. These changes are all due to phenazone (antipyrine), an inducing agent in its own right. Chloral hydrate is not an inducing agent in either man or animals, although, through its metabolite, trichloracetic acid, which is strongly bound to plasma albumin, it may be the cause of other types of drug interaction (Seller and Koch-Weser, 1971).

(3) *Phenazone* (antipyrine). Antipyrine is widely used as a tool in pharmacology as a test drug for studying interindividual differences in rates of drug oxidation. When given chronically it will stimulate the rate of oxidation of other drugs, e.g. warfarin (Breckenridge *et al.*, 1971) and bilirubin glucuronidation (Orme *et al.*, 1974). Antipyrine is considerably less potent as an inducing agent than phenobarbitone in the rat (Breckenridge, 1973b) (see Fig. 9.7).

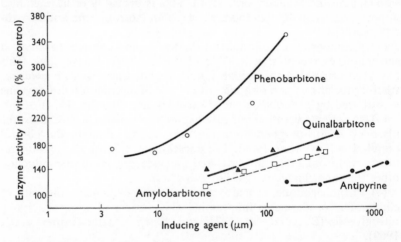

Fig. 9.7. Change in microsomal enzyme activity *in vitro* with liver microsomal preparations from rats pretreated with varying doses of four different inducing agents. (From, Breckenridge *et al.*, 1973b, reproduced by kind permission of the editor of *Clinical Pharmacology and Therapeutics*).

(4) *Ethchlorvynol*. This hypnotic has been shown to antagonize the hypoprothrombinaemic action of warfarin and bishydroxycoumarin in man (Johannson, 1968) but the mechanism of this antagonism is quite unclear. There is no evidence that it is due to enzyme induction.

(5) *Benzodiazepines*. Chlordiazepoxide may cause an increased urine excretion of 6β-hydroxycortisol with no change in the rates of warfarin elimination measured at the same time. It is interesting, however, that in the rat, chlordiazepoxide does increase rates of drug

oxidation, albeit at a dose considerably in excess of that given to man (Orme *et al.*, 1972). Nitrazepam and diazepam do not increase the rate of drug oxidation in man.

(6) *Phenothiazines.* Long-term administration of chlorpromazine results in a shortening of the plasma half-life of antipyrine in volunteers (S. H. Curry, personal communication). There are no comparable data for other phenothiazines but it would perhaps be surprising if the same effect was not seen with other congenors. In addition, chlorpromazine appears to stimulate its own metabolism (Curry *et al.*, 1972).

(7) *Antihistamines.* There is no good evidence in man or dogs (Hunninghake and Azarnoff, 1968) that antihistamines will stimulate rates of drug oxidation. Diphenhydramine, however, is a component of the hypnotic mandrax (methaqualone and diphenhydramine) and Ballinger *et al.*, (1972) have shown, in 3 patients addicted to mandrax, a shortening of the plasma half-life of antipyrine and an increased urine excretion of 6β-hydroxycortisol, both parameters changing appropriately when mandrax was withdrawn. That this effect was due to the antihistamine and not methaqualone is perhaps suggested by the studies of Orme (1974) where methaqualone, 250 mg nightly for 30 days, caused no change in either the steady state plasma concentration or pharmacological response of warfarin. Mandrax, however, was not used in these studies with warfarin; and the dose of mandrax taken by the addicts was not, of course, accurately known.

Other Drugs

(1) *Spironolactone.* This aldosterone antagonist has been shown to alter significantly the rates of oxidation of several drugs in animals, but results in man have been slightly inconsistent. In one study spironolactone, 50 mg three times daily for 7 days, caused a shortening of the plasma half-life of antipyrine in 4 of 9 subjects (Taylor *et al.*, 1972). For the group as a whole, however, there was no significant change in antipyrine half-life, or D-glucaric excretion, when given spironolactone. In a second study, 100 mg spironolactone per day for 14 days, caused a decrease in antipyrine half-life in 9 volunteers, associated with an increased urine excretion of 4-hydroxyantipyrine. There was also an increased urine excretion of 6β-hydroxycortisol in these subjects (Huffman *et al.*, 1973).

(2) *Hydrocortisone* (cortisol) in man will enhance the elimination of antipyrine and other drugs oxidized by liver microsomal enzymes (Breckenridge *et al.*, 1973a). This is a dose-dependent effect and change of elimination is immediate. It appears to be an effect on metabolism rather than any other pharmacokinetic parameter.

(3) *Diphenylhydantoin.* It has been known for many years that administration of diphenylhydantoin to man causes an increased excretion of 6β-hydroxycortisol in urine and an increase in the urine

ratio of 6β-hydroxycortisol to 17-oxogenic steroids (Werk *et al.*, 1964). More recently, however, it has been suggested that diphenylhydantoin may have more marked effects on steroid metabolism. It has also been suggested that diphenylhydantoin may increase tissue uptake of cortisol in the hepatic smooth endoplasmic reticulum. This is compatible with the observation that diphenylhydantoin causes an increase in urine 6β-hydroxycortisol, a shortening of the disappearance time of cortisol from the plasma, an increase in its metabolic clearance rate, its production rate, and thus its secretion rate (Choi *et al.*, 1971). Diphenylhydantoin administration has also been shown to increase urine excretion of D-glucaric acid (Hunter, 1971a). There is no good evidence that it causes increases in rates of drug oxidation in man, although in rats it is an inducing agent resembling phenobarbitone (Eling *et al.*, 1970).

(4) *Phenylbutazone.* Phenylbutazone administration causes an increase in aminopyrine elimination in man (Chen *et al.*, 1962) and an increase in urine excretion of 6β-hydroxycortisol (Poland, 1970). The most important drug interactions with phenylbutazone in man, however, are either concerned with displacement of other drugs from protein binding or inhibition of rates of drug oxidation.

(5) *Ethanol.* There is a large and confused literature on the effects of ethanol on rates of drug oxidation in man. It has been shown that a diet containing 42% of its total calorific content as ethanol given for 12 days resulted in a significant increase in liver microsomal pentobarbital hydroxylase activity, but not benzpyrene hydroxylase activity in serial liver biopsies taken before and after the dietary treatment (Rubin and Lieber, 1968). In ten chronic alcoholics, the plasma clearance of tolbutamide was found to be greater than in controls. On the other hand, the acute infusion of ethanol caused an increase in the plasma half-life of tolbutamide in 6 volunteers, all of whom had normal liver function (Carulli *et al.*, 1971). Acute ethanol administration has also been shown to decrease the clearance of pentobarbitone, meprobamate (Rubin *et al.*, 1970) and warfarin (Breckenridge and Orme, 1971) in man.

Insecticides, Food Additives and Pollutants

In these days of heightened awareness of the importance of environmental pollution, the effect of insecticides and food additives on drug oxidation has not gone unnoticed.

Subjects exposed to DDT in their work have been shown to excrete more D-glucaric acid (Hunter *et al.*, 1971b) and 6β-hydroxycortisol (Poland *et al.*, 1970) in their urine than their non-exposed fellows, and to metabolize phenylbutazone (Poland *et al.*, 1970) and antipyrine (Kolmodin *et al.*, 1969) more rapidly too. DDT is stored in body fat and may remain in the body long after exposure. The inducing effect of DDT was put to therapeutic use at one stage to treat patients with unconjugated hyperbilirubinaemia (Hunter et al., 1971b). Likewise, the effect of o,p'-DDD in increasing urine excretion of 6β-hydroxy-cortisol has been used in clinical medicine in treating patients with Cushing's syndrome (Southern et al., 1966).

There is, of course another aspect of the environmental pollutant story. There is now good evidence that many pollutants encountered are chemical carcinogens (e.g. polycyclic hydrocarbons found in various smoked food, in cigarettes or in the air—(Conney, 1967). It has been shown in animals that administration of inducers of benz-pyrene hydroxylase will block the carcinogenic potency of these sub-strates (Conney et al., 1960). Substances which induce benzpyrene hydroxylase activity in the gastrointestinal tract, lung and skin (the portals of entry of the potential carcinogen) may thus lessen the toxicity (Wattenberg, 1966). The genetic differences in benzpyrene hydroxylase activity which exist in different individuals may also be important in determining susceptibility to environmental toxins. Cigarette smokers have been shown to have higher benzpyrene hydroxylase activity in the placenta, and the activity can be graded according to cigarette consumption (Welch et al., 1968). Whether this increase serves as a protective function or not is open to question, for it is at least theoretic-ally possible that some of the metabolites produced in increased amounts may show greater toxicity than the unchanged drug.

The metabolism of phenacetin has been studied in 9 non-smokers and 9 smokers (smoking more than 15 cigarettes/day). Plasma concen-trations of phenacetin were significantly lower in the smokers. More-over, the ratio of plasma concentrations of total p-acetamidophenol (the principal metabolite of phenacetin) to phenacetin was seven-fold greater in smokers. Whether this effect is due to changes in enzymes within the gastrointestinal mucosa or liver is uncertain (Pantuck et al., 1974).

Cigarette smoking has also been shown to enhance the metabolism of nicotine in man, and this may explain the smoker's tolerance and addiction to nicotine (Beckett and Triggs, 1967). Cigarette smokers also appear to require a larger dose of pentazocine for analgesia, but whether this can be explained by enzyme induction is not clear (Keeri-Szanto and Pomeroy, 1971). Long-term smokers of marihuana metab-olize Δ^9-tetrahydrocannabinol more rapidly than non-smokers (Lemberger et al., 1971).

Drugs which Stimulate their own Metabolism in Man

One of the causes of drug tolerance is enzyme induction. It was pointed out by Remmer in 1958 that long-term administration of barbiturates resulted in increased activity of the liver microsomal enzymes which inactivated the barbiturates themselves and thus their pharmacological effect would be lessened. This, of course, is only one cause of tolerance to barbiturates. Not all inducing agents stimulate their own rate of metabolism. For example, both phenylbutazone and diphenylhydantoin can be given for long periods to patients without causing a fall in their steady state plasma concentration, although in the dog and rat phenylbutazone does stimulate its own metabolism and its toxicity, as measured by the production of gastric ulcers, is lessened (Burns *et al.*, 1967). Tolbutamide, ethanol, antipyrine, glutethimide and meprobamate, on the other hand, stimulate their own metabolism with presumably a corresponding lessening of their pharmacological effect (Conney, 1967).

Enzyme Induction and Drug Toxicity

The ability of a drug to stimulate its own rate of metabolism is of importance in toxicity studies. If the drug produces a metabolite more toxic than the parent drug, then long-term administration may enhance its toxicity; if, on the other hand, the drug is more toxic than its metabolites, the reverse will be true. Acetanilide, phenacetin, chlorguanide and prontosil are transformed in man from inactive or relatively inactive drugs to metabolites of therapeutic value (Williams, 1963). Phenylbutazone, aspirin and primidone, active *per se*, also have active metabolites (Williams, 1963). Octamethylpyrophosphoramide (Schradan) is relatively inactive but it is metabolized to a potent choline esterase inhibitor; inducing agents increase the toxicity of such substances in animals and perhaps in man (Kato, 1961).

Recent studies implicate induction in the aetiology of tissue lesions caused by certain compounds (Brodie *et al.*, 1971). The liver and perhaps the kidney have the capacity to convert stable halogenated hydrocarbons such as bromobenzene to relatively unstable epoxides. These epoxides destroy the cells where they are formed by reacting covalently with nucleophilic groups in proteins and nucleic acids.

Inducing agents, by increasing the rate of production of metabolites, may lead to tissue lesions. This may well be the basis of the increased hepatic toxicity seen, e.g., with paracetamol, which is enhanced by the administration of inducing agents (Prescott *et al.*, 1971).

9.5. CHANGES IN NORMAL BODY CONSTITUENTS PRODUCED BY ADMINISTRATION OF INDUCING AGENTS

The effect of administration of inducing agents (usually phenobarbitone) on the turnover of several normal body constituents has been studied in man:

1. Steroids (cortisol, cholesterol, testosterone, vitamin D)
2. Bilirubin.
3. Porphyrins

Changes in other parameters have also been investigated:

4. Liver blood flow
5. Bile flow and bromsulphthalein retention

Steroids

(1) *Cortisol*. Effects on 6β-hydroxycortisol excretion and implications of this have been discussed previously.

(2) *Cholesterol*. Enzymes catalyzing the rate-limiting steps of both synthesis and catabolism of cholesterol are situated within the endoplasmic reticulum of the liver (Boyd and Percy-Robb, 1971). Administration of phenobarbitone to experimental animals changes the rate of cholesterol metabolism (Jones and Armstrong, 1965) and these studies have recently been extended to man (Miller and Nestel, 1973). Phenobarbitone, 240 mg/day, was given to four subjects for between 13 and 18 days; in three of these subjects (in whom phenobarbitone caused a significant shortening of the plasma half-life of antipyrine) there was an increased faecal excretion of bile acids, associated with an increased pool size and turnover of cholic acid. These subjects also developed significant increases in plasma cholesterol and triglyceride concentrations (Fig. 9.8). This interesting study also suggested that the cause of the increased plasma cholesterol was an increased cholesterol synthesis, which may have implications in the genesis of vascular disease in man.

(3) *Testosterone*. N-phenylbarbitone has been shown to decrease the urine excretion of androsterone and the major metabolites of testosterone and to increase the excretion of its more polar metabolites (Southern *et al.*, 1969). The clinical significance of this observation is uncertain.

(4) *Vitamin D*. In the last few years there has been a great advance in our understanding of the metabolism and action of vitamin D. Not only is it a vitamin but it also merits the description of a hormone (Kodicek, 1974).

Vitamin D is oxidized by liver microsomal enzymes to form 25-hydroxycholecalciferol (25-HCC) (Hahn *et al.*, 1972). This represents an activation process, and the liver's ability to produce 25-HCC is limited. This may well be due to a feedback-type control mechanism, or there may be some genetic factors controlling the 25-hydroxylation. Administration of barbiturates has been shown to accelerate the conversion of cholecalciferol to 25-HCC (Hahn *et al.*, 1972), and it is now

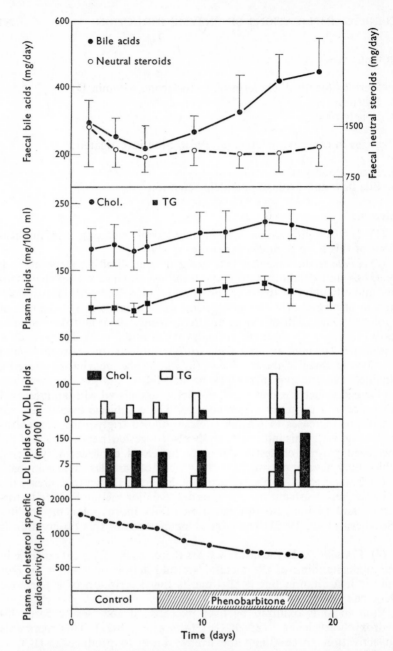

Fig. 9.8. Effects of phenobarbitone (240 mg/day for 13–18 days) on plasma cholesterol and triglyceride concentrations, plasma lipoproteins, plasma cholesterol specific radioactivity and faecal steroid excretion in three normal subjects. Chol., cholesterol; TG, triglyceride; VLDL, very-low-density lipoprotein; LDL, low-density lipoprotein. Each vertical bar represents the standard error of the mean. The specific radioactivity and lipoprotein results were obtained from one subject only. (From, Miller and Nestel, 1973, reproduced by kind permission of the editor of *Clinical Science and Molecular Medicine*).

widely recognized that epileptics given long-term anticonvulsants develop osteomalacia (Dent *et al.*, 1970) (this has not been reported with epileptics taking diphenylhydantoin alone and barbiturates would appear to be an essential drug). This obviously cannot be explained on the basis of the activation mechanism outlined above; the most probable explanation is that further conversion of 25-HCC to more polar, biliary-excreted, metabolites is also increased, thus leading to a depletion of body pools of vitamin D. The story of vitamin D does not end there, however. A series of experiments by Kodicek and his colleagues in Cambridge have clearly shown that the agent primarily responsible for the action of vitamin D is not 25-HCC, but 1,25-dihydroxychol-ecalciferol (1,25-DHCC) which is exclusively produced in the mito-chondrial fraction of the kidney and whose rate of formation and elimination is not influenced by the administration of inducing agents. 1,25-DHCC is a real hormone, because it is transported from its site of production to sites of action, namely, the gut and bone. For an excellent review of this fascinating field, readers are referred to Kodicek (1974).

Bilirubin

Bilirubin UDP-glucuronyl transferase is a hepatic microsomal enzyme which catalyses the conjugation of bilirubin—an essential stage in the elimination of the pigment. This enzyme is deficient in various congenital forms of unconjugated hyperbilirubinaemia (Gilbert's syndrome, Crigler-Najjar syndrome). Inducing agents have been successfully used to increase bilirubin elimination in affected infants (Yaffe *et al.*, 1966; Crigler and Gold, 1966), the agent usually being given to the mother in pregnancy if the deficiency was suspected, or to the infant itself. Other situations where inducing agents have been used to lower bilirubin concentrations are in unconjugated hyperbilirubinaemia in adult life (Black and Sherlock, 1970; Orme, Davies and Breckenridge, 1974) and in patients with chronic intrahepatic cholestasis (Thompson and Williams, 1967). Several agents have been used, including phenobarbitone (Black and Sherlock, 1970) (Fig. 9.9), phetharbital (Hunter *et al.*, 1971a), DDT (Hunter *et al.*, 1971b) and antipyrine (Orme *et al.*, 1974). Interestingly, the fall produced by phenobarbitone may not be due to enzyme induction (as defined by a specific increase in enzyme activity per unit of protein weight) as was originally suggested by Black and Sherlock (1970), whereas antipyrine does appear to cause an increase in specific enzyme activity. It is known that phenobarbitone administration may cause several other changes relevant to this situation, e.g. an increase in liver blood flow, an increase in organic anion binding proteins which transport bilirubin within the liver cells, and an increase in bile flow. From the therapeutic point of view, as well as the biochemical, antipyrine might appear a more logical form of treatment for patients with

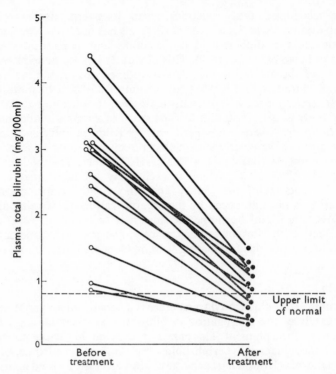

Fig. 9.9. Effect of phenobarbitone (180 mg/day for 14 days) on plasma
bilirubin concentration in patients with Gilbert's syndrome. (From,
Hunter *et al.*, 1971a, reproduced by kind permission of the editor of
the *Lancet*).

unconjugated hyperbilirubinaemia, since the side-effects of barbiturates
are unacceptable to many patients when these drugs are given over long
periods. DDT has also been used in the treatment of patients with
unconjugated hyperbilirubinaemia, but since this material or its active
metabolites remain in body fat stores for many months after administration, and the long-term effects of induction are not without potential
hazard, it appears unwise to use DDT in this situation.

Porphyrins

The porphyrias are metabolic disorders mostly of hereditary nature,
characterized by increased excretion of porphyrins and of their precursors. In this group of diseases the synthesis of these pigments is
greatly increased on account of the failure of the control mechanisms
that normally regulate the synthesis of haem (de Mattheis, 1967)

Porphyrins have been subdivided on biochemical and clinical bases into two main groups, erythropoetic and hepatic. It has been known for many years that certain drugs, especially sedatives, may cause derangement of liver porphyrin metabolism in man. These drugs include phenobarbitone, sulphonal, tolbutamide, chlorpropamide, oestrogens, griseofulvin and many others (de Matteis, 1967). The biochemical basis of the precipitation of acute porphyuria is still the subject of considerable debate and is reviewed elsewhere in this book (see Chapter 7). Increased activity of the enzyme δ-ALA synthetase caused by enzyme induction is perhaps the most likely basis.

Liver Blood Flow

It has been observed that, in rats, the same dose of phenobarbitone which causes a 12 to 15-fold decrease in the plasma half-life of drugs may only produce a 3 to 4-fold increase in liver enzyme activity measured *in vitro*. This suggests that factors other than specific enzyme activity may be important in controlling rates of drug oxidation *in vivo*. Using a heat exchange method in conscious rats, it has been shown that phenobarbitone and antipyrine both cause an increase in liver blood flow of between 33% and 175% above control values (Ohnhaus *et al.*, 1971). The mechanism of this increase is uncertain. In monkeys, too, administration of phenobarbitone increased total hepatic blood flow, as measured by a radioactive microsphere technique (Nies *et al.*, 1974). In these studies in monkeys it was shown that the clearance of both antipyrine and d-propranolol was increased, and the plasma half-life decreased, by administration of phenobarbitone. The mechanism of the increased clearance differed. With antipyrine, the increased clearance was due to enzyme induction, i.e. an effect on its rate of metabolism; whereas with d-propranolol (whose hepatic extraction ratio is higher than that of antipyrine) the increase in liver blood flow contributed as much to the enhanced clearance as did stimulation of drug metabolism.

Bile Flow and Bromsulphthalein (BSP) Retention

Increased bile flow after phenobarbitone treatment in rats has been clearly documented (Roberts and Plaa, 1967). The mechanism of this effect is uncertain; it may be related to an alteration of disposition of taurocholate or other bile salts, or it may be due to the increase in liver blood flow mentioned above. In man, Kreek and Sleisenger (1968) have shown that two adults with non-haemolytic unconjugated hyperbilirubinaemia treated with phenobarbitone showed increases in BSP clearance from the plasma; this may be due to an increased rate of bile flow, but other explanations are equally possible.

9.6. ENZYME INDUCTION IN TISSUES
OTHER THAN THE LIVER

Enzyme induction in man has been shown in tissues other than the liver.

Placenta

The enzymic hydroxylation of 3,4-benzpyrene was not detected in human placentae obtained after childbirth from women who were not cigarette smokers, whereas enzyme activity was present in the placentae obtained from all individuals who smoked cigarettes (Welch *et al.*, 1968). The induction of carcinogen metabolizing enzymes, such as benzpyrene hydroxylase, has been suggested as a mechanism through which women may protect themselves and their foetuses from polycyclic hydrocarbons found in their environment.

White Blood Cells

Benzpyrene hydroxylase is found in human lymphocytes, and its inducibility shows genetic variation in different individuals (Kellerman *et al.*, 1973b). It has been found possible to divide the white population into groups showing low, intermediate and high inducible benzpyrene hydroxylase activity. Most interestingly, it has also been shown that, in 50 patients with bronchogenic carcinoma, the proportion with highly inducible levels of enzyme activity is significantly greater than in the control population (Kellerman *et al.*, 1973a). All these patients with bronchogenic carcinoma were heavy smokers but whether subjects with initially high inducibility are more susceptible to exposure to environmental carcinogens is an intriguing possibility. Significantly, there is already a body of opinion who, on independent grounds, have suggested that the susceptibility to lung cancer is genetically determined.

Skin

Normal human skin contains benzpyrene hydroxylase, and samples of foreskin are readily available. When skins were incubated in a medium containing benzanthracene, increases ranging from 2- to 5-fold in benzpyrene hydroxyase activity were found (Alvarez *et al.*, 1973). Interestingly, there were considerable interindividual variations not only in the inducibility of the enzyme in various skins, but also in the basal enzyme activity. Whether this was due to exposure of the mother to different environmental influences or whether these differences were genetic, is uncertain. It is possible that individuals whose skin metabolizes carcinogens more readily may be more susceptible to the carcinogenic action of polycyclic hydrocarbons ingested orally, through inhaled smoke, or by absorption through the skin. The skin is one of the

largest organs in the body, its weight for an average man being estimated at 4 kg, or three times the weight of the liver. It is not known
whether benzpyrene hydroxylase activity in the skin correlates with
that found in other tissues, e.g. liver and lung.

Other Tissues

Cigarette smoking and phenothiazines have been shown to stimulate
benzpyrene hydroxylase activity in the rat lung (Wattenberg and
Leong, 1965); whether the same is true in human lung is uncertain.
Phenobarbitone, however, had no effect on either benzpyrene hydroxylase activity or pentobarbitone hydroxylation in homogenates of rat
lung.

In the small intestine in the rat, several phenothiazines, including
chlorpromazine, promazine and pyrathiazine, have been shown to
induce benzpyrene hydroxylase activity (Wattenberg et al., 1962).
Again, there are no data on this in man.

9.7. THERAPEUTIC USES OF ENZYME INDUCTION

From the preceding account, it might appear that enzyme induction
should be considered an undesirable phenomenon. In many instances
this is so, and the practice of therapeutics is made correspondingly more
difficult. Some physicians have argued that if all patients were given
inducing agents, the great interindividual differences in the rates of
drug oxidation, which complicate prescribing, would be minimized!

Inducing agents have been given to benefit the patient in situations
where diminished enzyme activity, due to inborn errors of metabolism,
result in the accumulation of potentially toxic substances. It should
always be kept in mind, however, that enzyme induction has many
effects other than those for which the drug is given and, secondly,
inducing agents, like any other drug, may have other undesirable side-
effects for the patient. There are three principal therapeutic areas
where inducing agents have been prescribed, as follows.

Patients with Unconjugated Hyperbilirubinaemia

In many such patients, administration of inducing agents is largely
a cosmetic exercise, but where the defect is severe, or intrahepatic
cholestasis is the underlying disease, itching and abdominal pain may
be diminished by administration of inducing agents.

Patients with Cushing's Syndrome

Werk et al. (1966) administered diphenylhydantoin chronically to
two patients with non-tumerous Cushing's syndrome and observed
improvement both clinically and biochemically. In these patients
there was an increased urine excretion of 6β-hydroxycortisol but no

11

increase in cortisol secretion (see section above, however). A similar effect has been seen with o,p'-DDD (Southern *et al.*, 1966), although there is considerable debate as to whether this effect is due to induction of hepatic steroid metabolism or to a direct toxic effect on the turnover. These forms of drug treatment are in no sense first-line forms of therapy for Cushing's syndrome and at the present, o,p'-DDD is kept for patients with inoperable adrenal cortical tumours, frequently with metastases. Moreover, the side-effects of o,p'-DDD are unpleasant and the drug is expensive.

Patients with Hereditary Disorders of Carbohydrate Metabolism

Many forms of glycogen storage disease are now known where certain essential enzymes are either absent or present in a partially inactive form. Some of these defects have been partially corrected by administration of inducing agents. For example, Pesch *et al.* (1960) reported that the treatment of galactosaemic patients with progesterone enhanced the metabolism of galactose. More recently, Moses *et al.* (1966) have stimulated hepatic glucose-6-phosphatase activity with triamcinolone, in a patient with glycogen storage disease. Whether or not this manoeuvre, and others like it, are in the overall interest of the patient, is uncertain.

REFERENCES

Aarts, E. M. (1965). *Biochem. Pharmacol.*, **14**, 359
Aggeler, P. M. and O'Reilly, R. A. (1969). *J. Lab. Clin. Med.*, **74**, 229
Alvarez, A. P., Kappas, A., Levin, W. and Conney, A. H. (1973). *Clin. Pharmacol. Therap.*, **14**, 30
Antlintz, A. M., Tolentino, M. and Kosai, M. F. (1968). *Curr. Ther. Res.*, **10**, 70
Ballinger, B., Browning, M., O'Malley, K. and Stevenson, I. H. (1972). *Brit. J. Pharmacol.*, **45**, 638
Beckett, A. H. and Triggs, E. J. (1967). *Nature*, **216**, 587
Black, M. and Sherlock, S. (1970). *Lancet*, **1**, 1359
Bledsoe, T., Island, D. P., Ney, R. L. and Liddle, G. W. (1964). *J. Clin. Endocrinol.*, **24**, 1303
Bogaert, M., Rosseel, M. T. and Belpaire, F. M. (1971). *Arch. Int. Pharmacodyn. Therap.*, **192**, 198
Boyd, G. S. and Percy-Robb, I. W. (1971). *American J. Med.*, **51**, 580
Breckenridge, A. and Orme, M. L'E. (1971). *Ann. N.Y. Acad. Sci.*, **179**, 421
Breckenridge, A., Orme, M. L'E., Thorgeirsson, S., Davies, D. S. and Brooks, R. V. (1971). *Clin. Sci.*, **40**, 351
Breckenridge, A., Burke, C. W., Davies, D. S. and Orme, M. L'E. (1973a). *Brit. J. Pharmacol.*, **47**, 434
Breckenridge, A., Orme, M., Davies, L., Thorgiersson, S. and Davies, D. S. (1973b). *Clin. Pharmacol. Ther.*, **14**, 514
Breckenridge, A., Orme, M. L'E., Wesseling, H., Bending, M. and Gibbons, R. (1974). In *Proceedings of an International Meeting on Drug Metabolism* (Garratini, S., ed.), to be published by Raven Press; N.Y.

Brodie, B. B., Cho, A. K., Krisna, G. and Reid, W. D. (1971). *Ann. N.Y. Acad. Sci.*, **179,** 11

Burns, J. J., Welch, R. M. and Conney, A. H. (1967). *Animal and Clinical Pharmacological Techniques in Drug Evaluation*, Vol. 2, p. 67, Year Book Medical Publishers; Chicago

Burstein, S. and Klaiber, E. L. (1965). *J. Clin. Endocrinol.*, **25,** 293

Calne, D. (1973). *Brit. J. Hospital Med.*, **9,** 171

Carulli, N., Manenti, F., Gallo, M. and Salvioli, G. F. (1971). *Europ. J. Clin. Invest*, **1,** 421

Chen, W., Vrindten, P. A., Dayton, P. G. and Burns, J. . (1962). *Life Sciences*, **1,** 35

Choi, Y., Thrasher, K., Werk, E. E., Sholiton, L. J. and Olinger, C. (1971). *J. Pharmacol. Exp. Ther.*, **176,** 27

Conney, A. H. (1967). *Pharmacol. Rev.*, **19,** 317

Conney, A. H., Davison, C., Gastel, R. and Burns, J. J. (1960). *J. Pharmacol. Exp. Therap.*, **130,** 1

Corn, M. (1966). *Thromb. Diath. Haemorrh.*, **16,** 606

Crigler, J. F. and Gold, N. I. (1966). *J. Clin. Invest.*, **45,** 998

Cucinell, S. A., Conney, A. H., Sansur, M. and Burns, J. J. (1965). *Clin. Pharmacol. Ther.*, **6,** 420

Curry, S. H., Lader, M. H., Mould, G. P. and Sararis, G. (1972). *Brit. J. Pharmacol.*, **44,** 370P

Dayton, P. G., Tarcan, Y., Chenkin, T. and Weiner, M. (1961). *J. Clin. Invest.*, **40,** 1797

de Matteis, F. (1967). *Pharmacol Rev.*, **19,** 523

Dent, C. E., Richens, A., Rowe, D. J. F. and Stamp, T. C. B. (1970). *Brit. Med. J.*, **4,** 69

Eling, T. E., Harbison, R. D., Becker, B. A. and Fouts, J. R. (1970). *J. Pharmacol. Exp. Therap.*, **171,** 127

Forrest, F. M., Forrest, I. S. and Serra, M. T. (1969). *Proc. West. Pharmacol. Soc.*, **12,** 31

Frantz, A. G., Katz, F. H. and Jailer, J. W. (1961). *J. Clin. Endocrinol.*, **21,** 1290

Hahn, T. J., Birge, S. J., Scharp, C. T. and Aviolo, L. V. (1972). *J. Clin. Invest.*, **51,** 741

Hammer, W. and Sjoqvist, F. (1967). *Life Sciences*, **6,** 1895

Hertz, R. and Tullner, W. W. (1968). *Proc. Soc. Exp. Biol. Med.*, **99,** 451

Huffman, D. H., Shoeman, D. W., Pentikainen, P. and Azarnoff, D. L. (1973). *Pharmacology*, **10,** 338

Hunninghake, D. and Azarnoff, D. L. (1968). *Arch. Int. Med.*, **121,** 349

Hunter, J., Maxwell, J. D., Stewart, D. A. and Williams, R. (1973). *Biochem. Pharmacol.*, **22,** 743

Hunter, J., Maxwell, J. D., Carrella, M., Stewart, D. A. and Williams, R. (1971a). *Lancet*, **1,** 572

Hunter, J., Maxwell, J. D., Stewart, D. A., Williams, R., Robinson, J. and Richardson, A. (1971b). *Gut*, **12,** 761

Hunter, J., Maxwell, J. D., Stewart, D. A., Williams, R., Robinson, J. and Richardson, A. (1972). *Nature, Lond.*, **237,** 339

Ideo, G., Morganti, A. and Dioguardi, N. (1972). *Digestion*, **5,** 326

Jelliffe, R. W. and Blankenhorn, D. H. (1966). *Clin. Res.*, **14,** 160

Johannson, S. A. (1968). *Acta Med. Scand.*, **184,** 297

Jones, A. L. and Armstrong, D. T. (1965). *Proc. Soc. Exp. Biol. Med.*, **119,** 1136

Kato, R. (1961). *Arzneim-Forsch.*, **11,** 797

Kawada, M., Yamada, K., Kagawa, Y., and Mano, Y. (1961). *Biochem. Tokyo*, **50,** 74

Keeri-Szanto, M. and Pomeroy, J. R. (1971). *Lancet*, **1,** 947

Kellerman, G., Shaw, C. R. and Luyten-Kellerman, M. (1973a) *New Engl. J. Med.*, **289,** 934

Kellerman, G., Luyten-Kellerman, M. and Shaw, C. R. (1973b). *Amer. J. Hum. Genet.*, **25,** 327

Kodicek, E. (1974). *Lancet*, **1,** 325

Kolmodin, B., Azarnoff, D. L. and Sjoqvist, F. (1969). *Clin. Pharmacol. Ther.*, **10,** 638

Kreek, M. J. and Sleisenger, M. H. (1968). *Lancet*, **2,** 73

Kuntzman, R., Jacobson, M. and Conney, A. H. (1966). *Pharmacologist*, **8,** 195

Kuntzman, R., Jacobson, M., Levin, W. and Conney, A. H. (1968). *Biochem. Pharmacol.*, **17,** 565

Lemberger, L., Tamarkin, N. R., Axelrod, J. and Kopin, I. J. (1971). *Science*, **173,** 72

Levi, A. J., Sherlock, S. and Walker, D. (1968). *Lancet*, **1,** 1275

McDonald, M. G., Robinson, D. S., Sylwester, D. and Jaffe, J. (1969). *Clin. Pharmacol. Ther.*, **10,** 80

Miller, N. E. and Nestel, P. J. (1973). *Clin. Sci. Mol. Med.*, **45,** 257

Morselli, P., Rizzo, M. and Garratini, S. (1971). *Ann. N.Y. Acad. Sci.*, **179,** 88

Moses, S. W., Levin, S., Chayoth, R. and Steinitz, K. (1966). *Pediatrics (N.Y.)*, **38,** 111

Mowat, A. P. (1968). *J. Endocr.*, **42,** 485

Neuvonen, P. J. and Pentilla, O. (1974). *Brit. Med. J.*, **i,** 535

Nies, A. S., Shand, D. G. and Branch, R. A. (1974). In *Proceedings of an International Meeting on Drug Metabolism* (Garratini, S., ed.), to be published by Raven Press; N.Y.

Ohnhaus, E. E., Thorgeirsson, S. S., Davies, D. S. and Breckenridge, A. (1971). *Biochem. Pharmacol.*, **20,** 2561

Orlowski, M. (1963). *Archivum Immunoliae et Therapiae Experimentalis*, **11,** 1

Orme, M. L'E., Breckenridge, A. and Brooks, R. V. (1972). *Brit. Med. J.*, **3,** 611

Orme, M. L'E., (1974). *MD Thesis*, p. 203, University of Cambridge

Orme, M. L'E., Davies, L. and Breckenridge, A. (1974). *Clin. Sci. Mol. Med.*, **46,** 511

Pantuck, E. J., Hsiao, K-C., Maggio, A., Nakamura, K., Kuntzman, R. and Conney, A. H. (1974). *Clin. Pharmacol. Ther.*, **15,** 9

Pelkonen, O. and Karki, N. T. (1973). *Chem.-Biol. Interactions*, **7,** 93

Pesch, L. A., Segal, S. and Topper, Y. J. (1960). *J. Clin. Invest.*, **39,** 178

Poland, A., Smith, D., Kuntzman, R. Jacobson, M. and Conney, A. H. (1970). *Clin. Pharmacol. Ther.*, **11,** 724

Prescott, L. F., Wright, N., Roscoe, P. and Brown, S. S. (1971). *Lancet*, **1,** 519

Remmer, H. (1958). *Naturwiss.*, **45,** 189

Remmer, H. (1962). In *Ciba Foundation Symposium on Enzymes and Drug Action*, p. 276, Little Brown; Boston

Riegelman, S., Rowland, M. and Epstein, W. L. (1971). *J.A.M.A.*, **213,** 426

Rizzo, M., Morselli, P. L. and Garratini, S. (1972). *Biochem. Pharmacol.*, **21,** 449

Roberts, R. J. and Plaa, G. L. (1967). *Biochem. Pharmacol.*, **16,** 827

Robinson, D. S. and McDonald, M. G. (1966). *J. Pharmacol. Exp. Ther.*, **153,** 250

Robinson, D. S. and Sylwester, D. (1970). *Ann. Int. Med.*, **72,** 853

Rosalki, S. B., Tarlow, D. and Rau, D. (1971). *Lancet*, **2,** 376

Rubin, E. and and Lieber, C. S. (1968). *Science*, **162,** 690

Rubin, E., Gang, H., Misra, P. S. and Lieber, C. S. (1970). *Amer. J. Med.*, **49,** 801

Schmid, K., Cornu, F., Imhof, P. and Keberle, H. (1964). *Med. Wochensch.*, **94,** 235

Sellers, E. M. and Koch-Weser, J. (1971). *Ann. N.Y. Acad. Sci.*, **179,** 213

Smith, S. E. and Rawlins, M. (1974). *Europ. J. Clin. Pharmacol.*, **7,** 71

Southern, A. L., Tochimoto, S., Strom, L., Ratuschni, A., Ross, H. and Gordon, G. (1966). *J. Clin. Endocrinol.*, **26**, 268

Southern, A. L., Gordon, G. G., Tochimoto, S., Krikun, E., Krieger, D., Jacobson, M. and Kuntzman, R. (1969). *J. Clin. Endocrinol.*, **29**, 251

Taylor, S. A., Rawlins, M. D. and Smith, S. E. (1972). *J. Pharm. Pharmacol.*, **24**, 578

Thompson, R. P. H. and Williams, R. (1967). *Lancet*, **2**, 646

Thrasher, K., Werk, E. E., Choi, Y., Sholiton, L. J., Meyer, W. and Olinger, C. (1969). *Steroids*, **14**, 455

Vesell, E. and Page, J. G. (1969). *J. Clin. Invest.*, **48**, 2202

Vajda, F., Williams, F. M., Davidson, S., Falkoner, M. and Breckenridge, A. (1974). *Clin. Pharmacol. Ther.*, **15**, 597

Van Dam, F. E. and Gribnau-Overkamp, M. J. H. (1967). *Folia Med. Neerl*, **10**, 141

Wagner, J. G., Northam, J. I., Alway, C. D. and Carpenter, O. S. (1965). *Nature (Lond.)*, **207**, 1301

Wattenberg, L. W. (1966). *Cancer Res.*, **26**, 1520

Wattenberg, L. W. and Leong, J. L. (1965). *Cancer Res.*, **25**, 365

Wattenberg, L. W., Leong, J. L. and Strand, P. J. (1962). *Cancer Res.*, **22**, 1120

Wattenberg, L. W., Leong, J. L. and Galbraith, A. R. (1968). *Proc. Soc. Exp. Biol. Med.*, **127**, 467

Welch, R. M., Harrison, Y. E., Conney, A. H., Poppers, P. J. and Finster, M. (1968). *Science*, **160**, 541

Werk, E. E., MacGee, J. and Sholiton, L. J. (1964). *J. Clin. Invest*, **43**, 1824

Werk, E. E., Sholiton, L. J. and Olinger, C. P. (1966). *2nd Int. Congress Hormonal Steroids*, Excerpt. Med. Found., Series 3, p. 301, N.Y.

Whitfield, J. B., Moss, D. W., Neale, G., Orme, M. and Breckenridge, A. (1973). *Brit. Med. J.*, **i**, 316

Williams, R. T. (1963). *Clin. Pharmacol. Ther.*, **4**, 234

Yaffe, S. J., Levy, G., Matsuzana, T. and Balish, T. (1966). *New Eng. J. Med.*, **275**, 1461

Chapter 10

Disruptions in Enzyme Regulation during Aging

Richard C. Adelman
Fels Research Institute and Department of Biochemistry,
Temple University School of Medicine
Philadelphia, Pennsylvania 19140, U.S.A.

10.1. INTRODUCTION

The single feature which probably characterizes all aging populations is a progressive impairment in the ability to adapt to environmental change. Adaptation may be expressed at a biochemical level by modifications in the rates of synthesis and degradation of enzymes, as well as by alterations in physiological activities. Thus, it was gratifying to discover that the susceptibility of at least certain enzymes to the regulation of their activities is impaired during aging. The present chapter will review the effects of aging on enzyme regulation; discuss the possible roles of alterations in appropriate hormonal interactions and hepatic gene expression; and suggest the manner in which these observations may provide the basis for eventual comprehension of the underlying biochemical mechanisms responsible for the physiological decline which accompanies old age.

10.2. PATTERNS OF ALTERED ENZYME REGULATION

The ability to stimulate an adaptive increase in the activity of a large number of enzymes is modified during aging, and is evident in various tissues of different species in response to a broad spectrum of stimuli. These phenomena have been reviewed previously (Adelman, 1971; Adelman *et al.*, 1972) and also are discussed below in conjunction with more recent examples.

The effects of aging on enzyme regulation or, for that matter, on all types of adaptive response, apparently may be subdivided into four

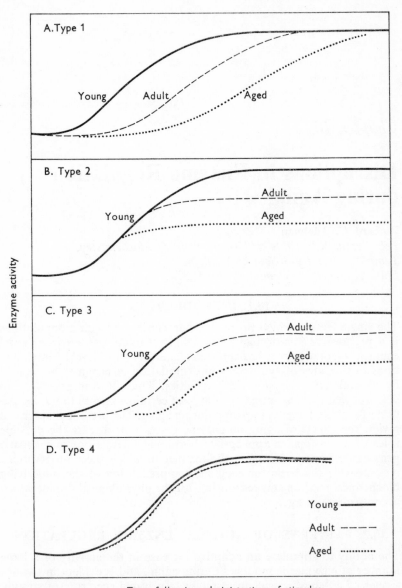

Time following administration of stimulus

Fig. 10.1. Patterns of Age-Dependent Enzyme Adaptation. The four major categories of effects of aging on the time course and magnitude of an hypothetical adaptive change in enzyme activity are illustrated generally. For the sake of convenience identical basal levels of enzyme activity are presented at the three indicated ages, changes in magnitude are presented as increases, and temporal modifications are presented as delays.

general categories. The first of these is expressed as a response (*Type 1*) whose initiation time (or adaptive latent period) is altered, but whose magnitude is not influenced during aging, as illustrated generally in Fig. 10.1(a). Examples of Type 1 response include increases in activity of glucokinase (Adelman, 1970a and 1971; Adelman *et al.*, 1972), tyrosine aminotransferase (Adelman, 1970b, and 1971; Adelman *et. al.*, 1972), microsomal NADPH-cytochrome c reductase (Adelman, 1971; Adelman *et al.*, 1972) and serine dehydratase (Rahman and Peraino, 1973) from rat liver following treatment *in vivo* with, respectively, glucose, ACTH, phenobarbital and cortisone; and also in hepatic tyrosine aminotransferase activity following exposure of mice to cold temperature (Finch *et al.*, 1969). Hypothetically, also included within the framework of this type of response are fluctuations of enzyme activity, including adaptive decreases of similar magnitude, whose initiation time occurs earlier in older animals. Although such changes remain to be discovered, consistent with the likelihood of such a possibility is the earlier initiation time for stimulated incorporation of ^{3}H-uridine into RNA in the isoproterenol-treated salivary gland of older rats (Roth *et al.*, 1974).

A second category of response (Type 2) reveals diminished or enhanced magnitude of adaptation in the absence of temporal change, as illustrated generally in Fig. 10.1(b). Examples of Type 2 response include the diminished increase in activities of ornithine aminotransferase and glucose 6-phosphatase in rat liver following treatment with cortisone and high-protein diet (Rahman and Peraino, 1973), and in activity of glutamine synthetase (Rao and Kanungo, 1972) in regenerating rat liver following treatment with cortisone. Administration of cortisone to adrenalectomized rats causes a greater increase in the glutamine synthetase activity of older rats (Rao and Kanungo, 1972). A large number of age-dependent enzyme adaptations apparently in this category (Adelman, 1971) were studied only at one time point.

A third category of response (Type 3) is modified temporally as well as in magnitude, as illustrated generally in Fig. 10.1(c). Examples of Type 3 response include delayed and inhibited increases in the activities of thymidine kinase and deoxythymidylate synthetase in rat salivary gland, following treatment *in vivo* with isoproterenol (Roth *et al.*, 1974); and also in the activity of catalase from mouse liver (Baird and Samis, 1971) or *Drosophila* (Nicolosi *et al.*, 1973) following treatment with aminotriazole.

A fourth category of enzyme adaptive responsiveness (Type 4) is apparently unaltered during aging, as illustrated generally in Fig. 10.1(d). Examples of Type 4 response include the increases in activities of glucokinase (Adelman and Freeman, 1972), tyrosine aminotransferase (Finch, *et al.*, 1969; Adelman and Freeman, 1972) and mitochondrial α-glycerolphosphate dehydrogenase (Bulos *et al.*, 1972) in response to treatment *in vivo* with, respectively, (*a*) insulin; (*b*) cortisol, corticosterone, insulin and glucagon, and (c) thyroxin. However, each of these reports is subject to criticism, as discussed in detail below.

Although divisible into distinct patterns of response, the influence of aging on enzyme regulation is extremely complex. Many of the observations referred to above were accomplished at ages not particularly old for the species in question (Adelman, 1971; Rahman and Peraino, 1973). Representative of the types of difficulties one might encounter are evident, for example, in the Type 1 response; the enzyme adaptations whose initiation time is delayed. The inductions of glucokinase, tyrosine aminotransferase and microsomal NADPH-cytochrome c reductase in rat liver following treatment *in vivo* with glucose, ACTH and phenobarbital, respectively, are characterized by an age-dependent adaptive latent period whose duration increases progressively from 2 to at least 24 months of age (Adelman, 1972). In contrast, the induction of tyrosine aminotransferase following exposure of mice to cold temperature is delayed only after 12 months of age (Finch *et al.*, 1969). Consistent with the former work done in rat liver, a delayed response by serine dehydratase to cortisone treatment is evident between 1 and 12 months of age (Rahman and Peraino, 1973), although older rats were not studied by these investigators.

The likelihood of species and strain differences, as well as the influence of the environment of animal maintenance, is also largely unknown. The inducibility of microsomal NADPH-cytochrome c reductase activity and synthesis by phenobarbital treatment is delayed by aging in Sprague-Dawley rats obtained from A. R. Schmidt (Adelman, 1971; Adelman *et al.*, 1972), whereas age-dependence is not detectable in Sprague-Dawley derived rats obtained from Charles River (Rotenberg *et al.*, 1974). That this enzyme may already be in a partially induced state in the Charles River rats probably accounts for the latter observation. Relative contributions by genetic and environmental factors remain to be determined.

Age-dependent enzyme adaptations are influenced also by sex differences. For example, the magnitude of the increase in serine dehydratase activity, as well as the loss of responsiveness in ornithine aminotransferase activity, in response to treatment with high-protein diet is more evident in male than in female rats of all ages tested (Rahman and Peraino, 1973). On the other hand, no sex differences were detected for the adaptation of rat liver glucokinase to glucose (Adelman, 1970a).

10.3. MECHANISMS OF AGE-DEPENDENT ENZYME ADAPTATION

The biochemical mechanisms responsible for these age-dependent alterations in the regulation of enzyme activity are not understood.

Perhaps the greatest difficulty lies in localization of the responsible lesions. For example, by allowing sufficient time for complete liver regeneration following partial hepatectomy of aged rats, there is created in essence, a population of aged rats in which are contained livers composed of primarily new, or 'young', liver cells. Both time course and magnitude of the age-dependent (Type 1) increase in activity of glucokinase and microsomal NADPH-cytochrome c reductase are identical in untreated and surgically treated rats of corresponding ages (Adelman, 1970c, 1971 and 1972). Thus, age-dependent alterations in hepatic enzyme regulation may be the consequence of genetic changes copied during the cell division which accompanies liver regeneration, and/or extrahepatic phenomena. General biochemical mechanisms which encompass these possibilities are alterations in: (a) control of hepatic gene expression; (b) sensitivity of liver to hormone or other extrahepatic effectors; and (c) availability to liver of hormone or other extrahepatic effectors.

Hepatic Gene Expression

Three types of experimental approaches have purported to examine the role of *de novo* protein synthesis in enzyme regulation during aging. In one approach, the direct stimulation of glucokinase by insulin treatment (Adelman and Freeman, 1972), of tyrosine aminotransferase by treatment with glucocorticoids, insulin or glucagon (Finch *et al.*, 1969; Adelman and Freeman, 1972; Gregerman, 1959), and of mitochondrial α-glycerolphosphate by thyroxin treatment (Bulos *et al.*, 1972) results in identical increases in catalytic activity in rats and mice of all ages examined. These data have been interpreted generally as support for the concept that the capacity of liver for hormonally stimulated protein synthesis is not impaired during aging. However, in one case (Gregerman, 1959) only a single time point in the induction was observed; in all cases only a single hormone dosage was employed (see more below); and as properly cautioned by Haining and Legan (1973), these data fail to distinguish between selective modifications in rates of enzyme synthesis and degradation, although the contribution by enhanced rates of synthesis generally is known in younger rats.

A second approach entails determination of changes *in vivo* in the synthetic and degradative rate constants of adaptive fluctuations in enzyme activities. Applied thus far to tryptophan oxygenase (Haining and Correll, 1969), xanthine oxidase (Haining *et al.*, 1970) and catalase (Haining and Legan, 1973) in rat liver, these kinetic analyses indicate generally a decrease in both the rate of synthesis and the rate of protein degradation in very old animals.

In an attempt to assess directly a rate of adaptive enzyme synthesis in aging rats, the rate of incorporation of amino acid into NADPH-cytochrome c reductase, purified to homogeneity from a total liver microsomal fraction following treatment with phenobarbital, was determined (Adelman *et al.*, 1972). The delayed initiation of increased enzyme activity in aged rats is accompanied by the delayed initiation of a stimulated rate of amino acid incorporation into purified enzyme. Changes in specific radioactivity of precursor pools do not account for this phenomenon, and a 30-minute pulse of radioactive precursor is probably not sufficient to permit a significant contribution by enzyme degrading reactions. Thus, it is tempting to suggest that this delayed enzyme adaptation is the consequence of delayed initiation of stimulated *de novo* enzyme synthesis in aged rats. However, based upon available data it is not yet possible to make distinctions between the effects of aging on the synthesis of new enzyme molecules on free or membrane-bound polysomes, the translocation of newly synthesized enzyme from polysome to endoplasmic reticulum, and the assembly of enzyme into proliferating endoplasmic reticulum membrane.

Based upon such data as the recent report of increased accumulation of inactive molecules of liver aldolase in aged mice (Gershon and Gershon, 1973), another alternative mechanism also must be given due consideration. NADPH-cytochrome c reductase prepared in homogeneous form from young and aged rats appears identical according to such criteria as kinetic constants, electrophoretic mobility and sedimentation behavior (R. C. Adelman, unpublished observations). However, the increased accumulation of either faulty molecules, synthetic and/or degradative intermediates, or other post-translational modifications could conceivably impair synthesis and/or activation of the native enzyme in response to phenobarbital treatment. Of course, properties of such a modified protein might preclude the likelihood of its co-purification with the native enzyme.

Liver Sensitivity

Frolkis (1970) observed that by varying the inducing dosages of cortisol or tyrosine, an apparent alteration in the time course and magnitude of response by hepatic tyrosine aminotransferase activity could be abolished in aged adrenalectomized rats. One possible interpretation of such data is an age-dependent change in the biochemical action of cortisol on liver. However, increases in activities of glucokinase in response to insulin (Adelman, 1970a; Adelman *et al.*, 1972), tyrosine aminotransferase in response to cortisol, corticosterone, insulin or glucagon (Finch *et al.*, 1969; Adelman and Freeman, 1972) microsomal NADPH cytochrome c reductase in response to phenobarbital (Adelman, 1971; Adelman *et al.*, 1972), serine dehydratase in response to cortisone (Rahman and Peraino, 1973), mitochondrial α-glycerolphosphate dehydrogenase in response to thyroxin (Bulos *et al.*, 1972), in addition to other examples (Adelman, 1971), are of nearly identical magnitude in intact animals of various ages. These data appear to be consistent with the absence of aging effects on hepatic responsiveness to hormonal stimulation.

However, both kinds of interpretation are susceptible to considerable criticism. For example, most of this work ignores the need for dose-response curves; the danger of which was demonstrated by Frolkis (1970). In several cases (Adelman, 1971; Rahman and Peraino, 1973) there is an inadequate sampling of ages, further complicating the difficulties in distinguishing between growth and aging effects. Furthermore, in none of these studies are distinctions made between relative contributions to the increase in enzyme activity by increased rates of enzyme synthesis and decreased rates of enzyme degradation. In other words, alterations in hormonally enhanced synthetic capacities may have been overlooked.

The most serious criticism of these apparent changes (or lack of them) in sensitivity to hormonal stimulation *in vivo*, however, is the total absence of any attempt to relate changes in enzyme responsiveness to the binding of hormone effectors to specific receptor molecules. It is now well established that certain tissues such as liver are endowed with the capacity to concentrate certain hormones by means of high affinity binding to specific carrier proteins; e.g., the glucocorticoid-binding proteins (Morey and Litwack, 1969; Singer and Litwack, 1971; Litwack *et al.*, 1971). Any meaningful evaluation of the effects of aging on hormonally stimulated enzyme induction must include data which relate to the specific binding of the hormone effector to the receptor molecule directly responsible for enhancement of *de novo* enzyme synthesis. In the case in which hormone receptor molecules are incorporated into the structure of the cellular plasma membrane (e.g., insulin, glucagon, epinephrine), the biochemical mechanism of message transmission between membrane-bound receptor and enhancement of *de novo* enzyme synthesis must be given due consideration as well.

Availability of Hormones

The only enzymes whose regulation is altered during aging, for which the physiological role of hormones *in vivo* has been examined in any detail, are glucokinase and tyrosine aminotransferase of rat liver (Adelman *et al.*, 1972; Adelman and Freeman, 1972). Although hardly doing justice to the complexity of the situation, suffice it to say that essential requirements for adaptive responsiveness of glucokinase to glucose include insulin and glucocorticoids, whereas effectors of negative influence include glucagon and catecholamines. In an apparently equally complex manner, the adaptive responsiveness of tyrosine aminotransferase to ACTH requires the generation and action of an intermediary hormone which is yet to be identified conclusively from among glucocorticoids, insulin, glucagon and epinephrine, each of which can enhance directly the rate of synthesis of this enzyme and each of whose serum levels may be increased by ACTH.

Effects of aging on hormone production were reviewed by Robertson *et al.* (1972) and by Bellamy (1967), among others and, to say the least, are quite contradictory and inconclusive. More to the point, however, the availability of hormones to the liver has not been studied at all as a function of aging. For example, the levels of hormones determined in peripheral blood, or of hormone metabolites determined in urine, may not relate even remotely to hormonal interaction with liver. The concentration of a circulating hormone is a steady state level, largely the resultant of two opposing processes: (*a*) synthesis and secretion into the circulation; and (*b*) removal and turnover which occurs primarily in liver and kidney. The concentration of a hormone in blood may differ enormously, depending upon the site of blood collection; e.g., prior or subsequent to passage through the liver. Thus, for example, newly secreted insulin which is immediately available to the liver can be assessed appropriately only in samples of blood collected from the portal vein. Similarly, free and serum-protein bound glucocorticoids which are available to the liver can be assessed appropriately only by giving due consideration to the dual delivery route, namely, the hepatic arterial and portal venous systems. Furthermore, it must ultimately be established whether impaired availability of hormones is a consequence of alterations in production by the appropriate tissue source, responsiveness of tissue source to pituitary or neural control mechanisms, and/or additional unknown factors still to be determined (Adelman *et al.*, 1972; Adelman and Freeman, 1972; Adelman *et al.*, 1974). Determination of relationships during aging between the availability of hormones to the liver, their specific receptor-molecule binding, and the initiation of enzyme synthesis is a major immediate objective of this author's laboratory.

10.4. SIGNIFICANCE

Alterations in the pattern of adaptive fluctuation in a large number of enzyme activities, characteristic of several tissues from different species, represent one biochemical manifestation of the progressive physiological decline which accompanies old age. Impairment of critical adaptive response may be viewed as one form of disease. Since the incidence of most diseases increases exponentially with age, the progressive impairment of adaptive response probably is intimately associated with many maladies of the aged. Thus, the causes of impaired enzyme adaptation may also be the basis for increased vulnerability and decreased performance which accompany old age.

Undoubtedly, the rapid proliferation of knowledge in the fields of enzyme regulation and hormone action will eventually provide the opportunity to determine the sequence of biochemical events that are responsible for regulatory impairments during aging. At present, the most promising direction in which to seek explanations of altered tissue responsiveness *in vivo* relates to availability and/or the action of appropriate hormone effectors. When these biochemical lesions are identified, it will be possible to initiate investigation of the truly fundamental issues, namely, the time in the lifespan at which these lesions first develop, and the nature and origin of those factors which are responsible for their expression.

ACKNOWLEDGEMENT

Preparation of this manuscript was supported in part by research grants HD-05874, HD-04382 and CA-12227 from the National Institutes of Health.

REFERENCES

Adelman, R. C. (1970a). *J. Biol. Chem.*, **245**, 1032
Adelman, R. C. (1970b). *Nature*, **228**, 1095
Adelman, R. C. (1970c). *Biochem. Biophys. Res. Commun.*, **38**, 1149
Adelman, R. C. (1971). *Exper. Gerontol.*, **6**, 75
Adelman, R. C. (1972). *Advances in Gerontological Research*, Vol. 4 (B. Strehler, ed.), 1
Adelman, R. C. and Freeman, C. (1972). *Endocrinol.*, **90**, 1551
Adelman, R. C., Freeman, C. and Cohen, B. S. (1972). *Advances in Enzyme Regulation* (G. Weber, ed.), **10**, 365
Adelman, R. C., Freeman, C. and Rotenberg, S. (1973). *Progress in Brain Research*, **40**, 509
Baird, M. B. and Samis, H. V. (1971). *Gerontologia*, **17**, 105
Bulos, B., Shukla, S. and Saktor, B. (1972). *Mech. Age. Devel.*, **1**, 227
Bellamy, D. (1967). *Symp. Soc. Exp. Biol.*, **21**, 427
Finch, C. E., Foster, J. R. and Mirsky, A. E. (1969). *J. Gen. Physiol.*, **54**, 690
Frolkis, V. V. (1970). *Exper. Gerontol.*, **5**, 37
Gershon, H. and Gershon, D. (1973). *Proc. Natl. Acad. Sci., U.S.*, **70**, 909
Gregerman, R. I. (1959). *Amer. J. Physiol.*, **197**, 63
Haining, J. L. and Correll, W. W. (1969). *J. Gerontol.*, **24**, 143
Haining, J. L. and Legan, J. S. (1973). *Exper. Gerontol.*, **8**, 85
Haining, J. L., Legan, J. S. and Lovell, W. J. (1970). *J. Gerontol.*, **25**, 205
Litwack, G., Ketterer, B. and Arias, I. M. (1971). *Nature*, **234**, 466
Morey, K. S. and Litwack, G. (1969). *Biochem.* **8**, 4813
Nicolosi, R. J., Bairs, M. B., Massie, H. R. and Samis, H. V. (1973). *Exper. Gerontol.*, **8**, 101
Rahman, Y. E., and Peraino, C. (1973). *Exper. Geront.* **8**, 93
Rao, S. S. and Kanungo, M. S. (1972). *Mech. Age. Devel.*, **1**, 61
Robertson, O. H. *et al.* (1972). In *Endocrines and Ageing*, MSS Information Corp.; New York, N.Y.
Rotenberg, S., Ceci, L. and Adelman, R. C., manuscript submitted
Roth, G. S., Karoly, K., Britton, V. and Adelman, R. C. (1974). *Exper. Gerontol.*, **9**, 1
Singer, S. and Litwack, G. (1971). *Endocrinol.*, **88**, 1448

Contributors

Richard C. Adelman — *Fels Research Institute and Department of Biochemistry, Temple University Medical School, 3420 North Broad Street, Philadelphia, Pa. 19140, U.S.A.*

T. K. Basu — *Department of Biochemistry, University of Surrey, Guildford, Surrey GU2 5XH*

Alasdair Breckenridge — *Department of Clinical Pharmacology, Royal Postgraduate Medical School, Hammersmith Hospital, Ducane Road, London, W.12.*

G. Curzon — *Department of Neurochemistry, Institute of Neurology, Queen Square, London, W.C.1.*

J. W. T. Dickerson — *Department of Biochemistry, University of Surrey, Guildford, Surrey GU2 5XH*

Francesco De Matteis — *Biochemical Mechanisms Section, M.R.C. Toxicology Unit, Carshalton, Surrey*

Dennis V. Parke — *Department of Biochemistry, University of Surrey, Guildford, Surrey GU2 5XH*

George A. Porter — *Department of Medicine, University of Oregon Medical School, 3181 S.W. Sam Jackson Park Road, Portland, Oregon, 97201, U.S.A.*

Mels Sluyser — *Department of Biochemistry, Antoni van Leeuwenhoekhuis het Nederlands Kankerinstituut, Sarphatistraat 108, Amsterdam C, The Netherlands*

M. Smethurst — *Research Department, The Marie Curie Memorial Foundation, The Chart, Oxted, Surrey*

D. C. Williams — *Research Department, The Marie Curie Memorial Foundation, The Chart, Oxted, Surrey*

Alan Wiseman — *Department of Biochemistry, University of Surrey, Guildford, Surrey GU2 5XH*

Index

RNA *contd.*
  synthesis, steroids and carcino-
    genesis, 147–153
  stabilization, 220
RNA-polymerase,
  effect of aldosterone, 124
  steroids and carcinogenesis, 147–
    153

Saccharomyces cerevisiae, 4–5, 17–22
Safrole, induction of microsomal
    enzymes, 213, 224, 236–237, 239,
    259, 261–262
Schradan, and enzyme induction, 290
Sequential induction, 6–7
Serine dehydratase,
  effect of age on induction, 305–
    306, 308
  perinatal development, 34
Sex differences, in the effect of age on
    enzyme induction, 306
Sex hormones, effects on hepatic
    microsomal enzyme induction,
    232
Sex steroid binding protein, 82
Short circuit current, in measurement
    of sodium transport, 108–117,
    129
SKF 525A, 209, 219, 231, 240
  effects on porphyrin synthesis,
    194–195
Skin,
  induction of benzpyrene hydroxy-
    lase in, 296–297
  microsomal enzyme induction, 249
Smoking,
  and induction of benzpyrene hy-
    droxylase, 249, 251, 255
  in microsomal enzyme induction,
    279, 289
Sodium transport,
  role of aldosterone, 105–137
  role of other corticosteroids, 110,
    121
Solubilization of hepatic microsomal
    haemoproteins, 225
Species differences, in microsomal
    enzyme induction, 215–217
Spin, of hepatic CO-binding cyto-
    chromes, 217

Spironolactone,
  an aldosterone antagonist, 118–
    119, 121
  effect on drug metabolism, 232–233
  in microsomal enzyme induction,
    283, 287
Staphylococcus aureus, 4
Stereoselective hydroxylations, 214
Steroid hormones,
  induction of enzyme synthesis,
    99–101
  mechanism of action, 79–101
  physiological effects, 80
  role in transcription, 100
Steroids,
  effect on development of enzymes,
    29–30, 33–34, 39, 41–43, 45–48,
    56–57, 60–61
  hydroxylation of, 210
  induction of hepatic microsomal
    enzymes, 228–235
  metabolism, 154–156
  and carbohydrate metabolism, 157
  microsomal enzyme induction by,
    282–283
Steroid receptors,
  in cytosol, 82–85, 94–95
  in nucleus, 85–86, 94–95
Stilboestrol,
  and RNA-polymerase in prostate,
    154
  transplacental carcinogenesis, 156
Strain differences, in microsomal
    enzyme induction, 215–217
Streptomyces albus, 17
Stress, effects on tryptophan metab-
    olism, 171–173
Sub-optimal inducers, 121
Substrate specificities, of different
    microsomal CO-binding cyto-
    chromes, 225, 230
Sucrase, perinatal development of, 29
Superhelical structure, of DNA, 88,
    92, 95–99
Synthesis of enzymes, induced by
    steroid hormones, 99–101

Terpenoids, induction of hepatic
    microsomal enzymes, 236–237